Fundamentals of Systems Analysis and Modeling of Biosystems and Metabolism

Authored By

Kazuyuki Shimizu

Institute of Advanced Biosciences
Keio University
Tsuruoka
Yamagata 997-0017
Japan

&

Yu Matsuoka

Kyushu Institute of Technology
Iizuka
Fukuoka 820-8502
Japan

CONTENTS

CHAPTERS

About the Authors

Prof. Kazuyuki Shimizu is based at the Kyushu Institute of Technology (KIT) and the Institute of Advanced Biosciences (IAB), Keio University, Japan. Following a BS and an MS in Chemical Engineering at Nagoya University, and a PhD in Chemical Engineering at Northwestern University, USA, he started his career in 1981 as a research associate at Nagoya University, Japan, and was promoted to associate professor in 1990. He then moved to the KIT as a professor in 1991. In 2000, he became an adjunct professor at IAB, Keio University, Japan. He has long been involved in research into ^{13}C-metabolic flux analysis (^{13}C–MFA), and studies on modeling and systems biology with the aim of gaining insight into the basic principles governing living cell systems. He recognizes the importance of uncovering the metabolic regulation mechanism of a cell system, based on both experimental (wet) and computational (dry) approaches. He has also organized a UK–Japan collaboration project on microbial systems biology toward developing virtual microbes by integration of the different levels of hierarchical omics information, such as transcriptomics, proteomics, metabolomics, and fluxomics data.

Dr.Yu Matsuoka is based at Kyushu Institute of Technology (KIT). Following a BS and PhD in Bioscience and Bioinformatics at KIT, she started her career in 2009 as a postdoctoral fellow at KIT. Her research interest is the systems biology, modeling, and computer simulation of a cell system.

FOREWORD

I am honored to be able to contribute to this exciting eBook about microbial systems biology, modeling, and metabolic regulation of the main metabolism of a cell. The eBook has excellent credentials for describing important knowledge and approaches for the understanding of metabolic regulation of a cell system in general, since the main metabolism is common and conserved in a variety of living organisms.

Professor Kazuyuki Shimizu and I have been the project leaders of UK-Japan microbial systems biology research for long time. He has been involved in ^{13}C-metabolic flux analysis and systems biology to clarify the metabolic regulation of *Esherichia coli*. He has published many articles in this area.

In summary, I am delighted to see such expertly produced and well referenced eBook which will contribute for both biochemical science and metabolic engineering as well as bioinformatics.

Johnjoe Mcfadden
Associate Dean (International)
Faculty of Health and Medical Sciences
University of Surrey
Guildford, Surrey
UK

PREFACE

The ultimate goal of systems biology is to reconstruct the cell system into the computer which can predict observable phenotypes. If this could be attained, the effects of culture environment and/or the specific genes knockout on the metabolism can be predicted without many exhaustive experiments, and thus more efficient metabolic engineering can be made. For this, it is quite important to understand the metabolic regulation mechanism and properly express it in the computer. Construction of a "virtual cell" will be an ambitious but realistic target that will provide significant benefits in the variety of practical applications from both scientific and metabolic engineering points of view.

The present book gives fundamentals of systems analysis and modeling as well as model identification, optimization and dynamics together with computer simulation toward the above ultimate goal. It is also critical to understand the metabolic regulation for the modeling, and thus some of the metabolic regulation of the main metabolism is also explained.

ACKNOWLEDGEMENTS

Declared None.

CONFLICTION OF AUTHORS

The authors confirm that this ebook contents have no conflict of interest.

Kazuyuki Shimizu
Institute of Advanced Biosciences
Keio University
Tsuruoka
Yamagata 997-0017
Japan
E-mail: shimi@bio.kyutech.ac.jp

&

Yu Matsuoka
Kyushu Institute of Technology
Iizuka
Fukuoka 820-8502
Japan

Fundamentals of Systems Analysis and Modeling of Biosystems and Metabolism

CHAPTER 1

Background

Abstract: A brief overview is given for the current status of systems biology and modeling. Systems biology focuses on the profiling of the whole cellular metabolism using high-throughput data of different levels of information to understand and unraveling the underlying principles of the living organisms. The systems biology allows the development of mathematical models that can be computationally simulated. Various modeling approaches can be classified into two such as flux balance analysis (FBA) based on the stoichiometric constraints, and the kinetic modeling based on the enzymatic kinetic expressions. Although the former approach can be extended to large genome-scale, it is difficult to incorporate the metabolic regulation mechanism and to express the dynamics, while the latter approach can reasonable incorporate the metabolic regulation mechanism. The problem for the kinetic modeling is the increase in the model parameters as the system size becomes large. It is important for the modeling of a cell system to properly understand and express how the environmental stimuli are detected, how those are transduced, and how the cell metabolism is regulated. It is quite useful from science and metabolic engineering points of view to develop quantitative models toward whole cell modeling.

Keywords: Virtual microbe, systems biology, flux balance analysis, genome-scale, kinetic modeling, metabolic regulation, metabolic engineering.

INTRODUCTION

In living organisms, metabolic network plays an essential role, where the metabolism comprises thousands of reactions that are involved in the degradation of available nutrient sources for biosynthesis of cellular constituents. Those cellular constituents are formed from several key building blocks such as amino acids for protein synthesis, fatty acids for lipids synthesis, nucleotides for DNA and RNA synthesis, and sugar moieties for carbohydrate synthesis. Those building brocks are formed from the precursor metabolites generated in the main metabolism [1, 2]. These characteristics are common to any living organisms. Namely, such biosynthesis nature is conserved among all living organisms.

For the cell metabolism to function, and for the cell to grow, several cofactors such as ATP, NAD(P)H *etc.,* as well as other nutrient sources such as nitrogen, phosphate, oxygen, metal ion *etc.,* are required. Since the central carbon metabolism is tightly connected with overall cell function, it is quite important to understand its regulation mechanism for metabolic engineering [3].

Kazuyuki Shimizu and Yu Matsuoka

Recently, significant progress has been made on molecular biology to understand the cellular metabolism. However, molecular knowledge alone is limited to clarify the cell system's behavior, where the systems behavior emerges from the interactions between the characterized molecules [4-6]. Thus, the systems biology has attracted much attention. The ultimate goal of systems biology is to reconstruct a cell system into a computer that can predict the observable phenotypes. If this could be attained, the effects of culture environment and/ or the specific genetic mutation on the cell metabolism can be predicted without conducting many exhaustive experiments. Thus, metabolic engineering may be made more efficiently, where the small set of selected experiments may be conducted for verification and to see the preferred phenotypes. In this way, the development of the appropriate model is useful for the efficient design of cell factories in practice. Moreover, the appropriate model can contribute for understanding the metabolic regulation mechanism.

In the present Chapter, an overview is given for a variety of modeling approaches [7]. The detailed methods for the modeling as well as the related fundamentals are explained in the following Chapters.

RECENT TRENDS OF SYSTEMS BIOLOGY

Systems biology focuses on the profiling of the whole cellular metabolism using high-throughput data of different levels of information to understand and unraveling the underlying principles of the metabolism. Functional genomics are important for characterizing the molecular constituents of a cell system, where systems biology addresses the missing links between molecules and physiology [8, 9]. In this approach, the integration of different levels of "omics" data is important (Fig. 1), where the data set containing transcriptomics, proteomics, metabolomics, and fluxes of the cell in response to the specific pathway mutation and the different growth rates is useful [10] for computational systems biology [11]. Moreover, the systems biology allows the development of mathematical models that can be computationally simulated.

In relation to systems biology approach, where it may be a rather top-down approach, another bottom-up approach called "synthetic biology" has emerged gaining some attention, where it is an approach for designing, synthesizing, and assemblying mega-base pair and its transplantation into such recepient cell as *Mycoplasm mycoides* [12], and the mega-genome transfer between different organisms [13]. It may be also of practical interest to reconstruct small, artificial biology systems such as a synthetic feedback loop for improving biofuels production [14]. As such, synthetic biology aims at creating novel functional parts, modules, circuits *etc.*, and has been shown to be useful in biotechnological applications [15, 16]. The *in vitro* construction of the metabolic pathways may be also considered as the so-called synthetic metabolic engineering [17].

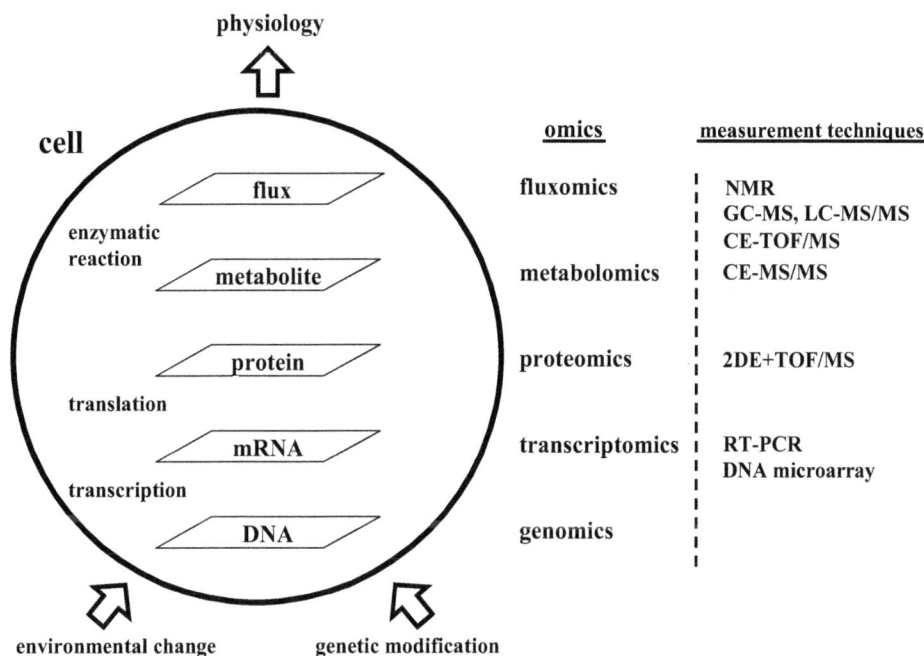

Figure 1: Integration of different levels of information.

In relation to synthetic biology, genome engineering has been made some progress, where the attempts have been made for minimization of genome (so-called Minimum Genome Project) to construct useful microbes by identifying minimal gene set, thus reducing the burden for the cell, resulting in improved cell growth rate [18-23].

Recent metabolic engineering has thus been upgraded from local to global or systems level by the systems biology and synthetic biology, possibly with evolutionary engineering [24]. The next advances may be expected at the intersection of the top-down systems approach and the bottom-up synthetic biology approach [25] with synergies between them [26, 27].

TYPICAL MODELING APPROACHES

A variety of models have been developed so far, where they are discriminated from others depending on the underlying assumptions, the data they require, and the accuracy of the model prediction [28, 29]. The modeling may start by considering the model structure, model (parameter) identification, and model validation [30]. The modeling approaches may be roughly classified into two such as **flux balance analysis** (FBA) method and **kinetic modeling** approach. The former is based on the linear constraints of stoichiometric equations, while the latter is based on the kinetic rate equations for the metabolic pathway reactions, where it is not straightforward to develop a kinetic model due to difficulty in identifying the underlying mechanism [31]. After the model structure was determined, model (parameter) identification, sensitivity analysis, identifiability, experimental design, and optimization may be considered in relation to the development of appropriate models [30, 32, 33]. Moreover, some model reduction may be considered depending on the available experimental data.

In either approach, the information on the metabolic flux distribution (MFD) is the most important, where the metabolic fluxes are located on top of different levels of the cell system (Fig. **1**). This information can be used for the analysis of metabolic regulation in response to culture environment and the specific pathway mutation [34-39]. This information can be obtained by the conventional **metabolic flux analysis** (MFA) based on the stoichiometric equations, or ^{13}C-MFA based on isotope balance as well as stoichiometric equations. Although MFA is important, this is the analysis method for the physiological state based on the experimental data of the measured specific rates, and it does not have the predictability. It is highly desirable to predict the phenotypic characteristics in response to culture environment and/ or the specific pathway mutation, where it is the main purpose of developing the models.

Flux Balance Analysis and its Genome-Scale Extension

Flux balance analysis (FBA) and its genome-scale extension has been a widely used approach based on the knowledge of known metabolic reaction stoichiometry in an organism and/ or the genes that encode each pathway enzyme [40]. FBA calculates the flux distribution through this metabolic network, thereby making it possible to predict the cell growth rate and the target metabolite production rate.

Metabolic reaction can be represented as stoichiometric equations, where the steady state mass balance may be expressed as

$$Sv = 0 \; S \in R^m \times R^n, v \in R^n \tag{1}$$

where S is the stoichiometric matrix, and is constituted of the stoichiometric coefficients with size of $m \times n$, and **v** is the flux vector of size n. Every row of S corresponds to one compound or metabolite, while the column corresponds to the associated reaction. In convention, the sign of the entry in S is negative if the corresponding metabolite is consumed, while it is positive if it is produced. In the case of large genome-scale FBA, most of the stoichiometric coefficients are zero, giving S to be a sparse matrix. Any **v** that satisfies Eq.(1) is called as the **null space** of S.

In general, there are more reactions or fluxes than the compounds or metabolites ($n > m$), which indicates that there are more reactions than equations, so that there is an excess degrees of freedom, and thus the system is under-determined. Although the stoichiometric constraints define a range of feasible solutions, it is not possible to determine the single point in the feasible space. Thus FBA approach essentially requires an objective function or a vector-valued objective function such as

$$z = c^T v \tag{2}$$

where $c^T v$ implies the linear combination of fluxes, and c is a weight vector for the objective function z. The objective function may typically be chosen as the specific cell growth rate or the maximum ATP production rate *etc.*, or those may be considered as a component of a vector-valued objective function. The problem of optimization of (2) under the constraint of Eq.(1) can be formulated and solved by the linear programming problem. Namely, FBA gives the flux distribution that optimizes the specific objective function(s), and thus the result depends on the objective function(s) employed. Therefore, careful inspection on the objective function(s) is necessary depending on the purpose of developing the model [41].

FBA approach can be extended for the metabolic engineering application such as OptKnock [42], a bi-level programming framework for identifying the gene knockout for strain improvement. This framework has been extended for various applications such as OptReg [43], OptForce [44], OptFlux [45], Differential Bees FBA (DFFBA) [46], and OptStrain [47]. Stoichiometry-based design algorithms are formulated as bi-level mixed integer linear programming problems [42-44, 48, 49], where the outer level optimizes the objective function(s), while the inner level optimizes the cellular system in response to environmental perturbations [50, 51].

The problem of FBA and its extension to genome-scale is the difficulty in analyzing the dynamic behavior. Some attempts have been made by incorporating the kinetic expressions for the uptake of multiple carbon sources and other nutrients into the quasi steady-state [52-54]. The dynamic multi-species metabolic modeling (DMMM) has been considered by incorporating the metabolites uptake kinetics into stoichiometric models of a microbial consocium [55, 56]. The steady-state metabolic flux distributions obtained by FBA and stoichiometric information may be used to parameterize the genome-scale kinetic models in response to small perturbations [57-60]. Lin-log kinetic expression and thermodynamics may be incorporated to constrain FBA simulation [61].

Some attempts have been made for the hybrid type of stoichiometric and kinetics based modeling [56, 62, 63]. The dynamic flux balance analysis (dFBA) has been considered for the diauxic growth of *E.coli* consuming glucose and acetate in the batch culture [64]. The OptForce formalism has been extended as k-OptForce by bridging the gap between stoichiometric approach and kinetic approach, where the procedure seamlessly integrates the mechanistic detail given by kinetic models within a constraint-optimization framework tractable for genome-scale [65].

As mentioned before, one of the advantages of FBA approach [66] may be the easy extension to large genome-scale models [67-69]. Thus a whole cell model has been developed for *Mycoplasm genetarium*, a urogenital parasite [70], where this model constitutes of 28 processes of the cell system [71]. This may be the dawn of virtual cell biology [72], as also mentioned as a grand challenge of the 21st century [73].

As also mentioned above, although powerful and attractive for the possible extension to the whole cell modeling or the so-called virtual microbes, the main drawback of the FBA approach is the difficulty in incorporating the metabolic regulation mechanisms. Although some attempts have been made for incorporating the transcriptional regulation mechanism into FBA framework by Boolean rules [74-77], the regulatory rules may not be based on the metabolic regulation mechanism, but based on the available experimental data, which show part or snapshot of the real regulation mechanism [78]. This regulation mechanism may be reasonably incorporated by the kinetic modeling approach, where this is the next topic.

As such, due to complexities of such a system, the conventional approach has focused primarily on each level of network as module such as signaling system, metabolism, and regulation *etc.* Metabolic reactions can be represented by the stoichiometry, where the regulatory reactions may be incorporated by the rule base using for example a Boolean formalism. FBA can be implemented in practice assuming quasi-steady state for the typical batch culture, since the typical time constant of the metabolic transient is smaller than the time scale for the biomass growth and other state variables such as extracellular substrate and metabolite concentrations. A kinetic model accounting for signal transduction, metabolism, and regulation has been constructed to describes the response of *S.cerevisiae* to osmotic stress [79], where this model connected the specific outputs of one network such as signaling network with the inputs of another network such as metabolism in a sequential manner. The drawback is that it is difficult to incorporate all the interactions. There is thus a growing interest for the development of the framework to study such networks from an integrated perspective [80]. FBA-based framework may be considered by incorporating signaling, metabolic, and regulatory networks at the genome-scale [81].

Kinetic Modeling Approach

Kinetic models for the cell metabolism incorporate the quantitative expression that connects fluxes and metabolite concentrations, where the flux v_j is expressed as a function of metabolites such as

$$v_j(v_j^{max}, x_j, p_j)$$

where v_j^{max} is the maximum reaction rate, x_j is a vector of metabolites related to v_j, and p_j is the vector of model parameters appeared in the expression of v_j. The parameter v_j^{max} may be also included in p_j, but it is separated from p_j for convenience. The time-course of metabolite concentrations can be obtained by solving the following ordinary differential equations (ODEs):

$$\frac{dx_i}{dt} = \sum_{j=1}^{n_i} S_{ij} v_j(v_j^{max}, x_j, p_j) \quad (i=1,\ldots,m) \tag{3a}$$

where the typical mechanistic expression for v_j may be expressed as the Michaelis-Menten type equation such as

$$v_j = \frac{v_j^{max} S_j}{K_{sj} + S_j} \tag{3b}$$

or Hill type expression. In the above expression, K_{sj} is the saturation constant, and S_j is the substrate concentration for the j-th reaction. The kinetic expression is determined based on the pathway enzyme reaction mechanism and characterization [82, 83]. One of the drawback of the kinetic modeling approach is the increase in the model parameters. Thus, various approximate kinetic forms such as lin-log [84-86], and log-lin kinetics [87], power law kinetic expression such as S-system [88], and generalized mass action [57,58] and others [89-91] have been considered in the past.

The primary attempts of incorporating the metabolic regulation mechanism into the kinetic models have been made by **Cybernetic modeling** approach, in which the cells are assumed to utilize the available nutrient sources with maximum efficiency by the optimal strategy [92]. This has been extended to more structure model [93], and applied for metabolic engineering purpose [94, 95]. In this approach, an elementary mode was considered as a metabolic subunit, where the elementary modes are a set of metabolic pathways by which the metabolic routes can be completely described, and any feasible fluxes can be represented by their combinations at steady state [96]. The elementary modes consist of a minimal set of reactions that function at steady state, which implies that the elementary mode cannot be a functional unit if any reaction is removed [96]. The hybrid type modeling has also been developed by assuming quasi steady-state for the intracellular metabolites [97, 98], and several applications were made for *E.coli* [98] and for yeast [78].

The large-scale extension is limited for kinetic modeling approach due to the increased number of model parameters and unambiguous model parameterization [99]. Several attempts have, therefore, been made by postulating a generalized uniform kinetic equation [84, 89, 100, 101], S-system formalism [88, 102], or a combination of *in vitro*-based lumped and approximate rate equations [103, 104].

Recently, the **ensemble modelling** (EM) has been considered to cope with large-scale kinetic modeling by successively reducing the size of parameter space based on the available experimental data together with thermodynamic constraints for the direction of the net fluxes [105]. In EM approach, any type of pathway reaction mechanism can be considered as well as already known mechanism, where each reaction can be decomposed into elementary reaction steps by mass action kinetics [106] such as

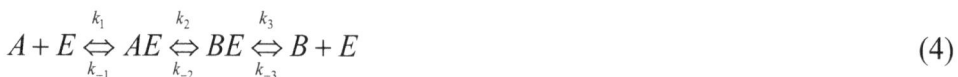

$$A + E \underset{k_{-1}}{\overset{k_1}{\Longleftrightarrow}} AE \underset{k_{-2}}{\overset{k_2}{\Longleftrightarrow}} BE \underset{k_{-3}}{\overset{k_3}{\Longleftrightarrow}} B + E \tag{4}$$

where A,B are the metabolites, E is the enzyme, and AE,BE are the complexes.

The EM procedure starts with initially assumed kinetic models that predict the experimentally observed phenotypic characteristics, and additional data such as those of the strain under environmental and/ or genetic perturbations are used to screen the model until a minimal set of kinetic models is obtained [106]. This modeling approach has been applied for lysine production [107], fatty acid production [108], aromatic production [105], robustness analysis for engineered non-native pathways [109], and modeling cancer cells [110]. Moreover, this approach has been applied for the modeling of *E.coli* that reasonable predicts the fluxes and metabolite concentrations of wild type and its single gene knockout mutants [111] based on the multi-omics data [10].

MODELING OF THE MAIN METABOLISM

Mathematical modeling for the main metabolic pathways is made for yeast [112, 113]. As for dynamics, *in vivo* kinetics with rapid perturbation experiments have been investigated with the so-called BioScope [114]. Fast quenching and efficient extraction, together with high throughput measurement of intracellular metabolite concentrations are essential for the prediction of the metabolite dynamics [115] (Fig. **2**). The modeling based on stimulus-response experiments has been made by several researchers [116].

The kinetic model equations for the glycolysis and pentose phosphate (PP) pathway have been developed for *E. coli* to simulate the transient data obtained by the fast sampling system [117]. The kinetic models for the TCA cycle and anaplerotic pathways as well as glycolysis and PP pathway were also considered to simulate the typical batch and continuous cultures with some rule based approach, where the cell growth rate was estimated based on the specific ATP production rate computed from the fluxes [118]. It is important to make modeling based on the integrated information from gene level to flux level by incorporating the roles of transcription factors [119, 120]. The phosphotransferase system (PTS) and catabolic regulation phenomena has been modeled by several researchers [121-126]. The important steps are how to incorporate (i) the effect of culture environment on global regulators, (ii) the effects of global regulators on the metabolic pathway genes, and (iii) the effects of metabolic pathway genes on the corresponding enzyme activities. It may be useful to incorporate the molecular mechanism as illustrated by several researchers [119, 120, 125, 127]. The detailed regulation mechanisms are given in Chapter 7.

The regulation mechanism of ammonia assimilation is also important. The emphasis rests on '*silicon-cell*' models based on experimental kinetic data for the enzymes involved that predict the flux of assimilation of extracellular ammonia into glutamate in *E. coli*. The comprehensive model [128] combined the metabolic regulation with signal transduction through the covalent modification of P_{II} and Glutamine synthetase by UTase and ATase. It shows that the regulation is distributed between the two modes of regulation. The systems biology analysis shows that only active import of ammonia or passive transport of ammonium ion is compatible with the concentration. The detailed mechanisms are explained in Chapter 7.

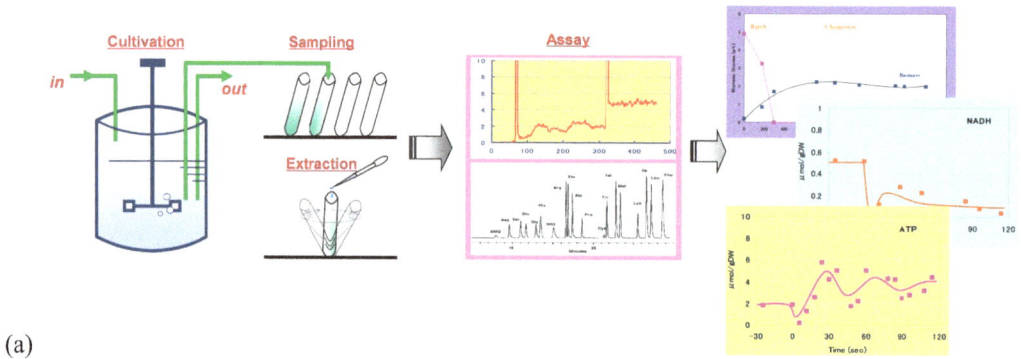

(a)

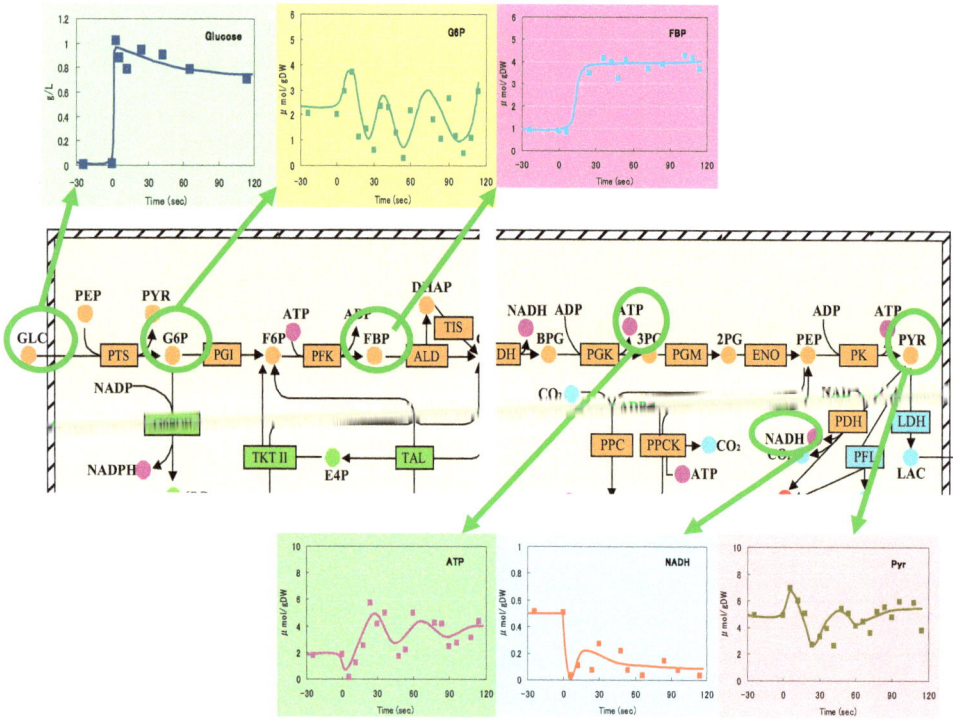

(b)

Figure 2 Fast sampling with extraction (a) and modeling for the dynamics of the intracellular metabolites (b) (Hua Q, private communication).

The above model may be incomplete in the sense that α-keto glutaric acid (αKG) pool size was assume to be constant, whereas it changes significantly during perturbation in nitrogen source availability or C/N ratio [129]. Moreover, it is important to take into account the interdependence of the metabolite pools and the cell growth, where αKG affects the glucose uptake by inhibiting enzyme I, EI of the phosphotransferase system (PTS) [130] as also mentioned in Chapter 7.

IMPORTANCE OF MODELING AND FUTURE PERSPECTIVES

The reactions of central metabolism are central to the cell synthesis and the hub on which nearly all catabolic and biosynthetic processes are built. They carry most of the carbon flux and are most amenable to experimentation and modeling. Gene regulatory control of central metabolism plays a key role in the adaptation of organisms to changes in their environment. Despite huge diversity of unicellular and multi-cellular organisms, the overall structure of central metabolic pathways is remarkably well conserved. The metabolic model for the central metabolism will provide a platform for further extension to peripheral metabolism and incorporation of gene regulation, protein synthesis, and cellular physiology. Completeness of the model may not be necessary for it to improve predictions or rationalizations.

The modelling will be of immense value to:

- biotechnologists aiming to improve fermentation performances such as the yield and productivity of the target metabolite.

- microbial engineers aiming to design novel microbes able to capture available carbon, degrade pollutants, vaccinate man and/or animals, build microbial fuel cells, produce next generation bio-fuels

- basic scientists aiming to understand metabolism in living cells, and its control for synthetic biology.

- systems biologists aiming to advance the science of modeling.

The product of this approach will greatly exceed the importance of the microbial genome sequencing projects, as it will be much closer to understanding biological function and will have widespread practical application.

OUTLOOK OF THE BOOK

In the present book, fundamentals of systems analysis are briefly explained in Chapter 2, followed by the explanation of the fundamentals of modeling in general, where the basic equations for mass balance, momentum balance, and heat balance are derived in Chapter 3. In Chapter 4, kinetic modeling of the main metabolic pathways together with fermentation and lysine synthetic pathways are explained. The kinetic models for ammonia assimilation pathways are also explained. In Chapter 5, model identification, sensitivity analysis, and the optimization techniques are briefly explained. In Chapter 6, after the explanation on the structure of steady state, the dynamics of the system are explained from the stability and bifurcation points of view. Then the effects of culture environments such as the availability of carbon sources and nitrogen sources as well as oxygen limitation on the metabolism in *E. coli* are explained in Chapter 7. Finally, the computer simulation of the main metabolism together with the estimation of the cell growth rate is considered, and compared with the experimental data in Chapter 8, where the catabolite regulation is considered in view of acetate overflow metabolism, and sequential and simultaneous assimilation of multiple sugars based on the kinetic models explained in Chapter 4 together with the incorporation of the effect of transcription factors. The computer simulation for ammonia assimilation is also considered, and some idea for the simulation of the metabolism switch from aerobic to anaerobic conditions is also given in Chapter 8.

ACKNOWLEDGEMENTS

It is acknowledged to Profs. H. Westerhoff of University of Manchester and J. McFadden of the University of Surrey in UK for the fruitful discussion on the development of virtual microbes.

REFERENCES

[1]   Nielsen J. It is all about metabolic fluxes. J Bacteriol 2003; 185: 7031-7035.
[2]   Varma A, Palsson BO. Metabolic capabilities of *Escherichia coli*: I. synthesis of biosynthetic precursors and cofactors. J Theor Biol 1993; 165: 477-502.
[3]   Nielsen J. Editorial: Industrial systems biology. Biotechnol J 2011; 6: 255.
[4]   Kitano H. Systems biology: a brief overview. Science 2002a; 295: 1662-1664.
[5]   Kitano H. Computational systems biology. Nature 2002b; 420: 206-210.
[6]   Stelling J. Mathematical models in microbial systems biology. Curr Opin Microbiol 2004; 7: 513-518.
[7]   Matsuoka Y, Shimizu K. Current status and future perspectives of kinetic modeling for the cell metabolism with incorporation of metabolic regulation. Bioresources and Bioprocessing 2015; 2:4
[8]   Westerhoff HV, Winder C, Messiha H, Simeonidis E, Adamczyk M, Verma M, Bruggeman FJ, Dunn W. Systems biology: the elements and principles of life. FEBS Lett 2009; 583: 3882-3890.

[9]   Bruggeman FJ, Westerhoff HV. Approaches to biosimulaiton of cellular processes. J Biol Phys 2006; 32: 273-288.

[10]   Ishii N, Nakahigashi K, Baba T, Robert M, Soga T, Kanai A *et al*. Multiple high-throughput analyses monitor the response of *E. coli* to perturbations. Science 2007; 316: 593-597.

[11]   Sauer U, Heinemann M, Zamboni N. Genetics. Getting closer to the whole picture. Science 2007; 316: 550-551.

[12]   Gibson DG, Glass JI, Lartigue C, Noskov VN, Chuang RY, Algire MA *et al*. Creation of a Bacterial Cell Controlled by a Chemically Synthesized Genome. Science 2010; 329: 52-56.

[13]   Itaya M, Tsuge K, Koizumi M, Fujita K. Combining two genomes in one cell: stable cloning of the Synechocystis PCC6803 genome in the *Bacillus subtilis* 168 genome. PNAS USA 2005; 102: 15971-15976.

[14]   Dunlop MJ, Keasling JD, Mukhopadhyay A. A model for improving microbial biofuel production using a synthetic feedback loop. Syst Synth Biol 2010; 4: 95-104.

[15]   Prather KJL, Martin CH. *De novo* biosynthetic pathways: rational design of microbial chemical factories. Curr Opin in Biotechnol 2008; 19: 468-474.

[16]   Na D, Lee KC. Construction and optimization of synthetic pathways in metabolic engineering. Curr Opin Microbiol 2010; 13: 363-370.

[17]   Ye X, Honda K, Sakai T, Okano K, Omasa T, Hirota R *et al*. Synthetic metabolic engineering- a novel, simple technology for designing a chimeric metabolic pathway. Microb Cell Fact 2012; 11: 120.

[18]   Kolisnychenko V, Plunket G III, Herring CD, Fehér T, Pósfai J, Blattner FR, Pósfai G. Engineering a reduced *Escherichia coli* genome. Genome Res 2002; 12: 640-647.

[19]   Goryshin, IY, Naumann TA, Apodaca J, Reznikoff WS. Chromosomal deletion formation system based on Tn5 double transposition: Use for making minimal genomes and essential gene analysis. Genome Res 2003; 13: 644-653.

[20]   Yu BJ, Sung BH, Koob MD, Lee CH, Lee JH, Lee WS, Kim MS, Kim SH. Minimization of the *Escherichia coli* genome using a Tn5-targeted Cre/*loxP* excision system. Nature Biotech 2002; 20: 1018-1023.

[21]   Hashimoto M, Ichimura T, Mizoguchi H, Tanaka K, Fujimitsu K, Keyamura K, Ote T, Yamakawa T, Yamazaki Y, Mori H, Katayama T, Kato J. Cell size and nucleotide organization of engineered *Escherichia coli* cells with a reduced genome. Mol Microbiol 2005; 55: 137-149.

[22]   Mizoguchi H, Mori H, Fjio T. *Escherichia coli* minimum genome factory. Biotechnol Appl Biochem 2007; 46: 157-167.

[23]   Kühner S, van Noort V, Betts MJ, Leo-Macias A, Batisse C, Rode M *et al*. Proteome organization in a genome-reduced bacterium. Science 2009; 326: 1235-1240.

[24]   Lee JW, Kim TY, Jang Y-S, Choi S, Lee SY. Systems metabolic engineering for chemicals and materials. Trends in Biotechnol 2011; 29: 370-378.

[25]   Lanza AM, Crook NC, Alper HS. Innovation at the intersection of synthetic and systems biology. Curr Opin in Biotech 2012; 23: 1-6.

[26]   Smolk CD, Silver PA. Informing biological design by integration of systems and synthetic biology. Cell 2011; 144: 855-859.

[27]   Nielsen J, Keasling JD. Synergies between synthetic biology and metabolic engineering. Nature Biotechnol 2011; 29: 693-695.

[28]   Selinger DW, Wright MA, Church GM. On the complete determination of biological systems. Trends Biotechnol 2003; 21:251-254

[29]   Machado D, Costa R, Rocha M, Ferreira E, Tidor B, Rocha I. Modeling formalisms in Systems Biology. AMP Expre 2011; 1:1-34

[30]   Almquist J, Cvijovic M, Hatzimanikatis V, Nielsen J, Jirstrand M. Kinetic models in industrial biotechnology-Improving cell factory performance. Metabolic Eng 2014; 24:38-60

[31]   Costa RS, Machado D, Rocha I, Pereira EC. Critical perspective on the consequences of the limited availability of kinetic data in metabolic dynamic modeling. IET Syst Biol 2011; 5:157-163

[32]   Ashyraliyev M, Fomekong-Nanfack Y, Kaandorp JA, Blom JG. Systems biology: parameter estimation for biochemical models. FEBS J 2009; 276:886-902

[33]   Cvijovic M, Bordel S, Nielsen J. Mathematical models of cell factories: moving towards the core of industrial biotechnology. Microb Biotechnol 2011; 4:572-584

[34]   Shimizu K. Metabolic flux analysis based on ^{13}C-labeling experiments and integration of the information with gene and protein expression patterns. Adv Biochem Eng Biotechnol 2004; 91:1-49

[35]   Sauer U. Metabolic networks in motion: ^{13}C-based flux analysis. Mol Systems Biol 2006; 2:62

[36]   Wittman C. Fluxome analysis using GC-MS. Microb Cell Fact 2007; 6:6

[37]   Shimizu K. Bacterial cellular metabolic systems. Woodhead Publ Ltd. Oxford 2013

[38]   Long CP, Antoniewicz MR. Metabolic flux analysis of *Escherichia coli* knockouts: lessons from the Keio collection and future outlook, Curr Opin Biotechnol 2014; 28: 127-133

[39]   Kromer JO, Nielsen LK, Blank LM (eds.). Metabolic flux analysis, Human press 2014

[40]   Orth JD, Thiele I, Palsson BO. What is flux balance analysis ?, Nat Biotechnol 2010; 28 (3): 245-248

[41]   Schuetz R, Kuepfer, Sauer U. Systematic evaluation of objective functions for predicting intracellular fluxes in *Escherichia coli*. Mol Syst Biol 2007; 3:119

[42]   Burgard AP, Pharkya P, Maranas CD. Optknock: a bilevel programming framework for identifying gene knockout strategies for microbial strain optimization. Biotechnol Bioeng 2003; 84:647–657

[43]   Pharkya P, Maranas CD. An optimization framework for identifying reaction activation/inhibition or elimination candidates for overproduction in microbial systems. Metab Eng 2006; 8:1–13

[44]   Ranganathan S, Suthers PF, Maranas CD. OptForce: an optimization procedure for identifying all genetic manipulations leading to targeted overproductions. PLOS Computational Biology 2010; 6: e1000744

[45]   Rocha I, Maia P, Evangelista P, Vilaca P, Soares S, Pinto JP *et al*. OptFlux: an open-source software platform for *in silico* metabolic engineering. BMC Syst Biol 2010; 4:45

[46]   Choon YW, Mohamad MS, Deris S, Illias RM, Chong CK, Chai LE *et al*. Differential bees flux balance analysis with OptKnock for *in silico* microbial strains optimization. PLoS One 2014; 9:e102744

[47]   Pharkya P, Burgard AP, Maranas CD. OptStrain: A computational framework for redesign of microbial production systems, Genom Res 2014; 14: 2367-2376

[48]   Yang L, Cluett WR, Mahadevan R. EMILiO: a fast algorithm for genome-scale strain design, Metab Eng 2011; 13: 272-281

[49]   Cotten C, Reed JL. Constraint-based strain design using continuous modifications (CosMos) of flux bounds finds new strategies for metabolic engineering. Biotechnol J 2013; 8: 595–604

[50]   Ibarra RU, Edwards JS, Palsson BO. *Escherichia coli* K-12 undergoes adaptive evolution to achieve *in silico* predicted optimal growth. Nature 2002; 420:186–189

[51]   Segrè D, Vitkup D, Church GM. Analysis of optimality in natural and perturbed metabolic networks. PNAS USA 2002; 99:15112–15117

[52]   Covert MW, Xiao N, Chen TJ, Karr JR. Integrating metabolic, transcriptional regulatory and signal transduction models in *Escherichia coli*. Bioinformatics 2008; 24: 2044–2050

[53]   Meadows AL, Karnik R, Lam H, Forestell S, Snedecor B. Application of dynamic flux balance analysis to an industrial *Escherichia coli* fermentation. Metab Eng 2010; 12:150–160

[54]   Feng X, Xu Y, Chen Y, Tang YJ. MicrobesFlux: a web platform for drafting metabolic models from the KEGG database. BMC Syst Biol 2012; 6:94

[55]   Zhuang K, Izallalen M, Mouser P, Richter H, Risso C, Mahadevan R *et al*. Genome-scale dynamic modeling of the competition between Rhodoferax and Geobacter in anoxic subsurface environments. ISME J 2011; 5:305–316

[56]   Salimi F, Zhuang K, Mahadevan R. Genome-scale metabolic modeling of a clostridial co-culture for consolidated bioprocessing. Biotechnol J 2010; 5:726–738

[57]   Jamshidi N, Palsson BØ. Formulating genome-scale kinetic models in the post-genome era. Mol Syst Biol 2008; 4:171

[58]   Jamshidi N, Palsson BØ. Mass action stoichiometric simulation models: incorporating kinetics and regulation into stoichiometric models. Biophysical Journal 2010; 98:175–185

[59]   Smallbone K, Simeonidis E, Broomhead DS, Kell DB. Something from nothing - bridging the gap between constraint-based and kinetic modelling. FEBS J 2007; 274:5576–5585

[60]   Smallbone K, Simeonidis E, Swainston N, Mendes P. Towards a genome-scale kinetic model of cellular metabolism. BMC Syst Biol 2010; 4:6

[61]   Fleming RM, Thiele I, Provan G, Nasheuer HP. Integrated stoichiometric, thermodynamic and kinetic modelling of steady state metabolism. J Theor Biol 2010; 264:683–692

[62]   Antoniewicz MR. Dynamic metabolic flux analysis—tools for probing transient states of metabolic networks. Curr Opin Biotechnol 2013; 24:973-978

[63]   Hoffner K, Harwood SM, Barton PI. A reliable simulator for dynamic flux balance analysis. Biotechnol & Bioeng 2013; 110:792-802

[64]   Mahadevan R, Edwards JS, Doyle FJ. Dynamic flux balance analysis of diauxic growth in *Escherichia coli*. Biophys J 2002; 83:1331–1340

[65]   Chowdhury A, Zomorrodi AR, Maranas CD. k-OptForce: integrating kinetics with flux balance analysis for strain design. PLoS Comput Biol 2014; 10:e1003487

[66]   Edwards JS, Covert MW, Palsson BØ. Metabolic modelling of microbes: the flux-balance approach. Environ Microbiol 2002; 4: 133-40.

[67]   Palsson B. Metabolic systems biology. FEBS Lett 2009; 583: 3900-3904.

[68]   Feist AM, Palsson B. The growing scope of applications of genome-scale metabolic reconstructions using *Escherichia coli*. Nat Biotechnol 2008; 26: 659-667.

[69]   Herrgard MJ, Swainston N, Dobson P, Dunn WB, Arga KY, Arvas M *et al*. A consensus yeast metabolic network reconstruction obtained from a community approach to systems biology. Nat Biotechnol 2008; 26: 1155-1160.

[70]   Karr JR, Sanghvi JC, Macklin DN, Gutschow MW, Jacobs JM, Bolival Jr B, Assad-Garcia N, Glass JI, Covert MW. A whole-cell computational model predicts phenotype from genotype. Cell 2012; 150: 389-401.

[71]   Gunawardera J. Silicon dreams of cells into symbols. Nature 2012; 30: 838-840.

[72]   Freddolino PL, Tavazoie S. The dawn of virtual cell biology. Cell 2012; 150: 248-250.

[73]   Tomita M. Whole-cell simulation: a grand challenge of the 21st century. Trends in Biotech 2001; 19: 205-210.

[74]   Covert MW, Palsson BØ. Transcriptional regulation in constraints-based metabolic models of *Escherichia coli*. J Biol Chem 2002; 277:28058–28064

[75]   Covert MW, Palsson BØ. Constraints-based models: regulation of gene expression reduces the steady-state solution space. J Theor Biol 2003; 221:309–325

[76]   Covert MW, Schilling CH, Famili I, Edwards JS, Goryanin II, Selkov E *et al*. Metabolic modeling of microbial strains *in silico*. Trends in Biochem Sci 2001; 26:179–186

[77]   Herrgård MJ, Fong SS, Palsson BØ. Identification of genome-scale metabolic network models using experimentally measured flux profiles. PLoS Comp Biol 2006; 2:676–686

[78]   Song HS, Morgan JA, Ramkrishna D. Systematic development of hybrid cybernetic models: application to recombinant yeast co-consuming glucose and xylose. Biotechnol & Bioeng 2009; 103:984–1002

[79]   Klipp E, Nordlander B, Kruger R, Gennemark P, Hohmann S. Integrative model of the response of yeast to osmotic shock, Nat Biotechnol 2005; 23, 975-982

[80]   Papin JA, Hunter T, Palsson BO. Reconstruction of cellular signaling networks and analysis of their properties. Nat Rev Mol Cell Biol 2005; 6, 99-111

[81]   Lee JM, Gianchandani EP, Eddy JA, Papin JA. Dynamic analysis of integrated signaling, metabolic, and regulatory networks. PLoS Comp Biol 2008; 4 (5) e1000086

[82]   Heinrich R, Rapoport TA. A linear steady-state treatment of enzymatic chains. General properties, control and effector strength. Eur J Biochem 1974; 42:89–95

[83]   van Riel NA. Dynamic modelling and analysis of biochemical networks: mechanism-based models and model-based experiments. Brief Bioinform 2006; 7:364–374

[84]   Heijnen JJ. Approximative kinetic formats used in metabolic network modeling. Biotechnol & Bioeng 2005; 91:534–545

[85]   Wu L, Wang WM, van Winden WA, van Gulik WM, Heijnen JJ. A new framework for the estimation of control parameters in metabolic pathways using lin-log kinetics. Europ J of Biochem 2004; 271:3348–3359

[86]   del Rosario RCH, Mendoza E, Voit EO. Challenges in lin-log modelling of glycolysis in *Lactococcus lactis*. Iet Systems Biology 2008; 2:136–149

[87]   Hatzimanikatis V, Emmerling M, Sauer U, Bailey JE. Application of mathematical tools for metabolic design of microbial ethanol production. Biotechnol & Bioeng 1998; 58:154–161

[88]   Voit Eberhard O Biochemical Systems Theory: A Review. ISRN Biomathematics 2013, 2013: 897658

[89]   Pozo C, Marín-Sanguino A, Alves R, Guillén-Gosálbez G, Jiménez L, Sorribas A. Steady-state global optimization of metabolic non-linear dynamic models through recasting into power-law canonical models. BMC Syst Biol 2011; 5:137

[90]   Sorribas A, Hernandez-Bermejo B, Vilaprinyo E, Alves R. Cooperativity and saturation in biochemical networks: A saturable formalism using Taylor series approximations. Biotechnol & Bioeng 2007; 97:1259–1277

[91]   Liebermeister W, Klipp E. Bringing metabolic networks to life: convenience rate law and thermodynamic constraints. Theor Biol Med Model 2006; 3:41

[92]   Ramkrishna D, Kompala DS, Tsao GT. Are Microbes Optimal Strategists. Biotechnol Prog 1987; 3:121–126

[93]   Varner J, Ramkrishna D. Metabolic engineering from a cybernetic perspective. 1. Theoretical preliminaries. Biotechnol Prog 1999; 15:407–425

[94]   Young JD. A system-level mathematical description of metabolic regulation combining aspects of elementary mode analysis with cybernetic control laws. PhD thesis, Purdue University 2005

[95]   Young JD, Henne KL, Morgan JA, Konopka AE, Ramkrishna D. Integrating cybernetic modeling with pathway analysis provides a dynamic, systems-level description of metabolic control. Biotechnol & Bioeng 2008; 100:542–559

[96]   Schuster S, Fell DA, Dandekar T. A general definition of metabolic pathways useful for systematic organization and analysis of complex metabolic networks. Nat Biotechnol 2000; 18:326–332

[97]   Kim JW, Dang CV. Multifaceted roles of glycolytic enzymes, Trends Biocem Sci 2005; 30: 142-150

[98]   Kim JI, Varner JD, Ramkrishna D. A hybrid model of anaerobic *E. coli* GJT001: combination of elementary flux modes and cybernetic variables. Biotechnol Prog 2008; 24:993–1006

[99]   Teusink B, Passarge J, Reijenga CA, Esgalhado E, van der Weijden CC, Schepper M *et al*. Can yeast glycolysis be understood in terms of *in vitro* kinetics of the constituent enzymes? Testing biochemistry. Eur J Biochem 2000; 267:5313–5329

[100]   Chakrabarti A, Miskovic L, Soh KC, Hatzimanikatis V. Towards kinetic modeling of genome-scale metabolic networks without sacrificing stoichiometric, thermodynamic and physiological constraints. Biotechnol J 2013; 8:1043–1057

[101]   Stanford NJ, Lubitz T, Smallbone K, Klipp E, Mendes P, Liebermeister W. Systematic construction of kinetic models from genome-scale metabolic networks. PLoS One 2013; 8:e79195

[102]   Savageau MA. Biochemical systems analysis. 3. Dynamic solutions using a power-law approximation. J Theor Biol 1970; 26:215–226

[103]   Dräger A, Kronfeld M, Ziller MJ, Supper J, Planatscher H, Magnus JB *et al*. Modeling metabolic networks in *C. glutamicum*: a comparison of rate laws in combination with various parameter optimization strategies. BMC Syst Biol 2009; 3:5

[104]   Costa RS, Machado D, Rocha I, Ferreira EC. Hybrid dynamic modeling of *Escherichia coli* central metabolic network combining Michaelis–Menten and approximate kinetic equations. Biosyst 2010; 100:150–157

[105]   Rizk ML, Liao JC. Ensemble modeling for aromatic production in *Escherichia coli*, PLoS One 2009; 4:e6903

[106]   Tan YK, Liao JC. Metabolic ensemble modeling for strain engineers. Biotechnol J 2012; 7:343–353

[107]   Contador CA, Rizk ML, Asenjo JA, Liao JC. Ensemble modeling for strain development of L-lysine-producing Escherichia coli, Metab Eng 2009; 11 (4-5) 221-233

[108]   Dean JT, Rizk ML, TanY, Dipple KM, Liao JC. Ensemble modeling of hepatic fatty acid metabolism with a synthetic glyoxylate shunt. Biophys J 2010; 98:1385–1395

[109]   Lee Y, Lafontaine Rivera JG, Liao JC. Ensemble modeling for robustness analysis in engineering non-native metabolic pathways. Metab Eng 2014; 25:63–71

[110]   Khazaei T, McGuigan A, Mahadevan R. Ensemble modeling of cancer metabolism. Front Physiol 2012; 3:135

[111]   Khodayari A, Zomorrodi AR, Liao JC, Maranas CD. A kinetic model of *Escherichia coli* core metabolism satisfying multiple sets of mutant flux data. Metab Eng 2014; 25:50–62

[112]   Rizzi M, Theobald U, Querfurth E, Rohrhirsch T, Baltes M, Reuss M. *In vivo* investigations of glucose transport in *Saccharomyces cerevisiae*. Biotechnol Bioeng 1996; 49: 316-327.

[113]   Rizzi M, Baltes M, Theobald U, Reuss M. *In vivo* analysis of metabolic dynamics in *Saccharomyces cerevisiae* Ⅱ. Mathematical model. Biotechnol Bioeng 1997; 55: 592-608.

[114]   Mashego MR, van Gulik WM, Vinke JL, Visser D, Heijnen J. *In vivo* kinetics with rapid perturbation experiments in *Saccharomyces cerevisiae* using a second-generation BioScope. Metab Eng 2006; 8: 370-383.

[115]   Hoque MA, Ushiyama H, Tomita M, Shimizu K: Dynamic responses of the intracellular metabolite concentrations of the wild type and *pykA* mutant *Escherichia* coli against pulse addition of glucose or NH_3 under those limiting continuous cultures. Biochem Eng J 2005; 26: 38-49.

[116]   Oldiys M, Takor R. Applying metabolic engineering techniques for stimulus-response experiments: chances and pitfalls. Adv Biochem Eng Biotechnol 2005; 92: 173-196.

[117]   Chassagnole C, Noisommit-Rizzi N, Schmid JW, Mauch K, Reuss M. Dynamic modeling of the central carbon metabolism of *Escherichia coli*. Biotechnol Bioeng 2002; 79: 53-73.

[118]   Kadir TA, Mannan AA, Kierzek AM, McFadden J, Shimizu K. Modeling and simulation of the main metabolism in *Escherichia coli* and its several single-gene knockout mutants with experimental verification. Microb Cell Fact 2010; 9: 88.

[119]   Kotte O, Zaugg JB, Heinemann M. Bacterial adaptation through distributed sensing of metabolic fluxes. Mol Sys Biol 2010; 6: 355.

[120]   Usuda Y, Nishio Y, Iwatani S, Van Dien SJ, Imaizumi A, Shimbo K, Kageyama N, Iwahata D, Miyano H, Matsui K. Dynamic modeling of *Escherichia coli* metabolic and regulatory systems for amino-acid production. J Biotechnol 2010; 147: 17-30.

[121]   Kremling A, Jahreis K, Lengeler JW, Gilles ED. The organization of metabolic reaction networks: A signal-oriented approach to cellular models. Metab Eng 2000; 2: 190-200.

[122]   Kremling A, Gilles ED. The organization of metabolic reaction networks. II. Signal processing in hierarchical structured functional units. Metab Eng 2001; 3: 138-150.

[123]   Kremlng A, Fischer S, Sauter T, Bettenbrock K, Gilles ED. Time hierarchies in the *Escherichia coli* carbohydrate uptake and metabolism. BioSystems 2004; 73: 57-71.

[124]   Sauter T, Gilles ED. Modeling and experimental validation of the signal transduction *via* the *Escherichia coli* sucrose phospho transferase system. J Biotech 2004; 110: 181-199.

[125]   Bettenbrock K, Fischer S, Kremling A, Jahreis K, Sauter T, Gilles ED. A quantitative approach to catabolite repression in *Escherichia coli*. J Biol Chem 2006; 281: 2578-2584.

[126]   Nishio Y, Usuda Y, Matsui K, Kurata H. Computer-aided rational design of the phosphotransferase system for enhanced glucose uptake in *Escherichia coli*. Mol Syst Biol 2008; 4: 160.

[127]   Kremling A, Bettenbrock K, Gilles ED. Analysis of global control of *Escherichia coli* carbohydrate uptake. BMC Syst Biol 2007; 1: 42.

[128]   Bruggeman FJ, Boogerd FC, Westerhoff HV, The multifarious short-term regulation of ammonium assimilation of *Escherichia coli:* dissection using an *in silico* replica, FEBS J 2005; 272:1965-1985

[129]   Yuan J, Doucette CD, Fowler WU, Feng XJ, Piazza M, Rabitz HA *et al.* Metabolomics-driven quantitative analysis of ammonia assimilation in *E. coli*. Mol Syst Biol 2009; 5:302

[130]   Doucette CD, Schwab DJ, Wingreen NS, Rabinowitz JD. Alpha-ketoglutarate coordinates carbon and nitrogen utilization *via* enzyme I inhibition. Nat Chem Biol 2011; 7: 894–901

CHAPTER 2

Basis for Biosystems Analysis

Abstract: Basic notion and systems analysis method is briefly explained for the preparation to the understanding of the later chapters. Non-linear and linear systems equations are explained, where the standard formulation and its representation is useful for the basis of various types of modeling. Transfer function is explained for the linear system as the input-output representation. Basic graph theory is explained with its applications, where it is important to analyze large-scale metabolic reaction networks. Data analysis such as regression analysis and the principal component analysis (PCA) are also briefly explained, where it is important for analyzing the experimental data and experimental design in relation to modeling.

Keywords: System equation, transfer function, graph theory, signal flow diagram, feedback control, regression analysis, principal component analysis (PCA).

INTRODUCTION

Before going into the explanation on the modeling, dynamics, optimization, and computer simulation, basic notion and fundamentals for systems analysis is explained for understanding the later chapters. Namely, here we consider mainly the linear systems such as the system representation, block diagram, transfer function, graph theory and its application, step and impulse responses, frequency response, and data analysis such as regression analysis and principal component analysis [1-7].

SYSTEMS REPRESENTATION

Process Equation and Transfer Function

As shown in Fig. **1**, let u be the **input variable**, **operation variable**, or **control variable**, x be the **state variable**, y be the **output variable** or **measurement variable**. In general, the dynamics of the process system may be expressed as

$$\frac{dx}{dt} = f_i(x_1, x_2, ..., x_n, u) \qquad (i = 1, 2, ..., n) \tag{1a}$$

$$y = g(x_1, x_2, ..., x_n) + d \tag{1b}$$

where these are called as **system equations**, where x_i is the state variable. In general only some of the state variables can be observed or measured.

Kazuyuki Shimizu and Yu Matsuoka

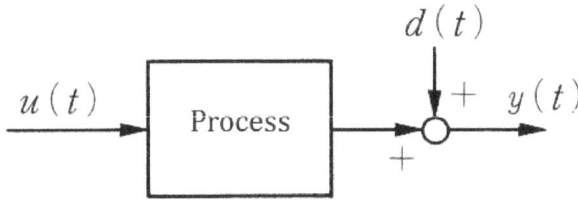

Figure 1: System representation.

Modeling is in some sense to find the appropriate function f_i in Eq.(1). However, the modeling is not easy in the sense that the model inherently has the uncertainty due to neglecting part of the real world. Therefore, the analysis based on the model developed must be careful by taking into account such uncertainty, and the applicability or robustness must be properly evaluated in practice. Moreover, in many cases, the system is **non-linear** and **time-variant**, and thus the analysis must be careful.

Consider the simple linear system as expressed as

$$\frac{dx}{dt} = a_1 x_1 + a_2 x_2 + \bullet \bullet \bullet + a_n x_n + b_1 u \qquad (i = 1,2,...,n)$$

$$\tag{2a}$$

$$y = c_1 x_1 + c_2 x_2 + \bullet \bullet \bullet + c_n x_n + d \tag{2b}$$

where this is the **single input** (u) and **single output** (y) (**SISO**) system. In the case where **multiple-inputs and multiple outputs** (**MIMO**), the interaction must be considered.

The above equations can be expressed in vector notation as

$$\frac{dx}{dt} = Ax + bu \tag{3a}$$

$$y = cx \tag{3b}$$

where

$$x^T = [x_1, x_2, ..., x_n], \qquad b^T = [b_1, b_2, ..., b_n], \qquad c = [c_1, c_2, ..., c_n]$$

$$A = \begin{bmatrix} a_{11} & . & . & a_{1n} \\ . & . & . & . \\ . & . & . & . \\ a_{n1} & . & . & a_{nn} \end{bmatrix}$$

where "^{T}" indicates the transpose such that $(a_{ij})^T=(a_{ji})$, where (a_{ij}) means the matrix with (i,j)th component of matrix A to be a_{ij}. Here, the disturbance d in Eq.(2b) was set to 0 for simplicity.

The system representation as shown in Eqs.(1), (2), and (3) is called as the **state space description**.

Consider the Laplace transformation (see **Appendix A**) of Eq.(3). The input-output relationship can be expressed as

$$\tilde{y}(s) = G(s)\tilde{u}(s) \tag{4a}$$

where

$$G(s) \equiv c(sI - A)^{-1}b \tag{4b}$$

where $\tilde{y}(s)$ and $\tilde{u}(s)$ are the Laplace transformed variable of y(t) and u(t), respectively. In Eq.(4), I is the unit matrix, and s is the Laplace variable, and "$^{-1}$" represents the inverse (matrix). In Eq.(4b), G(s) is called as **transfer function**.

Transfer Function of CSTF

Consider the continuous stirred tank fermenter as shown in Fig. **2**, where F is the flow rate [l/h], V is the volume [l], C_0 and C are the concentrations of the component of concern [g/l] in the feed and in the fermenter, respectively.

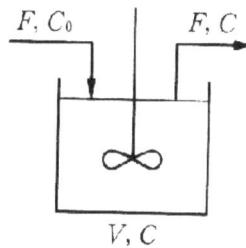

Figure 2: Continuous stirred tank fermentor.

The system equation is expressed by the mass balance as

$$V\frac{dC}{dt} = F(C_0 - C) + rV \tag{5}$$

where r is the reaction rate [g/l.h]. Consider the 1st order reaction as expressed by r=kC. By dividing Eq.(5) by F, we have

$$\theta\frac{dC}{dt} = (-1 + k\theta)C + C_0 \tag{6}$$

where k is the reaction rate constant, θ is defined as V/F, and is called as **residence time** [h]. If k and θ are constant, Eq.(6) becomes a linear equation, and it can be solved by the Laplace transformation as

$$s\tilde{C}(s) - C(0) = (-1 + k\theta)\tilde{C}(s) + \tilde{C}_0(s) \tag{7a}$$

where $\tilde{C}(s)$ and $\tilde{C}_0(s)$ are the transformed variables of C and C_0, respectively. Suppose that C(0)=0 without loss of generality, then the transfer function can be obtained as

$$G(s) \equiv \frac{\tilde{C}(s)}{\tilde{C}_0(s)} = \frac{1}{s\theta + 1 - k\theta} \tag{7b}$$

where this system is called as the 1st order delay system, since the denominator of Eq.(7b) is the 1st order with respect to s.

In the similar way, the transfer function of the system as shown in Fig. 3 can be derived, where n CSTRs are connected in series.

$$G(s) = \frac{1}{(s\theta_1 + 1 - k\theta_1)(s\theta_2 + 1 - k\theta_2)........(s\theta_n + 1 - k\theta_n)} \tag{8}$$

where $\theta_i \equiv V_i/F$ (i=1,2,...,n).

In general, the system as expressed by the following ordinary differential equation is the n-th order delay system:

$$\frac{d^{(n)}y}{dt^n} + a_{n-1}\frac{d^{(n-1)}y}{dt^{n-1}} + ... + a_0 y = b_m\frac{d^{(m)}u}{dt^m} + ... + b_0 u \qquad (n > m) \tag{9}$$

where "$^{(i)}$" means the i-th derivative. Based on the assumption that

$$\frac{d^{(n-1)}y(0)}{dt^{n-1}} = \frac{d^{(n-2)}u}{dt^{n-2}} = ... = y(0) = 0$$

$$\frac{d^{(m-1)}u(0)}{dt^{m-1}} = \frac{d^{(m-2)}u}{dt^{m-2}} = ... = u(0) = 0$$

the transfer function can be obtained as

$$G(s) \equiv \frac{\tilde{y}(s)}{\tilde{u}(s)} = \frac{b_m s^m + ...+ b_0}{s^n + a_{n-1}s^{n-1} + ...+ a_0}$$

(10)

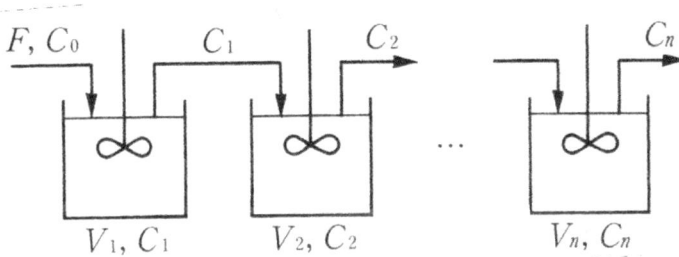

Figure 3: Multiple continuous stirred tank fermentor.

Note that this system can be also expressed by the state space description as

$$A = \begin{bmatrix} 0 & 1 & 0 & & 0 \\ 0 & 0 & 1 & & 0 \\ & & & & \\ 0 & 0 & 0 & & 1 \\ -a_0 & -a_1 & -a_2 & & -a_{n-1} \end{bmatrix}$$

where this form of A-matrix is called as **companion matrix**.

2.3. Block Diagram

The process system as expressed by Eq.(8) can be expressed as shown in Fig. **4**, where this figure is called as **block diagram**.

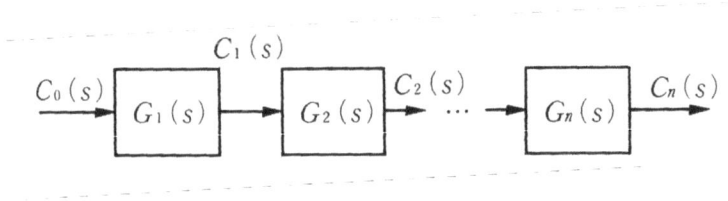

Figure 4: Block diagram for multiple CSTFs.

The block diagram is consisted of three elements as given in Fig. **5**. For example, consider the typical feedback system as shown in Fig. **6**, and apply the rules of Fig. **5**.

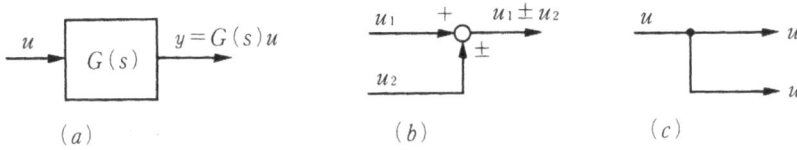

Figure 5: Three elements for the block diagram.

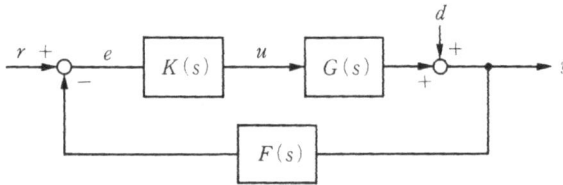

Figure 6: Feedback control system.

Then we have the following relationship:

$$y = G(s)u + d \tag{11a}$$

$$u = K(s)e \tag{11b}$$

$$e = r - F(s)y \tag{11c}$$

By substituting Eq.(11c) into Eq.(11b), and then substituting the resulting equation into Eq.(11a), we have

$$y = G(s)K(s)\{r - F(s)y\} + d \tag{12a}$$

There are two inputs r and d, and one output y. Then we have the relationships between r and y, and also d and y, and those are expressed as

$$G_r(s) = \frac{G(s)K(s)}{1 + G(s)K(s)F(s)} \tag{12b}$$

$$G_d(s) = \frac{1}{1 + G(s)K(s)F(s)} \tag{12c}$$

In Fig. **6**, r is called as the **set point** for y to be controlled at r, and d is called as **disturbance**. The overall transfer functions as expressed by Eq.(12b) and (12c) are called as the **closed-loop transfer functions**.

GRAPH THEORETIC APPROACH

The graph representation of the system is useful for analyzing the network such as the metabolic pathway network, gene network *etc.* often appear in bio-systems. Consider here the basis of graph theory and its applications [8-10].

Basic Notion of the Graph

As shown in Fig. **7**, graph G consists of the set V of **vertexes** or **nodes**, and the set E of **edges**. Namely, $G=(V,E)$, where $V=\{v_1, v_2, \ldots, v_n\}$, and $E=\{e_1, e_2, \ldots, e_m\}$. In Fig. **7**, v_1 and v_2 are called as **adjacent** by e_1. If the adjacent nodes are the same such as e_4, it is called as **self-loop**. The number of edges that connect to the node is called as the **degree** of the node, and is expressed as $d(\cdot)$. For example, in the case of Fig. **7**, $d(v_1)=1$ and $d(v_2)=3$ and so on.

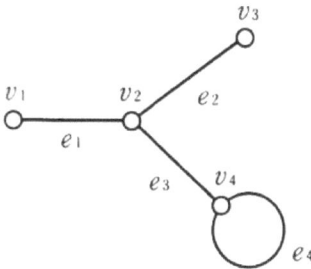

Figure 7: Typical graph.   **Figure 8:** Path and circuit.

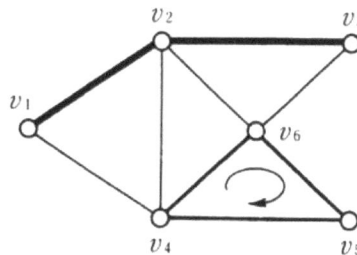

Consider the pathway from v_1 to v_3 in Fig. **8**, where there are several routes such as the one starting with v_1 *via* v_2 to v_3, or from v_1 *via* v_4 and v_6 and reach to v_3 and so on. Let us call the **path** for the route that goes through nodes only once, and call **circuit** or cycle for the pathway such that both starting and terminal nodes are identical such that $v_4 \rightarrow v_6 \rightarrow v_5 \rightarrow v_4$. If all the nodes are connected, and any nodes can reach to any other nodes, then this is called as **connected graph**. As shown in Fig. **9**, the graph is not necessarily connected. In such a case, we call the **component** for each sub-graph. If $G'=(V',E')$ satisfies $V' \subset V$ and $E' \subset E$, then G' is the sub-graph of G, where G_1 and G_2 are the sub-graphs of G in the case of Fig. **9**.

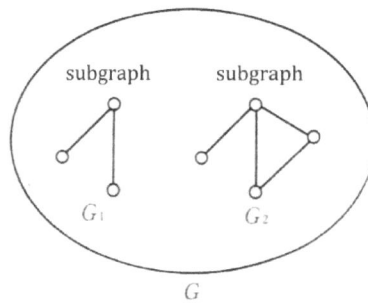

Figure 9: Subgraphs G_1 and G_2 contained in G.

Types of graph

Euler Graph

Let us call the **Euler graph** for the one that has the following characteristics: Namely, starting with any nodes, one can go through all the nodes once, and return to the original node. For example, Fig. **10a** is the Euler graph, while Fig. **10b** is not.

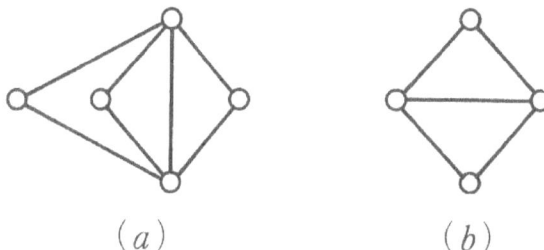

(a) (b)

Figure 10: Euler graph (a), and not Euler graph (b).

This graph is important in practice, where historically this application is known as **Konigsberg bridge problem** as shown in Fig. **11a**, where the problem is to find the possibility if starting with any place of A,B,C, or D, one can go through 7 bridges only once, and return to the original place. This problem was solved by Leonhard Euler in 1936 based on the graph theory. The following theorem is useful to check if the graph is the Euler graph or not:

Theorem 1: The necessary and sufficient condition for the graph to be the Euler graph is that the degrees of all the nodes must be even.

For example, the degrees of all the nodes are either 2 or 4 and even for Fig. **10a**, and thus this is the Euler graph, whereas the graph as shown in Fig. **10b** contains the nodes whose degree is 3, and thus this is not the Euler graph. In the case of Konigsberg bridge problem, Fig. **11a** can be represented by the graph as shown in Fig. **11b**, and it is easy to see that this is not the Euler graph by applying the above theorem.

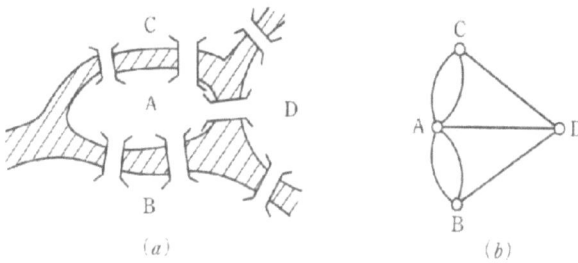

Figure 11: Konigsberg problem and its graph representation.

Regular Graph and Complete Graph

Regular graph is defined as the graph where the degrees of all the nodes are the same for all the nodes as shown in Fig. **12a**. If all the nodes are connected to all the other nodes, it is called as **complete graph** as shown in Fig. **12b**. Complete graph is also the regular graph by definition.

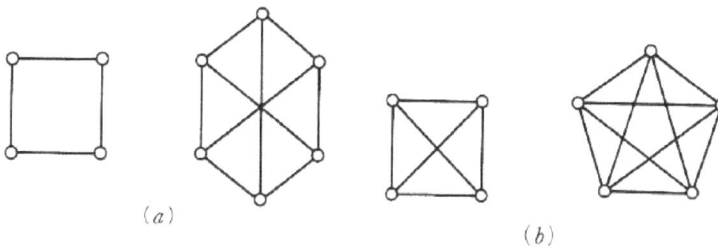

Figure 12: Regular graph (a) and complete graph (b).

Isomorphic Graph

As shown in Fig. **13**, two graphs are called as isomorphic if the connection characteristics or the graph properties are the same.

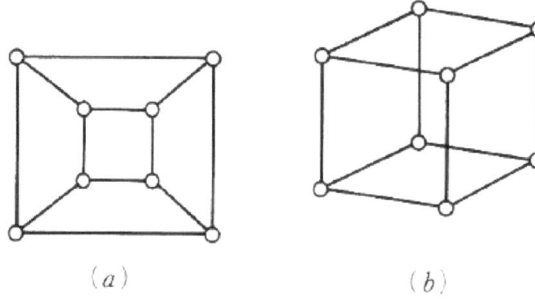

Figure 13: Isomorphic graphs.

Hamilton Circuit

As shown in Fig. **14**, the circuit that goes through all the nodes once and returns to the original node is called as **Hamilton circuit**, where there is no need to go through all the edges.

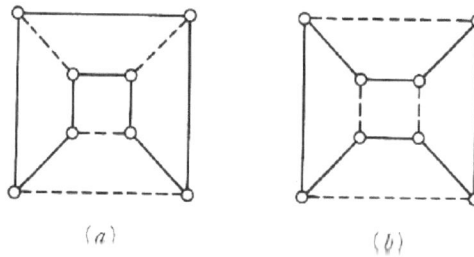

Figure 14: Hamilton circuit.

Tree Graph

The graph that does not contain circuit or closed loop as shown in Fig. **15a** is called as **tree**. Tree has the following properties:

☐   Connected graph

☐   Does not contain circuit

☐   The number of edges is N-1 with N nodes

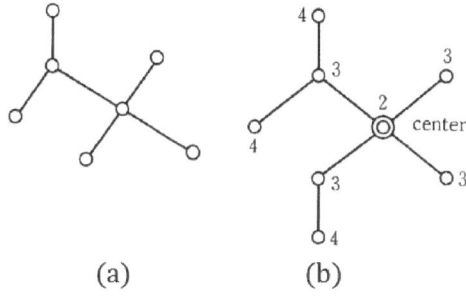

(a)   (b)

Figure 15: Tree graph (a) and eccentricity (b).

Let $d(v_i,v_j)$ be the distance between v_i and v_j, where it is defined as the minimum number of edges among the pathways connecting v_i and v_j. The distance must satisfy the following conditions:

☐ $d(v_i,v_j) > 0$ for $v_i \neq v_j$, $d(v_i,v_i) = 0$

☐ $d(v_i,v_j) = d(v_j,v_i)$

☐ $d(v_i,v_j) \leq d(v_i,v_k) + d(v_k,v_j)$

Consider the node v_i in the graph, and measure the distance to all the other nodes $d(v_i,v_j)$. Let the **eccentricity** of v be E(v), and is defined as the maximum distance as shown in Fig. **15b** for each node in the graph. Moreover, let the node(s) that has the minimum eccentricity be called as **center** (or bi-centers *etc.*) of the graph as shown in Fig. **15b**, and its eccentricity is called as **radius** of the graph.

As the special tree graph as shown in Fig. **16**, it is called as **binary tree**, where it contains a root with degree of 2 and the others with degree of either 1 or 3. This graph is often used for the decision making process.

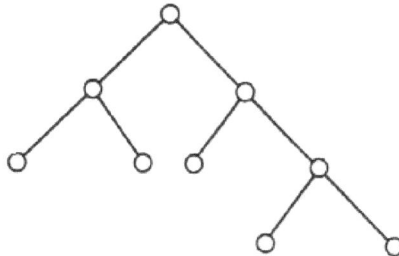

Figure 16: Binary tree.

Consider the graph as shown in Fig. **17**, where among the sub-graphs which contain all the nodes, the tree graph as shown by the bold solid line in the figure is called as **spanning tree** T. The edges belonging to the spanning tree is called as **branches**, while the other edges are called as **chords**. In the case of Fig. **17**, T={1,3,4,5,8}. There are many other spanning trees such as T={1,3,4,5,7} *etc*. Let M be the number of edges, and let N be the number of nodes of the graph, then the number of chord is M-N-1. In the case of Fig. **17**, 8-6+1=3, where chord ={2,6,7}.

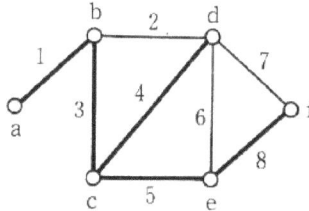

Figure 17: Spanning tree.

If we add one chord to the spanning tree, one circuit is formed, where it is called as **fundamental circuit**. In the example of Fig. **17**, {2,3,4},{4,5,6}, and {4,5,8,7} form the fundamental circuits, whereas {6,7,8} is not the fundamental circuit.

As mentioned before, there are several spanning trees for the given graph, and how can we found all the set of the spanning trees? For this, first find one spanning tree as shown for example in Fig. **18**. Then add one chord to form fundamental circuit, from which other spanning trees can be found by assigning each edge in the circuit as chord (**cyclic interchange**). In the above example, the fundamental circuit {4,5,6} as shown in Fig. **17** generates other spanning trees as shown in Fig. **18**. The similar operation can be made for the other fundamental circuits to find all the other spanning trees.

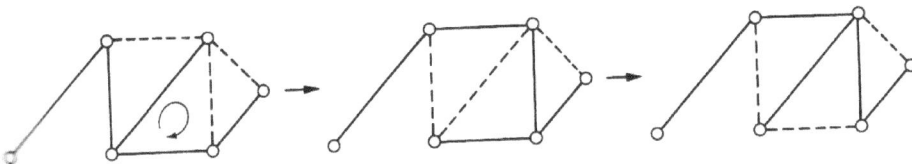

Figure 18: Cyclic interchange.

Cut-Set and Connectivity

The **cut-set** K_i of the connected graph G is the set of edges, where the removal of the set of edges makes the graph disconnected. In particular, the minimum set is called as the cut-set among the edges. For example, as shown in Fig. **19a**, {2,4,5}, {1}, {7,8}, {3,4,5} are the cult set, whereas {6,7,8} is not the cut-set, since this is not the minimum set, where the minimum set is {7,8}.

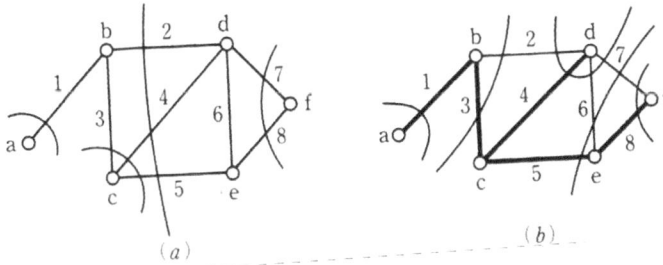

Figure 19: Cut set (a) and fundamental cut set.

Among the number of edges contained in the cut-sets, the minimum number is called as the **edge connectivity** of the graph. In the example as shown in Fig. **19a**, it is 1.

Consider the relationship between spanning tree and the cut-set as shown in Fig. **19b**, where the following theorem holds:

Theorem 2: All the cut-sets of G must have at least one branch for all the spanning trees.

All the branches of the spanning tree T are the cut-set, where each cut-set disconnects the tree into two connected graphs. For example as shown in Fig. **19b**, if {1} was removed, T is disconnected to {a} and {b,c,d,e,f}, if {4} was removed, T is disconnected to {a,b,c,e,f} and {d}, and if {5} was removed, T is disconnected to {a,b,c,d} and {e,f}.

Let the cut-set which contains only one branch of the spanning tree be called as **fundamental cut-set**. There exists N-1 fundamental cut-sets for the graph with N nodes. If we add one chord to the fundamental cut-set, one circuit is formed, where this is the fundamental circuit as mentioned before.

Theorem 3: For the given spanning tree T, the chord that forms the fundamental circuit C_i occurs only in the fundamental cut set associated with the branches belonging to C_i.

For example of Fig. **19b**, chord 6 for the fundamental circuit with the branches of T such as 4 and 5. In this case, the chord occurs in the two fundamental cut-sets such as {2,4,6,7} and {5,6,7}associated with the above two branches.

Theorem 4: For the given tree, the branch b_i that forms the fundamental cut-set K_i is contained in all the fundamental circuits associated with the chord for K_i.

For example of Fig. **19b**, edge 5 is the branch that determines the fundamental cut-set {5,6,7}, and it occurs only in the fundamental circuit {4,5,6} and {4,5,8,7} in association with chords 6 and 7, respectively.

Directed Graph

As shown in Fig. **20**, the graph such that all the edges have the directions is called as **directed graph**, **digraph**, or **oriented graph**. Here, the node v_i that the arrow of edge e_k is originated is called as **initial vertex** (or node), and the terminal node or directed node is called as the **terminal vertex** (or node).

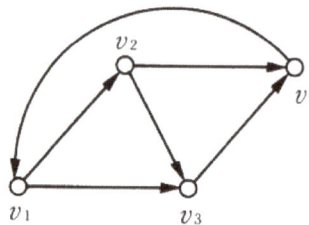

Figure 20: Directed graph.

Let the number of edges directing toward some node v_i is called as the **in-degree** (in-valence or inward demi-degree), and is expressed as $d^+(v_i)$, while the number of edges directing out of the node is called as **out-degree** (out-valence or outward demi-degree), and is expressed as $d^-(v_i)$. In the example of Fig. **20**, $d^+(v_1)=1$, $d^+(v_2)=1$, $d^+(v_3)=2$, $d^+(v_4)=2$, $d^-(v_1)=2$, $d^-(v_2)=2$, $d^-(v_3)=1$, $d^-(v_4)=1$.

Then the following relationship holds:

$$\sum_{i=1}^{N} d^+(v_i) = \sum_{i=1}^{N} d^-(v_i) \tag{13}$$

In the case of Fig. **20**, this value is 6.

Consider next the set X composed of x_i such that $X=\{x_1,x_2,\ldots,x_n\}$. Let the ordered pair x_i and x_j be defined as the magnitudes, or cause-effect relationship *etc.*, and be expressed as (x_i, x_j). Then the following set is called as direct product set:

$$X \times X = \{(x_i, x_j): x_i, x_j \in X\}$$

and the relationship between the two components in X is called as **binary relation** R. The binary relation R on X is the subset of $X \times X$ such as $R \subset X \times X$. If $(x_i, x_j) \in R$, then x_i has the relationship R with x_j, and express $x_i R x_j$. In the case where $(x_i, x_j) \notin R$, x_i does not have the relationship R with x_j, and is expressed as $x_i \overline{R} x_j$. For example, If R is the relationship "larger than", then this relationship can be expressed for $\{3,4,7,5,8\}$ as the graph as shown in Fig. **21**.

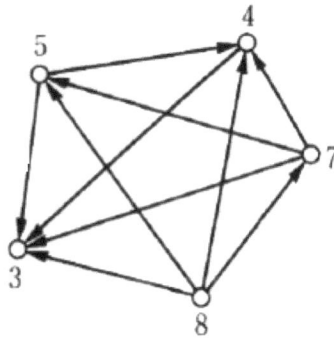

Figure 21: Graph representation for binary relation.

Graph and its Matrix Representation

The graph represents the connective structure of the system, and is useful for the visual inspection. However, computational treatment is inevitable as the graph becomes complicated often encountered in the metabolic network pathways. Let a_{ij} be the (ij)th component of the A matrix, where if there exists an edge from v_i to v_j, then $a_{ij}=1$, otherwise $a_{ij}=0$. For example as shown in Fig. **22**,

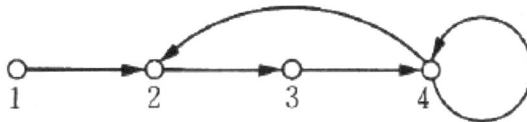

Figure 22: Directed graph.

A matrix becomes as

$$A = \begin{bmatrix} 0 & 1 & 0 & 0 \\ 0 & 0 & 1 & 0 \\ 0 & 0 & 0 & 1 \\ 0 & 1 & 0 & 1 \end{bmatrix}$$

where the raw indicates "from" and the column indicates "to". In the above example, $a_{12}=1$, which indicates that there is a directed edge from node v_1 to v_2. Moreover, the diagonal element such as $a_{44}=1$, which means that there is a self-loop at the node v_4. In this way, there is a one-to-one relationship between the graph and the matrix as shown above. This kind of matrix is called as **adjacency matrix** or **incidence matrix** (sometimes called also as associated matrix or relation matrix).

In order to find the properties of the network as represented by the graph, the matrix operation is required, where the Boolean algebra is used as follows:

$$x + y = \max \ (x, y), \qquad x - y = \min \ (x, y) \tag{14}$$

For example

$$0 + 1 = \max \ (0,1) = 1, \qquad 1 + 1 = \max \ (1,1) = 1, \qquad 0 \times 1 = \min \ (0,1) = 0$$

and so on, and the matrix operation is made as follows:

$$\begin{bmatrix} 0 & 1 & 1 \end{bmatrix} \begin{bmatrix} 1 \\ 0 \\ 1 \end{bmatrix} = 0 \times 1 + 1 \times 0 + 1 \times 1 = 0 + 0 + 1 = 1$$

Consider the graph as shown in Fig. **23a**.

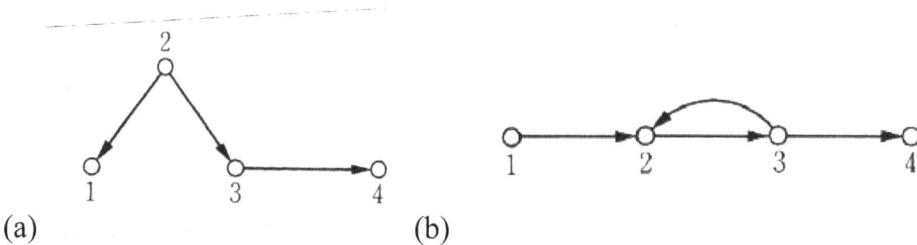

(a)   (b)

Figure 23: Directed graph without (a) and with (b) recycle loop.

The adjacency matrix becomes as

$$A = \begin{bmatrix} 0 & 0 & 0 & 0 \\ 1 & 0 & 1 & 0 \\ 0 & 0 & 0 & 1 \\ 0 & 0 & 0 & 0 \end{bmatrix}$$

Moreover, A^2 and A^3 become as the followings by applying Boolean algebra:

$$A^2 = \begin{bmatrix} 0 & 0 & 0 & 0 \\ 0 & 0 & 0 & 1 \\ 0 & 0 & 0 & 0 \\ 0 & 0 & 0 & 0 \end{bmatrix}, \quad A^3 = \begin{bmatrix} 0 & 0 & 0 & 0 \\ 0 & 0 & 0 & 0 \\ 0 & 0 & 0 & 0 \\ 0 & 0 & 0 & 0 \end{bmatrix}$$

Consider the relationship between A^m and the graph properties. If there exist edges going from node i to j in m steps, then $a_{ij}^{(m)}$ of A^m is 1, where $a_{ij}^{(m)}$ is the (i,j)the component of A^m. In the case of A^2, $a_{24}^{(2)}=1$, which corresponds to the fact that there exists a pathway from node 2 to 4 in two steps. In the case of A^3, all the components are 0, which indicates that there is no pathway with 3 steps. This means that if the graph does not contain recycle, A^n becomes 0 matrix as n increases.

Consider next the case where the recycle is contained as shown in Fig. **23b**, where the adjacency matrix becomes as

$$A = \begin{bmatrix} 0 & 1 & 0 & 0 \\ 0 & 0 & 1 & 0 \\ 0 & 1 & 0 & 1 \\ 0 & 0 & 0 & 0 \end{bmatrix}$$

In the similar way as done before, A^2 and A^3 become as

$$A^2 = \begin{bmatrix} 0 & 0 & 1 & 0 \\ 0 & 1 & 0 & 1 \\ 0 & 0 & 1 & 0 \\ 0 & 0 & 0 & 0 \end{bmatrix}, \quad A^3 = \begin{bmatrix} 0 & 1 & 0 & 1 \\ 0 & 0 & 1 & 0 \\ 0 & 1 & 0 & 1 \\ 0 & 0 & 0 & 0 \end{bmatrix}$$

where $a_{22}^{(2)}=a_{33}^{(2)}=1$, which indicates that the pathways that start from node 2 and 3 are returned to the original position in two steps.

Consider then the following operation: $R=A+A^2+\ldots+A^n$ until the matrix R does not change, where such matrix is shown as R^*, and is called as **reachability matrix**. As for the above example,

$$R_2 = A + A^2 = \begin{bmatrix} 0 & 1 & 1 & 0 \\ 0 & 1 & 1 & 1 \\ 0 & 1 & 1 & 1 \\ 0 & 0 & 0 & 0 \end{bmatrix}, \qquad R_3 = A + A^2 + A^3 = \begin{bmatrix} 0 & 1 & 1 & 1 \\ 0 & 1 & 1 & 1 \\ 0 & 1 & 1 & 1 \\ 0 & 0 & 0 & 0 \end{bmatrix}$$

$$R_4 = A + A^2 + A^3 + A^4 = \begin{bmatrix} 0 & 1 & 1 & 1 \\ 0 & 1 & 1 & 1 \\ 0 & 1 & 1 & 1 \\ 0 & 0 & 0 & 0 \end{bmatrix} \qquad \therefore R^* = \begin{bmatrix} 0 & 1 & 1 & 1 \\ 0 & 1 & 1 & 1 \\ 0 & 1 & 1 & 1 \\ 0 & 0 & 0 & 0 \end{bmatrix}$$

where if (i,j)th component of R^* is 1, then there exists a pathway from node i to j. In the above example, from node 1 to 2,3, and 4, from node 2 to2,3, and 4, and from node 3 to 2,3, and 4. Moreover, it indicates that there exist no pathway that can reach to node 1, and reach from node 4.

Consider Fig. **24a**, where there exist several recycle loops. The sub-graphs which contain such nodes as v_2 and v_3, v_4 and v_5, and v_4, v_5, and v_6 are called as **recycle net**. In particular, the recycle nets that contain such nodes as v_2 and v_3, and v_4, v_5, and v_6 are called as **maximal recycle net**. The maximal recycle net can be detected by manipulation of the reachability matrix R^* as follows:

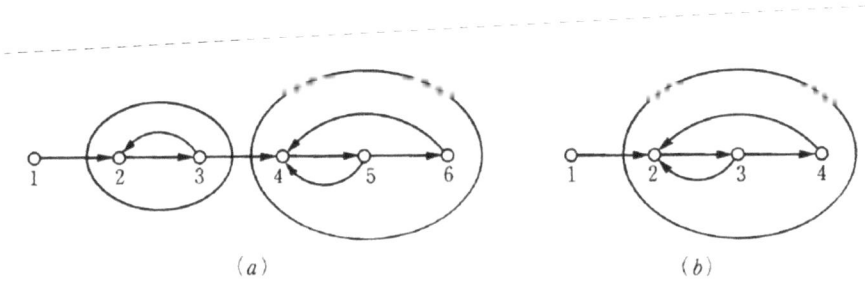

Figure 24: Recycle net and maximal recycle net.

Consider for example the graph as shown in Fig. **24b**, where the reachability matrix and its transpose are expressed as

$$R* = \begin{bmatrix} 0 & 1 & 1 & 1 \\ 0 & 1 & 1 & 1 \\ 0 & 1 & 1 & 1 \\ 0 & 0 & 0 & 0 \end{bmatrix}, \quad (R*)^T = \begin{bmatrix} 0 & 0 & 0 & 0 \\ 1 & 1 & 1 & 1 \\ 1 & 1 & 1 & 1 \\ 1 & 1 & 1 & 1 \end{bmatrix}$$

Consider next the operation of $R*\cap(R*)^T$, where the matrix manipulation for $\cap$ is defined as follows: Namely, the (i,j)th component of X for $X=B\cap C$ is expressed as $x_{ij}=\min(b_{ij},c_{ij})$, where b_{ij} and c_{ij} are the (i,j)th component of B and C, respectively. Thus $R*\cap(R*)^T$ can be computed as

$$R*\cap(R*)^T = \begin{bmatrix} 0 & 0 & 0 & 0 \\ 0 & 1 & 1 & 1 \\ 0 & 1 & 1 & 1 \\ 0 & 0 & 0 & 0 \end{bmatrix}$$

where the node v_i and v_j are included in the maximal recycle net if (i,j)th element is 1.

Consider another way of matrix representation for the graph as shown in Fig. **25**,

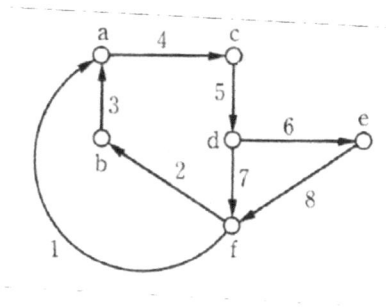

Figure 25: Directed graph with different labeling.

and consider the following matrix:

$$M'* = \begin{bmatrix} 1 & & 1 & -1 & & & \\ & 1 & -1 & & & & \\ & & 1 & -1 & & & \\ & & & 1 & -1 & -1 & \\ & & & & 1 & & -1 \\ -1 & -1 & & & & 1 & 1 \end{bmatrix}$$

where each raw corresponds to the node, while the column corresponds to the edge. In the matrix M', 1 indicates the "incoming to", while -1 indicates the "outgoing from". For example, edge 1 goes from node f to node a, and thus the 1st raw (node a) and the 1st column is 1, while the 6th raw (node f) and the 1st column is -1, and so on. This matrix M' is also called as incidence matrix, and this has the following properties:

- Each column contains only one "+1" and "-1"

- The number of +1 (or -1) in a raw corresponds to in-degree (our-degree).

- Self-loop cannot be represented.

- (M')^{T}**1**=0 and thus the raw is linearly dependent, and the rank of M' is N-1 or less (N being the number of nodes of the graph), where **1**≡[1, 1, ..., 1]^{T}.

Consider again the graph as shown in Fig. **25**, where the **circuit matrix** is given as

$$C = \begin{bmatrix} 1 & -1 & -1 & & & & & & & \\ & 1 & & 1 & 1 & 1 & & 1 & & \\ & & & & & 1 & -1 & -1 & & \\ 1 & & & & 1 & 1 & & 1 & & \\ 1 & & & & 1 & 1 & 1 & & 1 & \\ & 1 & & 1 & 1 & 1 & 1 & & & \end{bmatrix}$$

Here, if the jth edge is contained in the ith circuit, (i,j)th element is 1, while its direction is the reverse to that of the circuit, its element is -1.

As for the incidence matrix M and the circuit matrix C, the following relationships hold:

$$MC^{T} = 0 \tag{15a}$$

$$CM^{T} = 0 \tag{15b}$$

Let us define the fundamental circuit matrix as

$$C = \begin{bmatrix} T \mid I \end{bmatrix}$$

where

$$C = \left[\begin{array}{cccccc|ccc} -1 & -1 & & & & 1 & & \\ 1 & 1 & 1 & 1 & & & 1 & \\ 1 & 1 & 1 & 1 & 1 & & & 1 \end{array}\right]$$

The fundamental cut-set matrix is also defined as

$$K = \begin{bmatrix} I_{N-1} \mid B \end{bmatrix}$$

and

$$K = \begin{bmatrix} 1 & & & & & 1 & -1 & -1 \\ & 1 & & & & 1 & -1 & -1 \\ & & 1 & & & & -1 & -1 \\ & & & 1 & & & -1 & -1 \\ & & & & 1 & & & -1 \end{bmatrix}$$

where the following equations hold:

$$KC^T = 0, \qquad CK_T = 0, \qquad B = -T^T$$

Signal Flow Diagram and its Applications

Consider **signal flow diagram (SFD)** or **signal flow graph (SFG)** as shown in Fig. **26**, where this is essentially equivalent to the block diagram as mentioned before. As shown in Fig. **26**, the input x_1 is converted to x_2 by g as

$$x_1 = gx_2 \tag{16}$$

where g is called as **transmittance**, and corresponds to the transfer function in the block diagram.

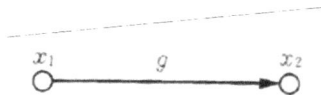

Figure 26: Signal flow diagram.

Fig. **27** shows the principles of operation. Namely, $x_2=g_2g_1x_1$ for the series connection, while $x_2=(g_1+g_2)x_1$ for the parallel connection. In the case of Fig. **27c**, we have the following equations:

$$z = g_1 x_1 + g_3 z$$

$$x_2 = g_2 z$$

from which we have

$$x_2 = \frac{g_2 g_1}{1-g_3} x_1$$

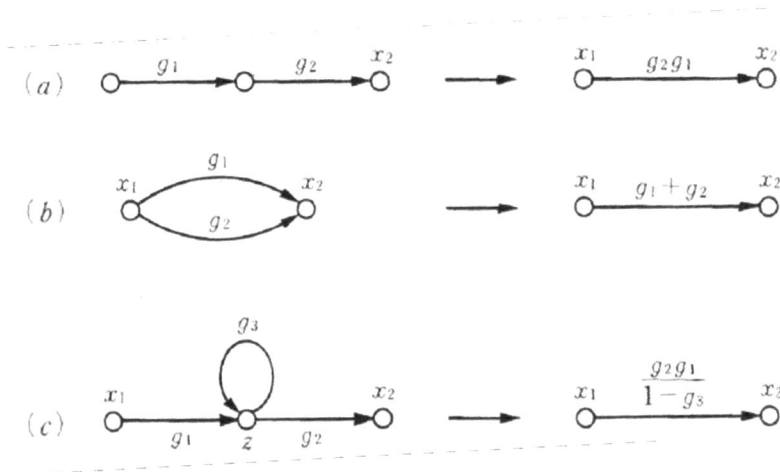

Figure 27: Transmittance of basic SFD.

Consider the SFD as shown in Fig. **28a**, and try to simplify it by applying the rules as given in Fig. **27**. For this, consider first removing the node ② as shown in Fig. **28b**, and in turn remove node ③, and finally obtain the simplified SFD as shown in Fig. **28d**.

Consider the following linear algebraic equations, and express as SFD to find the input-output relationship:

$$bx_1 - x_2 + fx_3 + ex_4 = 0 \tag{17a}$$

$$gx_1 + dx_3 - x_4 = 0 \tag{17b}$$

$$cx_2 - x_3 = 0 \tag{17c}$$

From Eq.(17c), $x_3 = cx_2$. This can be expressed as Fig. **29a**. Then $x_4 = gx_1 + dx_3$ holds from Eq.(17b), and this relationship can be expressed as Fig. **29b**. From Eq.(17a), $x_2 = bx_1 + fx_3 + ex_4$, and thus the SFD of the whole equations can be represented as shown in Fig. **29c**. From this SFD, the input-output relationship from x_1 to x_4 can be obtained by simplifying the SFD as mentioned above.

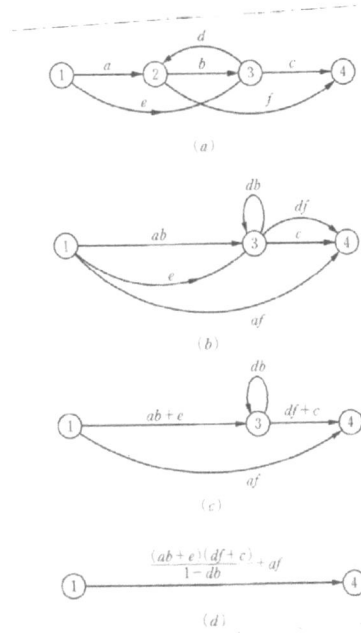

Figure 28: Reduction of SFD.

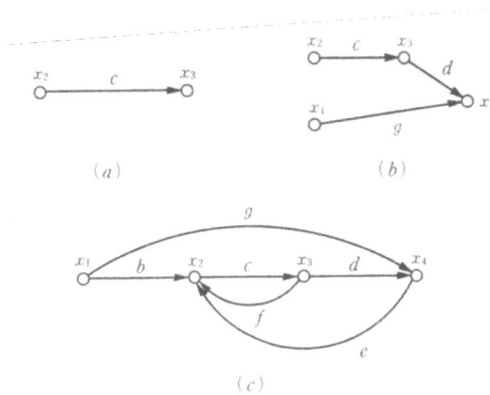

Figure 29: Graph representation for algebraic equations.

As the graph becomes complex, it becomes difficult to obtain the transmittance of SFD. For this, the following Mason's theorem is useful:

Theorem 5: The transmittance from one node to another node in SFD can be expressed as

$$T = \frac{\sum_k P_k \Delta_k}{\Delta} \tag{18}$$

where P_k is the gain of the forward path, and Δ is called as graph determinant defined as

$$\Delta \equiv 1 - \sum_i L_i + \sum_{ij} L_i L_j - \sum_{ijk} L_i L_j L_k + \dots \tag{19}$$

where L_i is the transmittance of the i-th closed-loop, $L_i L_j$ is the product of the transmittances of L_i and L_j but exclude the overlapped loops, and $L_i L_j L_k$ is the product of the transmittances of L_i, L_j, and L_k where those must not be overlapped among them. Moreover, Δ_k in Eq.(18) is the one subtracting the loop gain of the overlapped loop k from Δ. Consider the example as shown in Fig. **30**, where the above calculation procedure becomes as

$$L_1 = a_{22}, \qquad L_2 = a_{12} a_{23} a_{31}, \qquad L_3 = a_{13} a_{31}$$

$$\Delta = 1 - (L_1 + L_2 + L_3) + (L_1 L_2 + L_2 L_3 + L_1 L_3) - L_1 L_2 L_3 = 1 - (L_1 + L_2 + L_3) + L_1 L_3$$
$$= 1 - (a_{22} + a_{12} a_{23} a_{31} + a_{13} a_{31}) + a_{22} a_{13} a_{31}$$

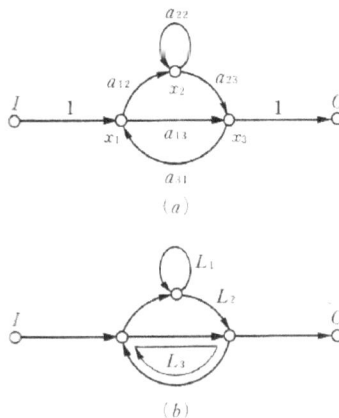

Figure 30: Computation of the transmittance by Mason's theorem.

For the forward path

$$P_1 = a_{12}a_{23}, \qquad P_2 = a_{13}$$

and

$$\Delta_1 = 1 - (L_1 + L_2 + L_3) + L_1 L_3 = 1$$

$$\Delta_2 = 1 - (L_1 + L_2 + L_3) + L_1 L_3 = 1 - L_1 = 1 - a_{22}$$

In the end, the transmittance can be obtained as

$$T = \frac{\sum_k P_k \Delta_k}{\Delta} = \frac{a_{12}a_{23} + a_{13}(1 - a_{22})}{1 - (a_{22} + a_{12}a_{23}a_{31} + a_{13}a_{31}) + a_{22}a_{13}a_{31}}$$

Consider next a little complex system as shown in Fig. **31a** for the multi-stage counter current system, where it is assumed that each state is completely mixed, and the vapor an the liquid leaving the stage are in equilibrium. In general, the following mass balance equation holds at the n-th stage:

$$L(x_{n-1} - x_n) = G(y_n - y_{n+1})$$

where L is the liquid flow rate and x is its concentration of the component of concern, and G is the gas flow rate and y is its concentration. This can be considered also for the liquid-liquid extraction process. The equilibrium relationship may be expressed as

$$y_n = mx_n$$

Although L,G, and m may change depending on the stages, here we assume those to be constant throughout the process without much loss of generality. Let

$$\lambda \equiv \frac{mG}{L}$$

Then the following equations are obtained by the mass balances:

$$x_n = \frac{1}{1+\lambda} x_{n-1} + \frac{\lambda/m}{1+\lambda} y_{n+1}$$

$$y_n = \frac{m}{1+\lambda} x_{n-1} + \frac{\lambda}{1+\lambda} y_{n+1}$$

These linear equations can be expressed by the SFD as shown in Fig. **31b**.

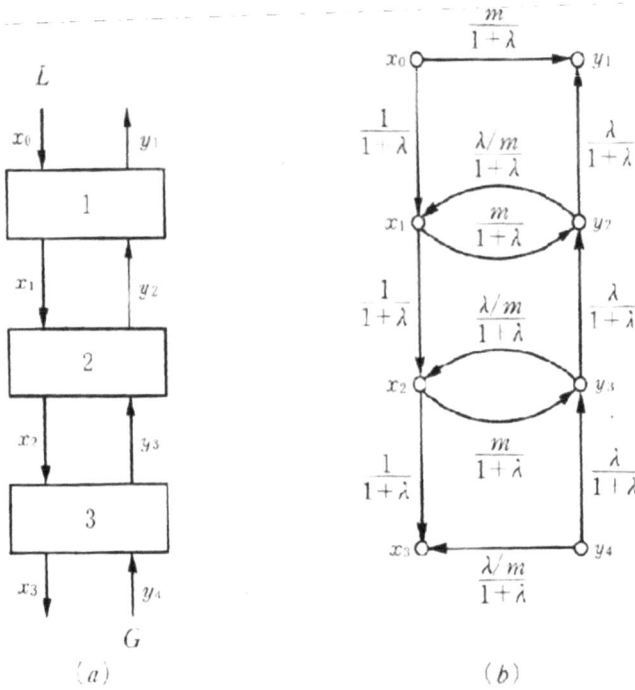

Figure 31: Multistage system and its SFD.

Consider obtaining the transmittance from x_0 (input to the system) to x_3 (output from the system) such that

$$\sum_i l_{ii} = \frac{\lambda}{(1+\lambda)^2} + \frac{\lambda^2}{(1+\lambda)^4} + \frac{\lambda}{(1+\lambda)^4}$$

$$\sum_{ij} L_i L_j = \frac{2\lambda}{(1+\lambda)^4}$$

$$\Delta = 1 - \frac{2\lambda}{(1+\lambda)^2}$$

$$P_k = P_1 = \frac{1}{(1+\lambda)^3}$$

$$\Delta_k = \Delta_1 = 1$$

Therefore, the transmittance is obtained as

$$\frac{x_3}{x_0} = \frac{\dfrac{1}{(1+\lambda)3}}{1 - \dfrac{2\lambda}{(1+\lambda)^2}} = \frac{1}{(1+\lambda)(1+\lambda^2)}$$

Computation of the Reaction Rate Based on Graph Theoretic Approach

Consider the following enzyme reaction scheme:

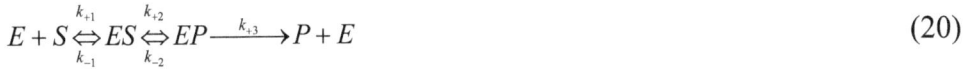

$$E + S \underset{k_{-1}}{\overset{k_{+1}}{\Longleftrightarrow}} ES \underset{k_{-2}}{\overset{k_{+2}}{\Longleftrightarrow}} EP \overset{k_{+3}}{\longrightarrow} P + E \tag{20}$$

where E, S, and P are the enzyme, substrate, and product, respectively. ES, EP are the complexes, and k_{+1}, k_{-1}, k_{+2}, k_{-2}, k_{+3} are the reaction rate constant. The reaction rate equation can be derived by graph representation [11].

For this first draw the graph as shown in Fig. **32a**, where each node represents the enzyme state such as E, ES, and EP in the present case. Then find all the routes that direct to each node (Fig. **32b**). The fraction of the existence of each enzyme state (E, ES, EP) may be expressed as

Existence of ES=(Summation of the product of reaction rate toward ES by referring Fig. **32b**)/(Total summation of the existence for all the enzyme states such as E, ES, EP).

This may be computed for ES as

$$\frac{[ES]}{[E_0]} = \frac{k_{+1}[S]k_{+3} + k_{+1}k_{+2}}{\Sigma} \tag{21a}$$

where [·] indicates the concentration, and E_0 is the total enzyme. In the similar way, the following equation can be derived for E:

$$\frac{[E]}{[E_0]} = \frac{k_{-1}k_{+3} + k_{-1}k_{-2} + k_{+2}[S]k_{+3}}{\Sigma} \tag{21b}$$

Moreover, the existence for EP can be expressed as

$$\frac{[EP]}{[E_0]} = \frac{k_{+1}[S]k_{+2}}{\Sigma} \tag{21c}$$

where Σ is the summation of the existence for all the enzyme state. Then the reaction rate can be expressed as

$$v = k_{+3}\,[EP] = \frac{k_{+3}[E_0][S]}{\dfrac{(k_{+3}+k_{-2}+k_{+2})[S]}{k_{+2}} + \dfrac{k_{-1}k_{+3}+k_{-1}k_{-2}}{k_{+1}k_{+2}}} \tag{22}$$

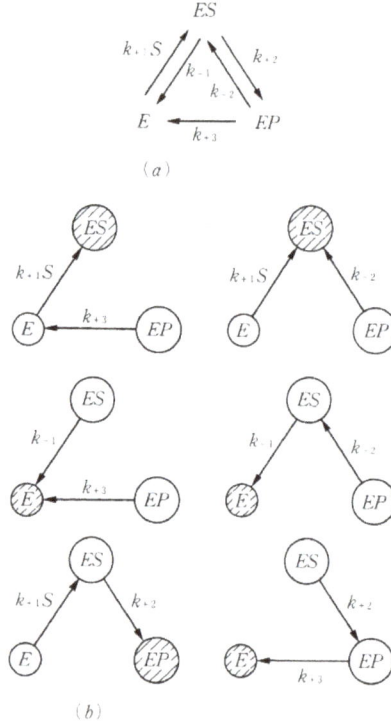

Flgure 32: Determination of reaction rate by King-Altman method.

SENSITIVITY ANALYSIS

The **parameter sensitivity** is defined as the change in the input and output relationship in response to the change in the parameter values such as

$$S_k = \left|\left(\frac{\Delta T}{\Delta T}\right)\Big/\left(\frac{\Delta k}{k}\right)\right| = \left(\frac{\Delta \ln T}{\Delta \ln k}\right) = \left(\frac{\Delta T}{\Delta k}\right)\Big/\left(\frac{T}{k}\right) \tag{23}$$

Where T is the transmittance of the system, and k is the parameter of concern. In the case of Fig. **27c**,

$$T = \frac{g_3 g_1}{1 - g_3} \tag{24}$$

as seen before. The parameter sensitivities for g_1, g_2, g_3 are obtained as

$$S_{g_1} = \left(\frac{g_2}{1 - g_3} \right) \Big/ \left(\frac{g_2}{1 - g_3} \right) = 1 \tag{25a}$$

$$S_{g_2} = \left(\frac{g_1}{1 - g_3} \right) \Big/ \left(\frac{g_1}{1 - g_3} \right) = 1 \tag{25b}$$

$$S_{g_3} = \frac{g_2 g_1}{(1 - g_3)^2} \Big/ \left(\frac{g_2 g_1}{g_3 (1 - g_3)} \right) = \frac{g_3}{1 - g_3} \tag{25c}$$

DYNAMICS OF THE LINEAR SYSTEM

Consider here the basic analysis of the linear system. More detailed analysis will be given in Chapter 5.

Step Response

Let the input-output relationship be expressed as

$$\tilde{y}(s) = G(s)\tilde{u}(s) \tag{26}$$

as mentioned before. Consider how the output y(t) changes in response to the step change in u(t) as shown in Fig. **33a**. Since the Laplace transformation of unit step function is 1/s such as $L[\mathbf{1}(t)] = 1/s$ (see Appendix A), the step response of the system is expresses as

$$y(t) = \mathcal{L}^{-1}\left[\frac{1}{s} G(s) \right] \tag{27}$$

where $\mathcal{L}^{-1}[\,\cdot\,]$ represents the inverse Laplace transformation.

Step Response of the 1ˢᵗ Order System

As seen before, the system equation for CSTF can be expressed as

$$G(s) = \frac{K}{\theta s + 1} \tag{28}$$

where K is the process gain, and θ is called as the **time constant**. The step response can be easily obtained as

$$y(t) = \mathcal{L}^{-1}\left[\frac{1}{s}\frac{K}{\theta s + 1}\right] = \mathcal{L}^{-1}\left[K\left(\frac{1}{s} - \frac{1}{s + 1/\theta}\right)\right] = K(1(t) - e^{-t/\theta}) \tag{29}$$

where $1(t)$ is the unit step function as seen in Fig. **33a**, and the time response of y(t) is given in Fig. **33b**.

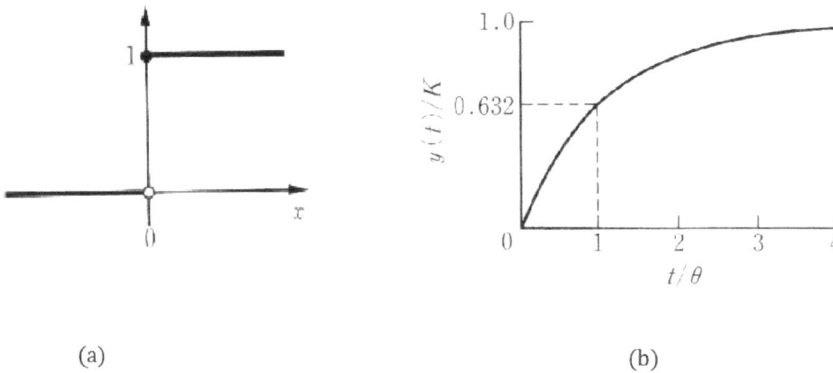

(a)   (b)

Figure 33: Step response.

Step Response of the 2ⁿᵈ Order System

Consider next the step response of the 2ⁿᵈ order system as expresses as

$$G(s) = \frac{K}{a_2 s^2 + a_1 s + a_0} \tag{30}$$

Let the **characteristic equation** be the denominator of G(s)=0 such that $a_2 s^2 + a_1 s + a_0 = 0$, and the **characteristic roots** to be the solution to this equation. Let D be the discriminant of this equation, and $D = a_1^2 - 4a_2$, then the step response depends on the value of D.

(a) The case where D>0 $(4a_2^2 < a_1^2)$

Consider the following example:

$$G(s) = \frac{2}{2s^2 + 3s + 1} = \frac{2}{(2s+1)(s+1)}$$

The step response is obtained as

$$y(t) = \mathcal{L}^{-1}\left[\frac{1}{s}\frac{2}{(2s+1)(s+1)}\right] = \mathcal{L}^{-1}\left[\frac{2}{s} + \frac{2}{s+1} - \frac{4}{s+1/2}\right] = 21(t) + 2e^{-t} - 4e^{-t/2}$$

As seen above, this can be expressed as the partial fraction (see **Appendix B**), and can be expressed as the sum of the 1st order system. The step response of the above example is given in Fig. **34a**. Note that the characteristic roots are -1 and -1/2 and both are negative, and the system converge to G(0), and it is stable, whereas if some of the roots are positive, the output y(t) becomes infinity as t→∞, and the system becomes unstable.

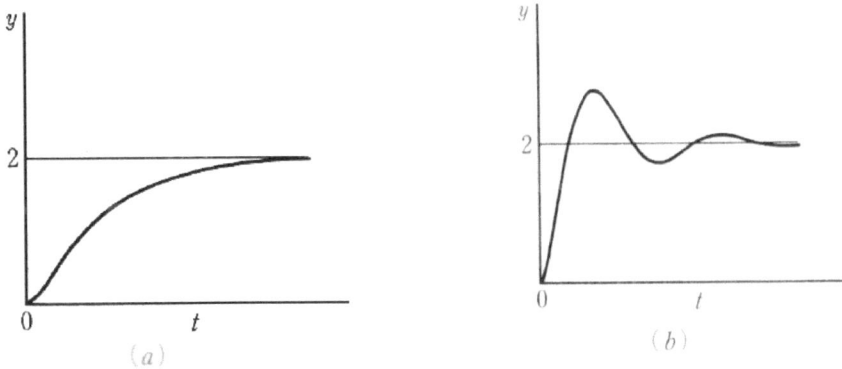

(a)

(b)

Figure 34: step response for the case of (a) and (c).

(a) The case where D=0
Consider for example

$$G(s) = \frac{2}{s^2 + 2s + 1} = \frac{2}{(s+1)^2}$$

The step response of the system becomes as

$$y(t) = \mathcal{L}^{-1}\left[\frac{1}{s}\frac{2}{(s+1)^2}\right] = \mathcal{L}^{-1}\left[\frac{2}{s} - \frac{2}{s+1} - \frac{2}{(s+2)^2}\right] = 2[1(t) - e^{-t} + te^{-t}]$$

This is the special case of (a).

(b) The case where D<0
Consider the following example:

$$G(s) = \frac{2}{s^2 + s + 1}$$

The step response can be obtained as (Appendix A):

$$y(t)=\mathcal{L}^{-1}\left[\frac{1}{s}-\frac{2}{s^2+s+1}\right]=\mathcal{L}^{-1}\left[\frac{2}{s}-\frac{2(s+1)}{s^2+s+1}\right]=\mathcal{L}^{-1}\left[\frac{2}{s}-\frac{2\{(s+1/2)+(1-1/2)\}}{(s+1/2)^2+1-1/4}\right]$$

$$=\mathcal{L}^{-1}\left[\frac{2}{s}-2\frac{(s+\frac{1}{2})+\frac{\sqrt{3}}{2}\frac{1}{\sqrt{3}}}{\left(s+\frac{1}{2}\right)^2+\left(\frac{\sqrt{3}}{2}\right)^2}\right]=2[1(t)-e^{-t/2}\cos\frac{\sqrt{3}}{2}t-e^{-t/2}\frac{1}{\sqrt{3}}\sin\frac{\sqrt{3}}{2}t]$$

The step response is given in Fig. **34b**, and it shows oscillatory behavior with convergence to G(0)=2.

Step Response of the Time-Delay System

Consider the time-delay system as shown in Fig. **35a**, where G(s) is expressed as

$$G(s) = e^{-t_d s} \tag{31}$$

The step response becomes as

$$y(t)=\mathcal{L}^{-1}\left[\frac{1}{s}u^{-t_d s}\right]=1(t-t_d)$$

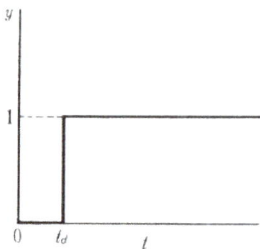
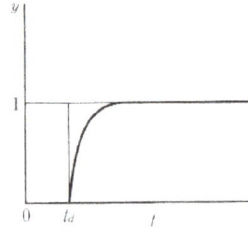

(a)   (b)

Figure 35: Step response of the time delay system.

Namely, the time-response is the delay by t_d as shown in Fig. **35a**. Therefore, the time-response for the following 1st order system with time delay becomes as the step response with just the time delay of the system

$$G(s) = G_1(s)e^{-t_d} = \frac{K}{\theta s + 1}e^{-t_d s} \tag{32}$$

Namely, as shown in Fig. **35b**, the step response is just the time delay by t_d for the step response of $G_1(s)$ such as $y(t-t_d)$.

Step Responses of the Higher Order Systems

Consider the n CSTFs connected in series as seen before. The transfer function of the system may be expressed as

$$G(s) = \frac{1}{\left(\dfrac{\theta}{n}s + 1\right)^n} \tag{33}$$

The step response can be expressed as

$$y(t) = 1(t) - \frac{1}{(n-1)!\,\theta}\frac{n}{}\left(\frac{nt}{\theta}\right)^{n-1}e^{-nt/\theta} \tag{34}$$

Fig. **36** shows the effect of n on the dynamics, where it implies that the step response tends to be the time-delay as $n \to \infty$.

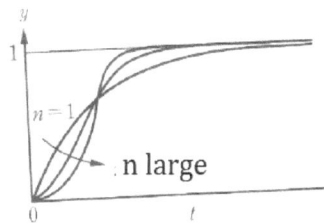

Figure 36: Step responses of the n-th order system.

Impulse Response

Consider the rectangular function as shown in Fig. **37** such that

$$y = \frac{1}{\tau} \qquad -\frac{\tau}{2} \le t \le \frac{\tau}{2}$$

$$y = 0 \qquad t < -\frac{\tau}{2}, \qquad t > \frac{\tau}{2}$$

The integration of the function from $-\infty$ to $+\infty$ gives as

$$\int_{-\infty}^{\infty} y(t)dt = \int_{-\tau/2}^{\tau/2} y(t)dt = 1 \tag{35}$$

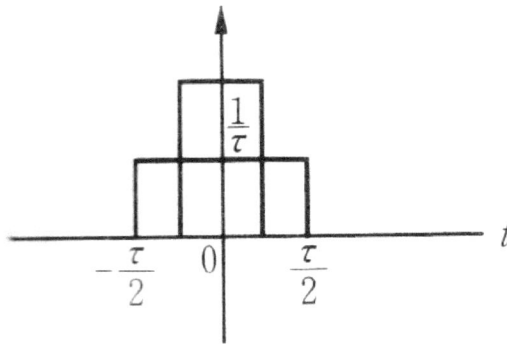

Figure 37: δ function.

Namely, the area is 1. If we keep the area to be constant, and reduce τ as $\tau \to 0$. Then the limiting function becomes as

$$\delta(t) = \begin{cases} \infty & t = 0 \\ 0 & t \ne 0 \end{cases} \tag{36a}$$

$$\int_{-\infty}^{\infty} \delta(t)dt = \int_{-0}^{+0} \delta(t)dt = 1 \tag{36b}$$

where this function is called as **delta function**. The **impulse response** is the response of the system when this delta function is imposed as input. Since $\mathcal{L}[\delta(t)]=1$, the impulse response is obtained by the inverse Laplace transformation of the process transfer function such that

$$g(t)=\mathcal{L}^{-1}[G(s)] \tag{37}$$

For example, the impulse response of the 1st order system as shown before becomes as

$$g(t)=\mathcal{L}^{-1}[G(s)]=\mathcal{L}^{-1}\left[\frac{K}{\theta s+1}\right]=\frac{K}{\theta}e^{-t/\theta} \tag{38}$$

and is expressed as Fig. **38**.

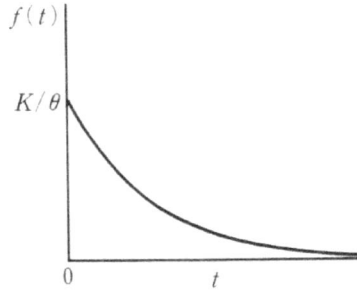

Figure 38: Impulse response of the 1st order system.

In the similar way, the impulse responses of the 2nd order system can be also expressed in the similar way as the step responses as

a)   D>0

$$g(t)=\mathcal{L}^{-1}\left[\frac{2}{2s^2+3s+1}\right]=\mathcal{L}^{-1}\left[\frac{4}{2s+1}-\frac{2}{s+1}\right]=2e^{-t/2}-2e^{-t}$$

b)   D=0

$$g(t)=\mathcal{L}^{-1}\left[\frac{2}{(s+1)^2}\right]=2te^{-t}$$

c)   D<0

$$g(t)=\mathcal{L}^{-1}\left[\frac{2}{s^2+s+1}\right]=\mathcal{L}^{-1}\left[\frac{\frac{\sqrt{3}}{2}\frac{4}{\sqrt{3}}}{\left(s+\frac{1}{2}\right)^2+\left(\frac{\sqrt{3}}{2}\right)^2}\right]=\frac{4}{\sqrt{3}}e^{-t/2}\sin\frac{\sqrt{3}}{2}t$$

where the impulse responses for the case of (a), (b), and (c) are shown in Fig. **39**.

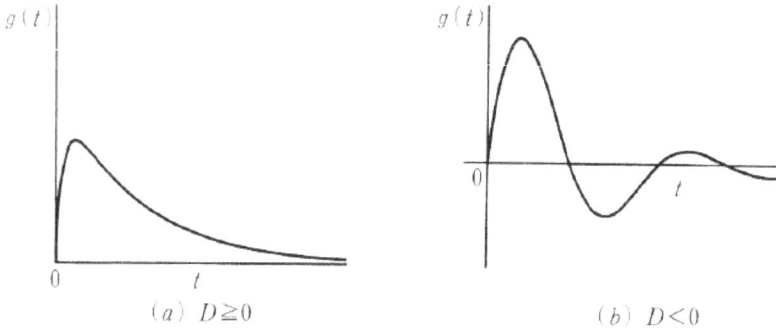

(a) $D \geqq 0$ (b) $D < 0$

Figure 39: Step responses of the 2nd order system.

In the case of time-delay system, the delta function is just delayed by t_d such that

$$g(t) = \mathcal{L}^{-1}[e^{-t_d s}] = \delta \ (t-t_d) \tag{39}$$

and the n-th order system with delay system, and the impulse responses can be expressed as

$$g_n(t) = \frac{1}{(n-1)!} \left(\frac{n}{\theta} \right) \left(\frac{nt}{\theta} \right)^{n-1} e^{-nt/\theta} \tag{40}$$

This response is shown in Fig. **40**.

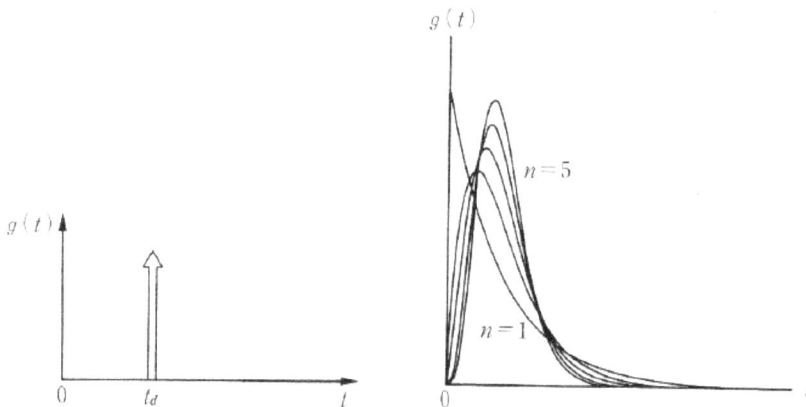

Figure 40: Impulse responses of the n-th order system.

Frequency Response

Basis for Frequency Response

Although it is not much seen in the bioprocesses, **frequency response** is also important to analyze the dynamics of the linear system, where input to the system is changed in sine wave fashion. The Laplace transformation of the sine wave becomes as

$$\mathcal{L}[A\sin \omega t] = \frac{A\omega}{s^2 + \omega^2}$$

where A is the gain or amplitude, and ω is the frequency of the sine. Thus the frequency response is obtained as

$$\phi\,(t) = \mathcal{L}^{-1}\left[\frac{A}{s^2 + \omega^2} G(s)\right] \tag{41}$$

For example, consider the frequency response of the 1st order system as

$$\phi\,(t) = \mathcal{L}^{-1}\left[\frac{A}{s^2 + \omega^2}\frac{K}{\theta s + 1}\right]$$

$$= \frac{AK}{\sqrt{1 + \omega^2\theta^2}}\sin(\omega t - \eta) + \frac{AK\omega\theta}{1 + \omega^2\theta^2}e^{-t/\theta} \tag{42}$$

where

$$\eta \equiv \tan^{-1}(\omega\ \theta)$$

The 2nd term of the RHS of the above equation becomes 0 as t→∞, and the 1st term shows the change in the gain by $AK/\sqrt{1 + \omega^2\theta^2}$, and the phase is delayed by η.

As such, the frequency response is defined as the process response after enough time was elapsed so that the 2nd term disappears. Thus the **gain** and the **phase lag** can be obtained by $|G(i\omega)|$ and $\angle G(i\omega)$, respectively. Let $G(i\omega)$ be the **frequency transfer function**. As shown in Fig. **41**, $G(i\omega)$ shows the locus in the complex plane from $\omega=0$ to $\omega\to\infty$. Let a and b be the real part and the imaginary part of $G(i\omega)$.

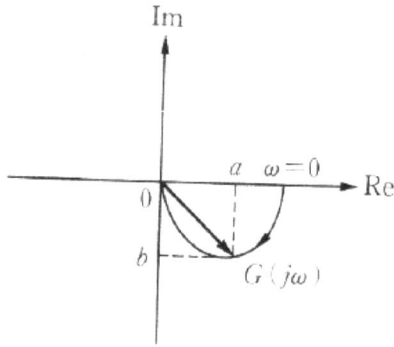

Figure 41: Frequency response in the complex plane.

Then

$$G(i\omega)=a(\omega)+b(\omega)i \tag{43}$$

where $a(\omega)\equiv\mathrm{Re}[G(i\omega)]$, and $b(\omega)\equiv\mathrm{Im}[G(i\omega)]$, and $|G(i\omega)|$ and $\angle G(i\omega)$ can be expressed as

$$|G(i\omega)| = \sqrt{a(\omega)^2 + b(\omega)^2} \tag{44a}$$

$$\angle G(i\omega) = \tan^{-1}\left(\frac{b(\omega)}{a(\omega)}\right) \tag{44b}$$

These are obtained for the 1st order, 2nd order, and time-delay systems as

a)   1st order system:

$$G(i\omega) = \frac{K}{\omega\theta i + 1} = \frac{K(1 - \omega\theta i)}{1 + \omega^2\theta^2}$$

$$|G(i\omega)| = \frac{K}{\sqrt{1 + \omega^2\theta^2}}, \qquad \angle G(i\omega) = \tan^{-1}(-\omega\ \theta)$$

b)   2nd order system

$$G(i\omega) = \frac{K}{1 - a_1\omega i - a_2\omega^2} = \frac{K\{(1 - a_2\omega^2) + a_1\omega i\}}{(1 - a^2\omega^2)^2 + a_1\omega^2}$$

$$|G(i\omega)| = \frac{K}{\sqrt{(1-a_2\omega^2)^2 + a_1^2\omega^2}}, \qquad \angle G(i\omega) = \tan^{-1}\left(\frac{a_1\omega}{1-a_2\omega^2}\right)$$

c) Pure time-delay system

$$G(i\omega) = e^{-t_d\omega i} = \cos(t_d\omega) - i\sin(t_d\omega)$$

$$|G(i\omega)| = \sqrt{\cos^2(t_d\omega) + \sin^2(t_d\omega)}, \qquad \angle G(i\omega) = \tan^{-1}[\tan(-t_d\omega)] = -t_d\omega$$

Consider the frequency response of the combined system such as

$$G(s) = G_1(s)G_2(s) \bullet \bullet \bullet G_n(s) \tag{45}$$

Since the followings hold

$$G(i\omega) = |G(i\omega)|\, e^{-\angle G(i\omega)} \tag{46a}$$

$$G_k(i\omega) = |G_k(i\omega)|\, e^{-\angle Gk(i\omega)} \tag{46b}$$

the following relationships hold.

$$|G(i\omega)| = |G_1(i\omega)| \; |G_2(i\omega)| \bullet \bullet \bullet |G_n(i\omega)| \tag{47a}$$

$$\angle G(i\omega) = \angle G_1(i\omega) + \angle G_2(i\omega) + \bullet \bullet \bullet + \angle G_n(i\omega) \tag{47b}$$

For example, consider the 1st order +pure time-delay system such as

$$G(i\omega) = \frac{K}{\theta s+1} e^{-t_d s} \tag{48}$$

Let $G_1(s) \equiv K/(\theta s+1)$ and $G_2(s) \equiv e^{-t_d s}$, then from the above relationship as shown in Eq.(46), the following equations are obtained:

$$|G(i\omega)| = |G_1(i\omega)| \; |G_2(i\omega)| = \frac{K}{\sqrt{1+\omega^2\theta^2}} \tag{49a}$$

$$\angle G(i\omega) = \angle G_1(i\omega) + \angle G_2(i\omega) = \tan^{-1}(-\omega\,\theta) - t_d\omega \tag{49b}$$

Representation of the Frequency Response

(a)Vector Locus

Vector locus is the plot of the locus of $G(i\omega)$ in the complex plane from $\omega=0$ to $\omega \to \infty$. Fig. **42a** shows the typical loci for the n-th order system for n=1,2, and 3. Fig. **42b** shows the vector locus of the pure time-delay system, and Fig. **42c** shows the vector locus of the 1st order system with pure time-delay. As can be seen, the vector locus starts from G(0) in the real axis, and converge to the origin as far as the order of the numerator of G(s) is smaller than that of the denominator.

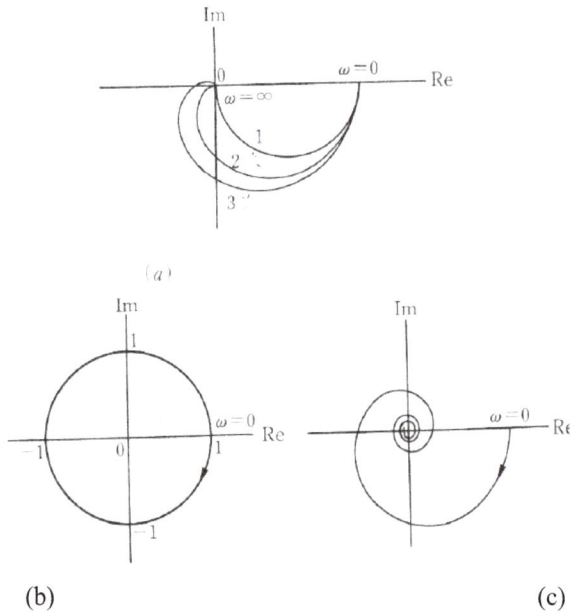

Figure 42: Vector loci: (a) n-th order system, (b) pure time-delay system, (c) 1st order system with pure time delay system.

(b) Bode diagram

Bode diagram is the plot of $\log|G(i\omega)|$ (or $-20\log|G(i\omega)|$ dB) and $\angle G(i\omega)$ with respect to ω. This diagram is often used for the control system design.

FEEDBACK SYSTEM

Consider the simple feedback system as shown in Fig. **43**, where the closed-loop transfer function is expressed as seen before as

$$G_{CL}(s) = \frac{G(s)K(s)}{1+G(s)K(s)} \tag{50}$$

where G(s) and K(s) are the transfer functions of the process and the controller, respectively. Consider for example the case of 1st order system with the process gain to be 1, and the controller with only the proportional gain K_c such as

$$G_{CL}(s) = \frac{\dfrac{K_c}{\theta s+1}}{1+\dfrac{K_c}{\theta s+1}} = \frac{K_c}{\theta s+1+K_c} = \frac{K_c}{\theta(s+\dfrac{1+K_c}{\theta})} \tag{51}$$

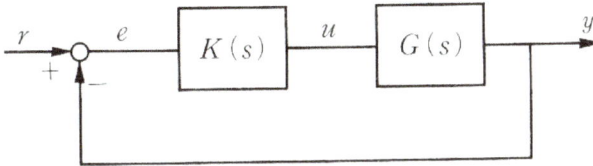

Figure 43: Feedback system.

The impulse response can then be expressed as

$$y_{CL}(t) = \mathcal{L}^{-1}[G_{CL}(s)] = \frac{K_c}{\theta} e^{-\frac{1+Kc}{\theta}t} \tag{52}$$

The characteristic root of G is $\lambda = -1/\theta$ (<0), while that of $G_{CL}(s)$ is $\lambda_{CL} = -(1+K_c)/\theta$. This means that the response speed can be increased by the proportional feedback system as illustrated in Fig. **44**.

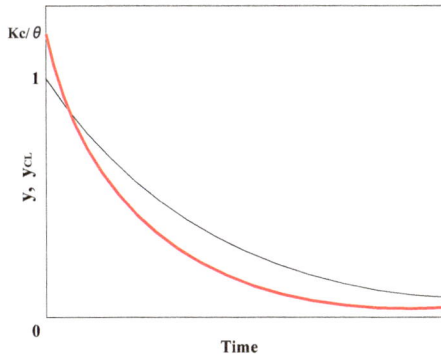

Figure 44: Effect of the feedback loop (red line) on the impulse response.

In the block diagram for the feedback system, the input to the system is expresses as u=K_ce. If the controller contains integral action such as $u(t) = K_i \int e(t)dt$ or $u(s) = K_i / s$, where it is also called as reset control, the offset will be minimized, while the response becomes slower.

DATA ANALYSIS

Linear Regression Analysis

Simple Regression Analysis

Consider fitting the n set of data such as $(x_1,y_1)(x_2,y_2) \cdot \cdot \cdot (x_n,y_n)$ to the line as shown in Fig. **45** such that

$$y_i = ax_i + b + e_i \qquad (i = 1,2,...,n)$$

(53)

where a and b are the model parameters, and e_i is the deviation or error of the ith data from the line. Consider fitting the values of a and b so that the sum of the distance between the data point and the line be minimized. Namely, find the values of a and b that minimize

$$J = \frac{1}{n}\sum_{i=1}^{n}(y - ax_i - b)^2$$

(54)

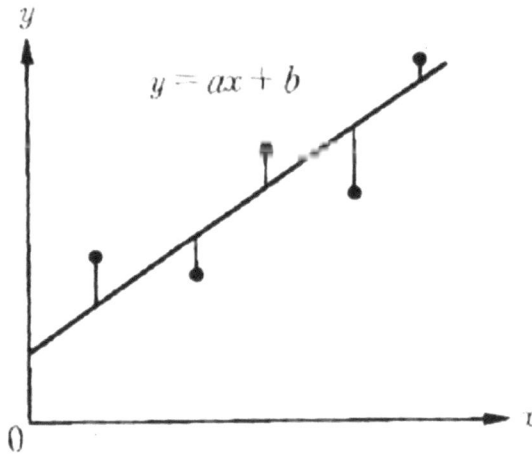

Figure 45: Linear regression.

This can be obtained by solving the following optimality condition:

$$\frac{\partial J}{\partial b} = \frac{1}{n}\sum_{i=1}^{n}(-2)(y_i - ax_i - b) = 0$$

(55a)

$$\frac{\partial J}{\partial a} = \frac{1}{n}\sum_{i=1}^{n}(-2x_i)(y_i - ax_i - b) = 0$$

(55b)

These equation can be re-expressed as

$$a\bar{x} + b = \bar{y}$$

(56a)

$$a\overline{X^2} + b\bar{x} = \overline{xy}$$

(56b)

where

$$\bar{x} \equiv \frac{1}{n}\sum_{i=1}^{n}x_i, \qquad \bar{y} \equiv \frac{1}{n}\sum_{i=1}^{n}y_i, \qquad \overline{x^2} \equiv \frac{1}{n}\sum_{i=1}^{n}x_i^{\,2}, \qquad \overline{xy} \equiv \frac{1}{n}\sum_{i=1}^{n}x_i y_i$$

Since the above equations are the linear equations, this can be solved as

$$\begin{bmatrix}\hat{a} \\ \hat{b}\end{bmatrix} = \begin{bmatrix}\bar{x} & 1 \\ \overline{x^2} & \bar{x}\end{bmatrix}^{-1}\begin{bmatrix}\bar{y} \\ \overline{xy}\end{bmatrix} = \frac{1}{(\bar{x})^2 - \overline{x^2}}\begin{bmatrix}\bar{x} & -1 \\ -\overline{x^2} & \bar{x}\end{bmatrix}\begin{bmatrix}\bar{y} \\ \overline{xy}\end{bmatrix} = \frac{1}{(\bar{x})^2 - \overline{x^2}}\begin{bmatrix}\bar{x}\bar{y} - \overline{xy} \\ -\overline{x^2}\bar{y} + \bar{x}(\overline{xy})\end{bmatrix}$$

(57)

where $\hat{a}$ and $\hat{b}$ are called as **least square estimates**.

Consider next finding the **regression** line for such data as $X_i = x_i - \bar{x}$ and $Y_i = y_i - \bar{y}$ (i=1,2,...,n), where $\bar{x}$ and $\bar{y}$ are the average values as defined above. Then the following relationships hold:

$$\sum_{i=1}^{n}X_i = \sum_{i=1}^{n}(x_i - \bar{x}) = \sum_{i=1}^{n}x_i - n\bar{x} = 0$$

(58a)

$$\sum_{i=1}^{n}Y_i = \sum_{i=1}^{n}(y_i - \bar{y}) = \sum_{i=1}^{n}y_i - n\bar{y} = 0$$

(58b)

Let the regression line be expressed as $Y=AX_i+B$, then the similar equation as (56a) can be expressed as

$$A\sum_{i=1}^{n} X_i + B = \sum_{i=1}^{n} Y_i \tag{59}$$

By substituting Eq.(58) into this equation, we have B=0. By the similar equation as Eq.(55b), we have

$$A = \frac{\sum_{i=1}^{n} X_i Y_i}{\sum_{i=1}^{n} x_i^2} = \frac{S_{xy}}{S_{xx}} \tag{60}$$

where S_{xx} is the variance of x, and S_{xy} is the co-variance of x and y. Namely, the regression line can be expressed as

$$y - \bar{y} = \frac{S_{xy}}{S_{xx}}(x - \bar{x}) \tag{61}$$

Let r_{xy} be the correlation coefficient defined as $r_{xy} \equiv S_{xy}/\sqrt{S_{xx}S_{yy}}$, then the following equation holds:

$$\frac{\sum_{i} e_i^2}{n} = S_{yy}(1 - r_{xy}^2) \tag{62}$$

Therefore, $|r_{xy}|$ shows the measure of how well the data fitted to the regression line.

Multiple Regression Analysis

Consider next the more general case such that y is not only a function of one variable, but also several other variables such as

$$y = a_0 + a_1 x_1 + a_2 x_2 + \ldots + a_d x_d \tag{63}$$

where this is called as **multiple regression equation**. The generalized linear regression model is expressed as

$$y = \Phi\theta + e \tag{64}$$

where

$$y \equiv [y_1, y_2, ..., y_n]^T, \qquad \theta \equiv [\theta_1, \theta_2, ..., \theta_d]^T, \qquad e \equiv [e_1, e_2, ..., e_n]^T$$

$$\Phi \equiv \begin{bmatrix} x_{11} & \cdot & \cdot & \cdot & x_{1d} \\ & \cdot & & & \cdot \\ \cdot & & & & \cdot \\ & \cdot & & & \cdot \\ x_{n1} & \cdot & \cdot & \cdot & x_{nd} \end{bmatrix}$$

The objective function to be minimized becomes as

$$J = (y - \Phi\theta)^T (y - \Phi\theta) = \|y - \Phi\theta\|^2 \tag{65}$$

where $\| \bullet \|$ indicates the norm. The least square estimate $\hat{\theta}$ can be obtained by

$$\frac{\partial J}{\partial \theta} = -2\Phi^T (y - \Phi\theta) = 0 \tag{66}$$

and the resulting equation may be expressed as

$$\Phi^T \Phi \hat{\theta} = \Phi^T y \tag{67}$$

This is called as the normal equation for the least square method, and this can be solved for θ as

$$\hat{\theta} = [\Phi^T \Phi]^{-1} \Phi^T y \tag{68}$$

as far as $[\Phi^T \Phi]$ is non-singular.

The above equation can be slightly extended using weight as weight matrix W such as

$$J = (y - \Phi\theta)^T W (y - \Phi\theta) \tag{69}$$

from which we have

$$\Phi^T W \Phi \hat{\theta} = \Phi^T W y \tag{70}$$

Then the least square estimate with weight can be obtained as

$$\hat{\theta} = [\Phi^T W \Phi]^{-1} \Phi^T W y \tag{71}$$

where $[\Phi^T \Phi]$ was assume to be nonsingular.

The normal equation as mentioned above is expressed as the multiple linear equations such as

$$Ax = b \tag{72}$$

Consider the perturbation of **b** such as **b**+ δ **b**, and thus the solution **x** also changes from **x** to **x**+ δ **x**. In such a case, since A(**x**+ δ **x**)=**b**+ δ **b**, the following equation can be obtained: δ **x**=A^{-1} δ **b**. Therefore, the following equation holds:

$$\|\delta x\| \le \|A^{-1}\| \|\delta b\| \tag{73}$$

Since $\|b\| \le \|A\| \|x\|$, the following equation holds:

$$\frac{\|\delta x\|}{\|x\|} \le \|A\| \|A^{-1}\| \frac{\|\delta b\|}{\|b\|} = c(A) \frac{\|\delta b\|}{\|b\|} \tag{74}$$

where $c(A)$ is called as the **condition number**, and is defined as

$$c(A) \equiv \frac{\overline{\sigma}(A)}{\underline{\sigma}(A)} \tag{75}$$

where $\overline{\sigma}$ and $\underline{\sigma}$ are the maximum and minimum **singular values**, respectively, and the singular values are defined as

$$\sigma_i(A) \equiv \lambda_i(A^T A) \tag{76}$$

where $\lambda_i(A^T A)$ is the i-th eigenvalue of A^{T}A. Eq.(74) implies that the relative error of the solution x may be amplified for the relative error of b. Namely, if the value of c(A) becomes large, the estimation accuracy becomes lower.

Principal Component Analysis

PCA for One Data Set

Consider n set of data plotted on the x-y plane as shown in Fig. **46**, where it implies that the data may be analyzed by a single variable by the appropriate change of coordinate (rotating the axis). Although this may be the special case, the data set as shown in Fig. **47a** may be better to analyze by the appropriate change of coordinate as shown in Fig. **47b**. Let X_i and Y_i be the deviation from the average values such as $X_i \equiv x_i - \bar{x}$ and $Y_i \equiv y_i - \bar{y}$, respectively. Then the new axis goes through the origin, and this can be expressed as the linear combination of X and Y such that

$$t_i \equiv aX_i + bY_i \tag{77}$$

where a and b are assumed to satisfy the following normalization condition:

$$a^2 + b^2 = 1 \tag{78}$$

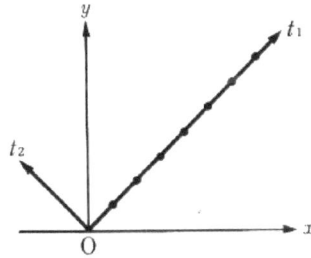

Figure 46: Principal component analysis by the rotation of axis.

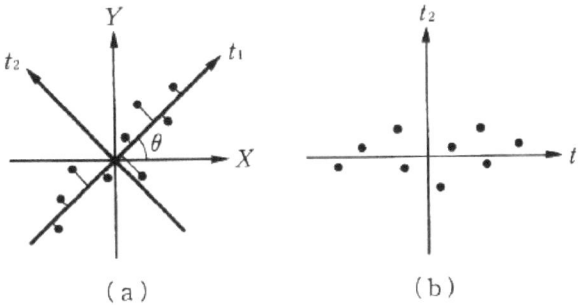

Figure 47: Principal component analysis for the scattered data.

Moreover, let t_i be the distance from the origin when the line from the data point in X-Y plane to the vertical line onto the t-axis. Then the new axis that minimizes the square summation S of t_i is called as the 1st or principal component axis. Thus the objective of the principal component analysis is to maximize such variance of the original data, and express as the principal component such that

$$S = \sum_{i=1}^{n} t_i^2 = \sum_{i=1}^{n} (aX_i + bY_i)^2 \tag{79}$$

The problem of finding a and b that maximize S as expressed by Eq.(79) under the constraint of Eq.(78) can be solved by using the Lagrange multiplier such that

$$J(a,b,\lambda) = \sum_{i=1}^{n} (aX_i + bY_i)^2 - \lambda(a^2 + b^2 - 1) \tag{80}$$

Taking the derivative with respect to a and b, we have

$$\left(\sum_{i=1}^{n} X_i^2 - \lambda \right) a + \left(\sum_{i=1}^{n} X_i Y_i \right) b = 0 \tag{81a}$$

$$\left(\sum_{i=1}^{n} X_i Y_i \right) a + \left(\sum_{i=1}^{n} Y_i^2 - \lambda \right) b = 0 \tag{81b}$$

For the solution of a and b to be non-zero, or nontrivial solution, the above equation must be linearly dependent, or the determinant of the coefficient matrix must be 0 such that

$$\begin{vmatrix} \sum_{i=1}^{n} X_i^2 - \lambda & \sum_{i=1}^{n} X_i Y_i \\ \sum_{i=1}^{n} X_i Y_i & \sum_{i=1}^{n} Y_i^2 - \lambda \end{vmatrix} = 0 \tag{82}$$

from which we have the following eigen equation:

$$\lambda^2 - \sum_{i=1}^{n} (X_i^2 + Y_i^2)\lambda + \sum_{i=1}^{n} X_i^2 \sum_{i=1}^{n} Y_i^2 - \left(\sum_{i=1}^{n} X_i Y_i \right)^2 = 0 \tag{83}$$

The values of a and b can be obtained by substituting the solution to this equation into Eq.(81), and using the normalization condition such as Eq.(78).

Consider such operation that $(81a) \times a + (81b) \times b$, and rearranging, we have

$$\sum_{i=1}^{n} X_i^2 a^2 + 2\sum_{i=1}^{n} X_i Y_i ab + \sum_{i=1}^{n} Y_i^2 b^2 = \lambda(a^2 + b^2) = \lambda \tag{84}$$

Substituting this equation into Eq.(79), we have

$$S = \sum_{i=1}^{n} t_i^2 = \|t\|^2 = \lambda \tag{85}$$

where $\| t \|$ is the norm of the vector $t = [t_1, t_2, \ldots, t_n]^T$. This equation implies that the maximum eigenvalue of Eq.(82) or (83) becomes the square summation of the 1st component t_i.

Let the data matrix D and the parameter vector **p** be defined as

$$D = [X \quad Y] = \begin{bmatrix} X_1 & Y_1 \\ X_2 & Y_2 \\ . & . \\ . & . \\ X_n & Y_n \end{bmatrix}, \qquad p = \begin{bmatrix} a \\ b \end{bmatrix}$$

Eq.(81) becomes as

$$[D^T D - \lambda]p = 0 \tag{86a}$$

or

$$[D^T D]p = \lambda p \tag{86b}$$

This equation indicates that the eigenvalue of $D^T D$ is λ, and **p** is the eigenvector. Namely, if the maximum eigenvalue is λ_{max}, then the eigenvector corresponding to λ_{max} such as a_1 and b_1 of **p** becomes the coefficient for the 1st component such that

$$t_1 = [X \quad Y] \begin{bmatrix} a_1 \\ b_1 \end{bmatrix} = Dp_1 \tag{87}$$

Let λ_{min} be the smaller eigenvalue, then the eigenvector for λ_{min} such as $[a_2, b_2]$ of $\mathbf{p}_2$ becomes the coefficient of the 2nd component such that

$$t_2 = [X\ \ Y]\begin{bmatrix} a_2 \\ b_2 \end{bmatrix} = Dp_2 \tag{88}$$

Note that two eigenvectors are orthogonal for the case of $\lambda_{max} \neq \lambda_{min}$.

Consider the following simple example to understand the above explanation:

$$D = [X\ \ Y] = \begin{bmatrix} X_1 & Y_1 \\ X_2 & Y_2 \\ X_3 & Y_3 \\ X_4 & Y_4 \end{bmatrix} = \begin{bmatrix} 2 & 2 \\ 1 & -1 \\ -1 & 1 \\ -2 & -2 \end{bmatrix}$$

In this case, $D^T D$ and Eq.(86b) become as

$$D^T D = \begin{bmatrix} 10 & 6 \\ 6 & 10 \end{bmatrix}, \qquad \begin{bmatrix} 10 & 6 \\ 6 & 10 \end{bmatrix}\begin{bmatrix} a \\ b \end{bmatrix} = \lambda \begin{bmatrix} a \\ b \end{bmatrix}$$

The eigen equation becomes as

$$\begin{vmatrix} 10-\lambda & 6 \\ 6 & 10-\lambda \end{vmatrix} = 0 \qquad \Rightarrow \qquad \lambda^2 - 20\lambda + 64 = (\lambda - 16)(\lambda - 4) = 0$$

and the eigenvalues are obtained as

$$\lambda_{max} = 16, \ \lambda_{min} = 4$$

The eigenvector for the 1st component becomes as

$$10a_1 + 6b_1 = 16a_1 \ \rightarrow \ b_1 = a_1$$

By normalization, we have

$$p_1^T = [a_1, b_1] = [1/\sqrt{2}, 1/\sqrt{2}]$$

In the same way, the eigenvector for the 2nd component becomes as

$$p_2^T = [a_2, b_2] = [-1/\sqrt{2}, 1/\sqrt{2}]$$

Therefore, the 1st component and the 2nd component are obtained as

$$t_1 = Dp_1 = \frac{1}{\sqrt{2}} X + \frac{1}{\sqrt{2}} Y, \qquad t_2 = -\frac{1}{\sqrt{2}} X + \frac{1}{\sqrt{2}} Y$$

The values of t_1 and t_2 for the 1st data $[X_1, X_2] = [2,2]$ becomes as $t_1 = 2\sqrt{2}$, $t_2 = 0$. Similarly, the other values are also obtained for the other data as

$$Dp = D[p_1, p_2] = \begin{bmatrix} 2\sqrt{2} & 0 \\ 0 & -\sqrt{2} \\ 0 & \sqrt{2} \\ -2\sqrt{2} & 0 \end{bmatrix}$$

If we plot these 4 points onto t_1-t_2 axis, these are the rotation of X-Y axis by $45°$. The relationship between (X,Y) and (t_1,t_2) becomes as

$$\begin{bmatrix} t_1 \\ t_2 \end{bmatrix} = \begin{bmatrix} -1/\sqrt{2} & 1/\sqrt{2} \\ -1/\sqrt{2} & 1/\sqrt{2} \end{bmatrix} \begin{bmatrix} X \\ Y \end{bmatrix} = \begin{bmatrix} p_1^T \\ p_2^T \end{bmatrix} \begin{bmatrix} X \\ Y \end{bmatrix} = p^T \begin{bmatrix} X \\ Y \end{bmatrix}$$

The conversion matrix for the rotation of axis P^T may be expressed as

$$P^T = \begin{bmatrix} \cos\theta & \sin\theta \\ -\sin\theta & \cos\theta \end{bmatrix}$$

where θ is the counter clockwise degree and $\theta = 45°$ in the above example.

As seen above, the principal component analysis is to find the new axis which gives the maximum variance and the axis are orthogonal. The main principal component analysis can be expressed as

$$D^T DP = D^T D[p_1, p_2] = [\lambda_1 p_1, \lambda_2 p_2] = [p_1, p_2] \begin{bmatrix} \lambda_1 & 0 \\ 0 & \lambda_2 \end{bmatrix} = P\Lambda \tag{89}$$

where this is considered to be the similarity transformation by $D^T D$. If we multiply P^{-1} from the left to the above equation, we have

$$P^{-1}D^T DP = \Lambda \tag{90}$$

where P is orthogonal as mentioned before such that $P^{-1}=P^T$. Then the above equation becomes as

$$P^{-1}D^T DP = P^T D^T DP = (DP)^T DP = \Lambda \tag{91}$$

For the above example,

$$(DP)^T DP = \begin{bmatrix} 2\sqrt{2} & 0 & 0 & -2\sqrt{2} \\ 0 & -\sqrt{2} & \sqrt{2} & 0 \end{bmatrix} \begin{bmatrix} 2\sqrt{2} & 0 \\ 0 & -\sqrt{2} \\ 0 & \sqrt{2} \\ -2\sqrt{2} & 0 \end{bmatrix} = \begin{bmatrix} 16 & 0 \\ 0 & 4 \end{bmatrix}$$

which indicates that $(DP)^T(DP)$ is the diagonal matrix, and its diagonal components are the eigenvalues.

PCA for Multiple Sets of Data

Consider the n set of data as expressed as

$$X = \begin{bmatrix} X_{11} & X_{12} & . & . & X_{1d} \\ . & . & . & . & . \\ . & . & . & . & . \\ . & . & . & . & . \\ X_{n1} & X_{n2} & . & . & X_{nd} \end{bmatrix} \tag{92}$$

The main component in this case is the d sets of data, and the 1st component is the one that maximizes the sum of square of this main component score such that

$$t_i = \sum_{k=1}^{d} X_{ik} p_k \tag{93}$$

where the normalization condition is expressed as

$$\sum_{k=1}^{d} p_k^2 = 1 \tag{94}$$

The square sum of the main component is expressed as

$$j = \sum_{k=1}^{n} t_k^2 = \sum_{i=1}^{n} \left(\sum_{k=1}^{d} X_{ik} p_k \right)^2 \tag{95}$$

The problem is to find $p_1, p_2, \ldots, p_d$ that maximize j as given above under the constraint of Eq.(94), and can be solved by using the Lagrange multiplier as before such that

$$J = j - \lambda \left(\sum_{k=1}^{d} p_k^2 - 1 \right) \tag{96}$$

By taking the derivative of this equation with respect to p_1, and setting the resulting equation as

$$\left(\sum_{k=1}^{d} X_{k1}^2 - \lambda \right) p_1 + \sum_{k=1}^{d} X_{k1} X_{k2} p_2 + \ldots + \sum_{k=1}^{d} X_{k1} X_{kd} p_d = 0 \tag{97}$$

Similar equations can be obtained for $p_2, p_2, \ldots, p_d$ as well, and these can be expressed by the matrix representation as

$$(X^T X - \lambda I) P = 0 \tag{98}$$

For this equation to have the nontrivial solution, the determinant of the coefficient matrix must be 0 such that

$$| X^T X - \lambda I | = 0 \tag{99}$$

Let $\lambda_1, \lambda_2, \ldots, \lambda_d$ be the solution or roots to this eigen equation. Let $\lambda_1 > \lambda_2 > \ldots > \lambda_d$ without loss of generality. Then, the 1st component t_1, the 2nd component t_2 *etc.* can be obtained as

$$t_1 = X^T p_1, \; t_2 = X^T p_2, \; \ldots, \; t_d = X^T p_d \tag{100}$$

where let the matrix be defined as

$$P \equiv [\mathbf{p}_1, \mathbf{p}_2, \ldots, \mathbf{p}_d]$$

Let us rewrite Eq.(98) as before as

$$X^T X \mathbf{p}_k = \lambda_k \mathbf{p}_k \ (k=1,2,\ldots,d) \tag{101}$$

Moreover, this can be further expressed as

$$X^T XP = X^T X[p_1, p_2, \ldots, p_d] = [\lambda_1 p_1, \lambda_2 p_2, \ldots, \lambda_d p_d]$$

$$= [p_1, p_2, \ldots, p_d] \begin{bmatrix} \lambda_1 & & & 0 \\ & \lambda_2 & & \\ & & \ddots & \\ 0 & & & \lambda_d \end{bmatrix} = P\Lambda \tag{102}$$

where P is the orthogonal matrix ($P^{-1}=P^T$), and Λ is the diagonal matrix with eigenvalues to be the diagonal elements. Multiplying P^{-1} from the left to the above equation, we have the following expression by the similarity transformation:

$$P^{-1} X^T XP = P^T X^T XP = \Lambda \tag{103}$$

The principal component analysis tries to express the data by the main components, and thus some of the original information will be lost. The extent to which extent the variance of the main components over the original information may be expressed as

$$\% \text{ Var of the 1}^{st} \text{ component} = \frac{\lambda_1}{\sum\limits_{k=1}^{d} \lambda_k} \times 100 \tag{104a}$$

$$\% \text{ Var of the 1}^{st} \text{ and 2}^{nd} \text{ components} = \frac{\lambda_1 + \lambda_2}{\sum\limits_{k=1}^{d} \lambda_k} \times 100 \tag{104b}$$

The equivalent information may be obtained by plotting the scores of the 1st and the 2nd components onto t_1, and t_2 axis. This plot is called as **Karhunen-Loove (K-L) plot**.

Singular Value Decomposition

Let the score matrix T be $[t_1, t_2, \ldots, t_d]$, then the following equation holds as mentioned before

$$T = XP \tag{105}$$

Since P is the orthogonal matrix, by multiplying P^{-1} from the right, we have

$$X = TP^{-1} = TP^{T} = [t_1, t_2, ..., t_d] \begin{bmatrix} p_1^{T} \\ p_2^{T} \\ . \\ . \\ p_d^{T} \end{bmatrix} \tag{106}$$

As seen before, the norm of the eigenvector **p** is normalized, and the norm of the score vector for a component ta is equal to the square root of the eigenvalue of λ_a of $X^{T}X$ such that

$$\|t_a\| = \sqrt{\lambda a} = \sigma_a \tag{107}$$

where σ_a is the singular value as defined before in Eq.(76).

Let the score vector **t** for the main component be normalized to 1, and let this be **u**, and the eigenvector **p** be expressed as **v**, and let Σ be the diagonal matrix whose diagonal elements are the singular values, then X can be expressed as

$$X = [u_1, u_2, ..., u_d] \begin{bmatrix} \sigma_1 & & & 0 \\ & \sigma_2 & & \\ & & . & \\ & & & . \\ 0 & & & \sigma_d \end{bmatrix} \begin{bmatrix} v_1^{T} \\ v_2^{T} \\ . \\ . \\ v_d^{T} \end{bmatrix} \tag{108}$$

This expression is called as **singular value decomposition**.

CONCLUDING REMARKS

Here, the basis for systems analysis was explained. Those are used in the later Chapters. Although most of the explanation is given for the linear system, this is also useful for the perturbation analysis of non-linear systems. The graph theoretic approach is useful for the metabolic network analysis, in particular as the network size becomes large. The data analysis is important for the analysis of experimental data and for the evaluation of the model.

APPENDIX A LAPLACE TRANSFORMATION

The integration of such function f(t) as $\sin t$ or t from 0 to ∞ becomes infinity, resulting in indetermination. However, the integration of $f(t)e^{-t}$ converges to the finite value. Thus, e^{-t} plays a role to converge the functions. Let us consider the following integration.

$$F(s) = \int_0^\infty f(t)e^{-st}dt \tag{A1}$$

For the $f(t)$ such as $\sin t$ and t, RHS of Eq. (A1) converges for the constant value of s. Although $f(t)e^{-st}$ is a function of time, it become a function of s after the integration. Transforming $f(t)$ into $F(s)$ by Eq. (A1) is called as Laplace transformation, and it is expressed as $\mathcal{L}$. Thus,

$$\mathcal{L}[f(t)] = \int_0^\infty f(t)e^{-st}dt = F(s) \tag{A2}$$

In addition, the following equation holds since Laplace transformation is the linear operation.

$$\mathcal{L}[af(t) + bg(t)] = a\mathcal{L}[f(t)] + b\mathcal{L}[g(t)] \tag{A3}$$

Laplace transformations of typical functions are shown in Table **A1** and Table **A2**. The basic properties of Laplace transformation are shown in Table **A3**.

The transformation of $F(s)$ into $f(t)$ is called as **inverse Laplace transformation**, and it is expressed as $\mathcal{L}^{-1}$.

Table A1: Laplace transformation of typical function

$f(t)$	$F(s) = \mathcal{L}[f(t)] = \int_0^\infty f(t)e^{-st}dt$
$u(t) = 1(t)$ (unit step function)	$\dfrac{1}{s}$
$\delta(t)$ (delta function)	1
t^n	$\dfrac{n!}{s^{n+1}}$ $(n = 0, 1, 2, \cdots)$
e^{at}	$\dfrac{1}{s-a}$
$\sin at$	$\dfrac{a}{s^2 + a^2}$
$\cos at$	$\dfrac{s}{s^2 + a^2}$

Table A2: Laplace transformation of complex function

$f(t)$	$F(s)$
te^{at}	$\dfrac{1}{(s-a)^2}$
$e^{at}\sin bt$	$\dfrac{b}{(s-a)^2+b^2}$
$e^{at}\cos bt$	$\dfrac{(s-a)}{(s-a)^2+b^2}$

Table A3: The basic properties of Laplace transformation

$\mathcal{L}\left[\dfrac{df(t)}{dt}\right]=sF(s)-f(0)$
$\mathcal{L}[f^n(t)]=s^nF(s)-s^{n-1}f(0)-\cdots-f^{(n-1)}(0)$
$\mathcal{L}[\int f(t)dt]=\dfrac{F(s)}{s}-\dfrac{f^{(n-1)}(0)}{s}$
$\mathcal{L}[f(t-L)]=e^{-Ls}F(s)$
$\lim\limits_{t\to 0}f(t)=\lim\limits_{s\to\infty}sF(s)$

APPENDIX B PARTIAL FRACTION

Let us consider the rational function expressed as.

$$f(x) = \frac{Q_m(x)}{P_n(x)} \equiv \frac{b_m x^m + \cdots + b_0}{x^n + a_{n-1} x^{n-1} + \cdots + a_1 x + a_0} \tag{B1}$$

where $P_n(x)$ and $Q_m(x)$ are the polynomials ($m < n$). In the case where $m \geq n$, f(x) can be re-expressed as

$$f(x) = R_{m-n}(x) + \frac{Q_{m'}(x)}{P_n(x)} \quad (n > m') \tag{B2}$$

where $R_{m-n}(x)$ is the (m-n)th order polynomial.

Eq. (B1) can be expresses as partial fractions as

$$f(x) = \frac{A_1}{x-a_1} + \frac{A_2}{x-a_2} + \cdots + \frac{A_n}{x-a_n} \tag{B3}$$

where $a_1, a_2, \cdots, a_n$ represent the solutions for the denominator of $f(x) = 0$. Namely, they are roots of $P_n(x) = 0$. By multiplying both sides of Eq.(B3) by $(x - a_i)$, we have

$$(x - a_i)f(x) = \frac{A_1(x-a_i)}{x-a_1} + \cdots + A_i + \cdots + \frac{A_n(x-a_i)}{x-a_n} \tag{B4}$$

Then, by the operation as $x \to a_i$, the following equation is obtained.

$$A_i = \lim_{x \to a_i} f(x)(x - a_i) \quad (i = 1, 2, \cdots, n) \tag{B5}$$

where it was assumed that $P_n(x) = 0$ does not have the same roots. In the case when $P_n(x) = 0$ has the same roots (for example, k-th term has the same root), $f(x)$ can be decomposed as follows:

$$f(x) = \frac{A_1}{x-a_1} + \cdots + \frac{A_{k2}}{(x-a_k)^2} + \frac{A_k}{x-a_k} + \cdots + \frac{A_n}{x-a_n} \tag{B6}$$

where A_i ($i = 1, 2, \cdots, n$) can be obtained by Eq.(B5) except for A_{k2} and A_k. A_{k2} can be obtained as follows:

$$A_{k2} = \lim_{x \to a_k} (x - a_k)^2 f(x) \tag{B7}$$

and A_k is obtained as follows:

$$A_k = \lim_{x \to a_k} \frac{d}{dx}\{(x - a_k)^2 f(x)\} \tag{B8}$$

DISCLOSURE

Part of this chapter has been previously published in [12].

REFERENCES

[1]   Tebbut P. Basic mathematics for chemist, John Wiley & Sons, Chichester 1994.
[2]   Hecht HG. Mathematics in chemistry-An introduction to modern method, Prentce-Hall, New York 1990.
[3]   Jenson VG, Jeffreys GV. Mathematical methods in chemical engineering, Academic Press 1977.
[4]   Wylie CR. Advanced engineering mathematics, McGraw-Hill 1975.
[5]   Bailey JE, Ollis DF. Biochemical engineering fundamentals. McGraw-Hill 1986.
[6]   Massart DL *et al.* (ed.) Chemometrics Tutorials, Elsevier 1991.
[7]   Amundsen NR. Mathematical methods in chemical engineering, Prentice-Hall, Englewood Cliffs, New Jersey 1966.
[8]   Deo N Graph theory with application to engineering and computer science, Prentice-Hall, Englewood Cliff, N.J., USA 1974
[9]   Mah RSH. Chemical process structures and information flows, Butterworth 1990.
[10]   Himmelblau DM. Bischoff KB, Process analysis and simulation, deterministic system, John Wiley, N.Y., USA 1968
[11]   King EL, Altman C. A schematic method of deriving the rate laws for enzyme-catalyzed reactions. J Phys Chem 1956; 60: 1375-1378
[12]   Shimizu K, Bioprocess systems analysis method, Corona Pub Co., Japan, 1997 (in Japanese).

CHAPTER 3

Fundamentals of Modeling of Biosystems

Abstract: The transport phenomena such as mass balance, momentum balance, and heat balance are briefly explained for the basis of modeling, where all the appropriate models must be constructed based on the principles which govern the organisms and their living environment. The model reduction by singular perturbation is explained for the simple enzyme reaction. The basic modeling approaches such as flux balance analysis (FBA), metabolic flux analysis (MFA), kinetic model, and their integration are briefly explained. Unstructure models are explained for the batch, fed-batch, and continuous cultures. For the modeling of Eco-systems, some population dynamic models are explained. As the data-driven modeling, the structure and the algorithm of artificial neural networks (ANNs) are explained together with simple examples. These modeling approaches may cover a variety of modeling approaches to the variety of cell systems.

Keywords: Flux balance analysis, metabolic flux analysis, kinetic modeling, mass balance, momentum balance, heat balance, unstructured model, cybernetic model, singular perturbation, population dynamics, age model, artificial neural networks (ANNs).

INTRODUCTION

The modeling is the extraction and reconstruction of the real world, and the model must contain the essential feature of the system of concern, but it is not necessary to incorporate all the details present in the real world. The purpose of modeling and computer simulation may be to understand the cell's behavior based on the predicted simulation result, and to optimize the culture condition. We may be able to deepen our knowledge about the metabolic regulation mechanism for further research progress for the dynamic metabolic analysis based on the computer simulation by the appropriate models. Moreover, with the reliable and valid models, the process performances may be improved, and can be validated *in silico*, thus saving time and resources, where the accurate metabolic models can contribute to the improvement of the cell factories in terms of yield, titer, and productivity of the desired product by the systematic metabolic engineering. The models range from microscopic intracellular cell machinery to macroscopic bioreactor processes. The typical dynamic modeling can be made based on the mass balances with stoichiometry or kinetics, giving a set of differential equations.

Many biological processes are dynamical or non-stationary in nature, and the conventional approach for the modeling of such systems is to set up mathematical

Kazuyuki Shimizu and Yu Matsuoka

expression by applying mass balances for each species in the system together with rate equations for biochemical reactions. Although the mechanistic modeling which describes the non-linear kinetic rate equations is straightforward, its extension to a large-scale is limited.

For the modeling of the cell's dynamical system, it must be careful about the time scales. Roughly speaking, the enzyme level regulation with allosteric modification of the enzyme with the metabolite is fast on the order of seconds or minutes, while the transcriptional reprogramming of the metabolism is slow on the order of minutes to hours. The enzyme level regulation includes kinase/ phosphatase reactions together with conformational changes of proteins or enzymes. The difference of the time scales poses computationally challenges for the quantitative analysis of the whole cell system, where the systems of having fast and slow dynamics are inherently "stiff" [1], and become sensitive to modeling errors [2]. Thus the model reduction [3], or singular perturbation [2] may be considered to overcome such a problem.

In the present chapter, transport phenomena is first explained for the notion of balance, which is the basis for modeling. Then, the simple modeling for enzymatic reaction for Michelis-Menten equation is explained, followed by the model reduction by singular perturbation. Then the modeling for the typical batch, fed-batch, and continuous cultures are explained with unstructured kinetic models. The modeling for population dynamics of eco-system is also explained. Moreover, the black-box type of modeling such as artificial neural networks (ANNs) are explained for the case of using only input-output data.

TRANSPORT PHENOMENA FOR THE BASIS OF MODELING

All the substances in the universe is composed of small particles, which are discrete in nature. In order to grasp the very basic nature of the substances, we have to understand the principles which govern the universe, or on earth. Here, we consider the **law of conservation** for substance, momentum, and heat, or in more generally **transport phenomena**. The notion of "**balance**" is the center for the modeling of natural phenomena and the engineering applications [4-6].

Mass Balance

Consider the small substance composed of the length Δx in x-axis direction, Δy in the y-axis, Δz in the z-axis direction in the 3-dimensional space as shown in Fig. **1**.

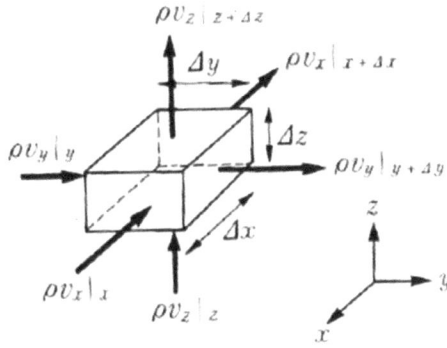

Figure 1: Mass balance for the rectangular object.

Consider the mass balance, or apply the law of conservation for this subject such that

[rate of mass accumulation] =[rate of mass in] – [rate of mass out]   (1)

Consider only x-axis direction first, where the mass entering at x per unit time and per unit area is ($\rho \, v_x$) | $_x$, and thus total mass of ($\rho \, v_x \Delta y \Delta z$) | $_x$ flows into the system, while ($\rho \, v_x \Delta y \Delta z$) | $_{x+\Delta x}$ flows out of the system, where ρ is the mass, and v_x is the velocity in the x-axis direction. In the similar way, by considering the mass balances for y-axis direction and z-axis direction, the following equation is obtained:

$$\Delta x \Delta y \Delta z \frac{\partial \rho}{\partial t} = \Delta y \Delta z [\rho v_x|_x - \rho v_x|_{x+\Delta x}] + \Delta z \Delta x [\rho v_y|_y - \rho v_y|_{y+\Delta y}] + \Delta x \Delta y [\rho v_z|_z - \rho v_z|_{z+\Delta z}] \quad (2)$$

Dividing this equation by $\Delta x \Delta y \Delta z$, we have

$$\frac{\partial \rho}{\partial t} = -[\frac{(\rho v_x)|_{x+\Delta x} - (\rho v_x)|_x}{\Delta x} + \frac{(\rho v_y)|_{y+\Delta y} - (\rho v_y)|_y}{\Delta y} + \frac{(\rho v_z)|_{z+\Delta z} - (\rho v_z)|_z}{\Delta z}] \quad (3)$$

Let Δx, Δy, Δz be infinitesimally small, and assume continuum, the following **equation of continuity** is obtained:

$$\frac{\partial \rho}{\partial t} = -(\frac{\partial}{\partial x} \rho v_x + \frac{\partial}{\partial y} \rho v_y + \frac{\partial}{\partial z} \rho v_z) \quad (4)$$

This may be expressed by the vector notation as

$$\frac{\partial \rho}{\partial t} = -\nabla \bullet \rho v \qquad (5)$$

where,

$$\nabla \equiv \left[\frac{\partial}{\partial x}, \frac{\partial}{\partial y}, \frac{\partial}{\partial z}\right], \quad v \equiv \left[v_x, v_y, v_z\right]^T$$

Eq.(4) is the representation in the Cartesian coordinate. In the case of cylindrical coordinate, Eq.(4) can be expressed as

$$\frac{\partial \rho}{\partial t} = -\left[\frac{1}{r}\frac{\partial}{\partial r}(\rho r v_r) + \frac{1}{r}\frac{\partial}{\partial \theta}(\rho v_\theta) + \frac{\partial}{\partial z}\rho v_z)\right] \qquad (5)'$$

where v_r and v_θ are the r-directional and θ-directional components of v. The relationship between the two coordinates is expressed as (Fig. **2**)

$$x = r\cos\theta, \quad y = r\sin\theta, \quad z = z \quad or \quad r = \sqrt{x^2 + y^2}, \quad \theta = \tan^{-1}\frac{y}{x}, \quad z = z$$

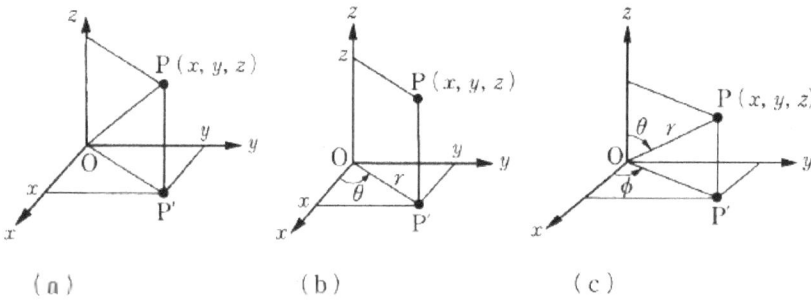

Figure 2: Change of coordinate: (a) Cartesian coordinate, (b) cylindrical coordinate, (c) spherical coordinate.

In the spherical coordinate, Eq.(4) can be expressed as

$$\frac{\partial \rho}{\partial t} = -\left[\frac{1}{r^2}\frac{\partial}{\partial r}(\rho r^2 v_r) + \frac{1}{r\sin\theta}\frac{\partial}{\partial \theta}(\rho v_\theta \sin\theta) + \frac{1}{r\sin\theta}\frac{\partial}{\partial \varphi}(\rho v_\varphi)\right] \qquad (5)''$$

where v_r, v_θ, and v_ϕ are the r-directional, θ-directional, and ϕ-directional components of v, respectively. The relationship between the Cartesian coordinate and the spherical coordinate is expresses as (Fig. **2**)

$$x = r\sin\theta\cos\varphi, \qquad y = r\sin\theta\sin\varphi, \qquad z = r\cos\theta$$

$$or \qquad r = \sqrt{x^2 + y^2 + z^2}, \qquad \theta = \tan^{-1}\left(\frac{\sqrt{x^2 + y^2}}{z}\right), \qquad \varphi = \tan^{-1}\left(\frac{y}{x}\right)$$

Now reconsider Eq.(4), and re-express as

$$\frac{\partial\rho}{\partial t} + v_x\frac{\partial\rho}{\partial x} + v_y\frac{\partial\rho}{\partial y} + v_z\frac{\partial\rho}{\partial z} = -\rho\left(\frac{\partial v_x}{\partial x} + \frac{\partial v_y}{\partial y} + \frac{\partial v_z}{\partial z}\right) \tag{6}$$

Let the total derivative be defined as

$$\frac{D}{Dt} \equiv \frac{\partial}{\partial t} + \frac{\partial}{\partial x}\frac{dx}{dt} + \frac{\partial}{\partial y}\frac{dy}{dt} + \frac{\partial}{\partial z}\frac{dz}{dt} = \frac{\partial}{\partial t} + v_x\frac{\partial v_x}{\partial x} + v_y\frac{\partial v_y}{\partial y} + v_z\frac{\partial v_z}{\partial z} \tag{7}$$

Using this notation, Eq.(6) can be expressed as

$$\frac{D\rho}{Dt} = -\rho\left(\frac{\partial v_x}{\partial x} + \frac{\partial v_y}{\partial y} + \frac{\partial v_z}{\partial z}\right) \tag{8a}$$

or in vector notation as

$$\frac{D\rho}{Dt} = -\rho(\nabla \bullet v) \tag{8b}$$

In the case of non-compressive fluid, ρ can be considered to be constant, and thus

$$D\rho/Dt=0.$$

So far, we considered only one component system. Consider the case of multi-component system, and also consider the case where the reaction is taking place in the system. In the similar way as considered above, the mass balance for the component A gives the following equation:

$$\frac{\partial\rho_A}{\partial t} = -\left(\frac{\partial}{\partial x}n_{Ax} + \frac{\partial}{\partial y}n_{Ay} + \frac{\partial}{\partial z}n_{Az}\right) + r_A \tag{9}$$

where n_{Ax}, n_{Ay}, n_{Az} are the mass of A which transports through the direction of x-axis, y-axis, and z-axis, respectively, per unit time and per unit area. r_A is the reaction rate of A.

If the concentration of A changes with respect to some direction such as x-direction, then the amount of A moves in the x-direction by diffusion. Let j_{Ax} be the flux by diffusion. Then the following equation holds:

$$j_{AX} = -cD\frac{dx_A}{dx} \tag{10}$$

where this is the **Fick's law of diffusion**, D is the diffusion coefficient, c is the molar density, and x_A is the molar fraction.

Consider the general system having volume V as shown in Fig. **3** instead of rectangular substance. Then the general form for mass balance may be expressed as

$$\int_V \left(\frac{\partial c}{\partial t}\right) dV + \int_s J \bullet n dS + \int_V r dV = 0 \tag{11}$$

where we consider general domain V instead of rectangular substance, and the first term in the left hand side of the above equation represents the change in the concentration of the component of interest. The 2nd term is the flux transports through the surface S per time.

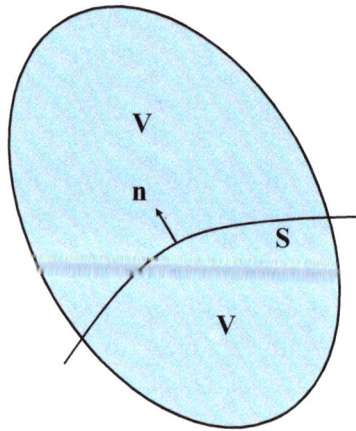

Figure 3: Mass balance for the general object.

The 3rd term is the rate of accumulation or disappearance of c by reaction. **J** is the flux, and **n** is the outward directed vector perpendicular to the surface of S (Fig. **3**). **r** is the reaction rate. Let us apply the **divergence theorem** to the 2nd term of the above equation such that

$$\int_S (J \bullet n) dS = \int_v (\nabla \bullet J) dv$$

Then, we have

$$\int_V \left(\frac{\partial c}{\partial t} + \nabla \bullet J + r \right) dV = 0 \tag{12}$$

where the flux by diffusion can be expressed as $J = -D\nabla c$. Then the **diffusion-reaction equation** for the x-axis direction can be expressed as

$$\frac{\partial c}{\partial t} = -r + D \frac{\partial^2 c}{\partial x^2} \tag{13}$$

where D was assumed to be constant. This equation is known as Belousov-Zhabotinsky (BZ) reaction equation, where this reaction system was first considered by Belousov in 1951, and modified by Zhabotinsky in 1970. This reaction system gives spacio-temporal oscillation by the simple chemical reaction with chemical wave.

Momentum Balance

Consider the momentum balance for the rectangular substances as shown in Fig. **1** again.

[Accumulated momentum per unit time]=

[momentum flow in]-[momentum flow out] +[power acted on the system] (14)

In order to derive **equation of motion** based on the above equation, we have to take into account the effects of ①convection, ②molecular transport, ③pressure, and ④gravity. In the similar way as considered for mass balance, consider the x-axis direction first, where the mass transported across the area $\Delta y \Delta z$ is $\rho v_x \Delta y \Delta z$ as seen before, where v_x is the velocity of the mass moved into the system. Therefore, the momentum accumulation is the difference between the amount of moved in at x and moved out at $x + \Delta x$ such that $\{(\rho v_x v_x) \mid x - (\rho v_x v_x \mid _{x+\Delta x})\} \Delta y \Delta z$. In the similar way, the momentum balance for the x-axis direction becomes as

$$\{(\rho v_x v_x)|_x - (\rho v_x v_x)|_{x+\Delta x}\}\Delta y \Delta z + \{(\rho v_y v_x)|_y - (\rho v_y v_x)|_{y+\Delta y}\}\Delta z \Delta x + \{(\rho v_z v_x)|_z - (\rho v_z v_x)|_{z+\Delta z}\}\Delta x \Delta y$$

where the 2nd block terms are the momentum transfer through $\Delta z \Delta x$ for x-axis direction, and the 3rd block terms are that through $\Delta x \Delta y$ for x-axis direction.

Consider next the molecular transport by diffusion. Let τ be the momentum transfer by molecular diffusion per unit time and per unit area. Then this can be expressed as

$$\tau_{yx} = -\eta \frac{\partial v_x}{\partial y}$$

(15)

where η is the coefficient. In the similar way as the momentum transfer by convection, the momentum balance by molecular diffusion may be expressed as

$$(\tau_{xx}|_x - \tau_{xx}|_{x+\Delta x})\Delta y \Delta z + (\tau_{yx}|_y - \tau_{yz}|_{y+\Delta y})\Delta z \Delta x + (\tau_{zx}|_z - \tau_{zx}|_{z+\Delta z})\Delta x \Delta y$$

The pressure balance can be expressed as $(p \mid _x - p \mid _{x+\Delta x}) \Delta y \Delta z$ in the x-axis direction. Moreover, let g_x be the x-component of the gravity g acting on the unit mass, then the gravity acting on the system in the x-axis direction is expressed as $\rho g_x \Delta x \Delta y \Delta z$.

In summary, the momentum balance equation can be expressed by considering all the factors and by assuming the continuum by infinitesimally reducing the magnitudes of Δx, Δy, and Δz as

$$\frac{\partial}{\partial t}(\rho v_x) = -[\frac{\partial}{\partial x}(\rho v_x v_x) + \frac{\partial}{\partial y}(\rho v_y v_x) + \frac{\partial}{\partial z}(\rho v_z v_x)] - (\frac{\partial \tau_{xx}}{\partial x} + \frac{\partial \tau_{yx}}{\partial y} + \frac{\partial \tau_{zx}}{\partial z}) - \frac{\partial p}{\partial x} + \rho g_x$$

(16)

where this is the equation for x-axis direction. The similar equations are obtained also for y-axis and z-axis directions, and the resulting equations can be expressed by vector-tensor representation as

$$\frac{\partial}{\partial t}(\rho v) = -[\nabla \bullet \rho vv] - [\nabla \bullet T] - \nabla p + \rho g$$

(17)

where vv is called dyadic and is computed as

$$vv = \begin{bmatrix} v_x \\ v_y \\ v_z \end{bmatrix} [v_x v_y v_z] = \begin{bmatrix} v_x v_x & v_x v_y & v_x v_z \\ v_y v_x & v_y v_y & v_y v_z \\ v_z v_x & v_z v_y & v_z v_z \end{bmatrix}$$

Then $\nabla \cdot vv$ is expressed as

$$\nabla \bullet vv = \begin{bmatrix} \dfrac{\partial v_x v_x}{\partial x} + \dfrac{\partial v_y v_x}{\partial y} + \dfrac{\partial v_z v_x}{\partial z} \\[2ex] \dfrac{\partial v_x v_y}{\partial x} + \dfrac{\partial v_y v_y}{\partial y} + \dfrac{\partial v_z v_y}{\partial z} \\[2ex] \dfrac{\partial v_x v_z}{\partial x} + \dfrac{\partial v_y v_z}{\partial y} + \dfrac{\partial v_z v_z}{\partial z} \end{bmatrix}$$

T is the matrix for τ such that

$$T \equiv \begin{bmatrix} \tau_{xx} & \tau_{xy} & \tau_{xz} \\ \tau_{yx} & \tau_{yy} & \tau_{yz} \\ \tau_{zx} & \tau_{zy} & \tau_{zz} \end{bmatrix}$$

Then $\nabla \cdot T$ becomes as

$$\nabla \bullet T = \begin{bmatrix} \dfrac{\partial \tau_{xx}}{\partial x} + \dfrac{\partial \tau_{yx}}{\partial y} + \dfrac{\partial \tau_{zx}}{\partial x} \\[2ex] \dfrac{\partial \tau_{xy}}{\partial x} + \dfrac{\partial \tau_{yy}}{\partial y} + \dfrac{\partial \tau_{zy}}{\partial z} \\[2ex] \dfrac{\partial \tau_{xz}}{\partial x} + \dfrac{\partial \tau_{yz}}{\partial y} + \dfrac{\partial \tau_{zz}}{\partial z} \end{bmatrix}$$

$\nabla \cdot p$ and ρg can be also expressed as

$$\nabla \bullet p = \begin{bmatrix} \dfrac{\partial p}{\partial x} \\[2ex] \dfrac{\partial p}{\partial y} \\[2ex] \dfrac{\partial p}{\partial z} \end{bmatrix}, \quad \rho g = \rho \begin{bmatrix} g_x \\ g_y \\ g_z \end{bmatrix}$$

Now Eq.(16) may be expressed as

$$\rho \frac{\partial v_x}{\partial t} + v_x \frac{\partial \rho}{\partial t} = -[\{v_x \frac{\partial \rho v_x}{\partial x} + \rho v_x \frac{\partial v_x}{\partial x}\} + \{v_x \frac{\partial \rho v_y}{\partial y} + \rho v_y \frac{\partial v_x}{\partial y}\} + \{v_x \frac{\partial \rho v_z}{\partial z}$$

$$+ \rho v_z \frac{\partial v_x}{\partial z}\}] - (\frac{\partial \tau_{xx}}{\partial x} + \frac{\partial \tau_{yx}}{\partial y} + \frac{\partial \tau_{zx}}{\partial z}) - \frac{\partial p}{\partial x} + \rho g_x \qquad (18)$$

Consider then multiplying v_x to the equation of continuity such as Eq.(4) as follows:

$$v_x \frac{\partial \rho}{\partial t} = -(v_x \frac{\partial}{\partial x}\rho v_x + v_x \frac{\partial}{\partial y}\rho v_y + v_x \frac{\partial}{\partial z}\rho v_z)$$

This equation can be used to rearrange Eq.(18) as

$$\rho \frac{\partial v_x}{\partial t} + \rho v_x \frac{\partial v_x}{\partial x} + \rho v_y \frac{\partial v_x}{\partial y} + \rho v_z \frac{\partial v_x}{\partial z} = -(\frac{\partial \tau_{xx}}{\partial x} + \frac{\partial \tau_{yx}}{\partial y} + \frac{\partial \tau_{zx}}{\partial z}) - \frac{\partial p}{\partial x} + \rho g_x \quad (19)$$

where the LHS of this equation can be expressed as total derivative such as $\rho \, Dv_x/Dt$, and also by considering all the directions for Eq.(18), the vector-tensor representation becomes as

$$\rho \frac{Dv}{Dt} = -[\nabla \bullet T] - \nabla p + \rho g \quad (29)$$

If we plot τ_{yx} *versus* $-dv_x/dy$ for various fluids, we can classify the fluids into several groups. Among them, the most typical fluid is the **Newtonian fluid** which shows the linear relationship between them, and important in engineering practice. Namely the following relationships hold:

$$\tau_{xx} = -2\eta \frac{\partial v_x}{\partial x} + 2\eta \frac{(\nabla \bullet v)}{3}, \; \tau_{yy} = -2\eta \frac{\partial v_y}{\partial y} + 2\eta \frac{(\nabla \bullet v)}{3}, \; \tau_{zz} = -2\eta \frac{\partial v_z}{\partial z} + 2\eta \frac{(\nabla \bullet v)}{3}$$

$$\tau_{xy} = \tau_{yx} = -\eta \left(\frac{\partial v_x}{\partial y} + \frac{\partial v_y}{\partial x} \right), \; \tau_{yz} = \tau_{zy} = -\eta \left(\frac{\partial v_y}{\partial z} + \frac{\partial v_z}{\partial y} \right), \; \tau_{zx} = \tau_{xz} = -\eta \left(\frac{\partial v_z}{\partial x} + \frac{\partial v_x}{\partial z} \right)$$

Substituting these relationships into Eq.(29), we have the following equation:

$$\rho \frac{\partial v_x}{\partial t} = \frac{\partial}{\partial x}[2\eta \frac{\partial v_x}{\partial x} - \frac{2}{3}\eta(\nabla \bullet v)] + \frac{\partial}{\partial y}[\eta(\frac{\partial v_x}{\partial y} + \frac{\partial v_y}{\partial x})] + \frac{\partial}{\partial z}[\eta(\frac{\partial v_z}{\partial x} + \frac{\partial v_x}{\partial x})] - \frac{\partial p}{\partial x} + \rho g_x$$

$$(30)$$

The similar equations can be obtained in the y and z directions, and these are called **Navier-Stokes** (NS) **equations**. In the case where ρ and η are constant, the NS equations become as

$$\rho \frac{Dv}{Dt} = \eta \nabla^2 v - \nabla p + \rho g \tag{31}$$

where ∇^2 is called as Laplacian, and is defined as

$$\nabla^2 \equiv \frac{\partial^2}{\partial x^2} + \frac{\partial^2}{\partial y^2} + \frac{\partial^2}{\partial z^2}$$

In the case when $\eta = 0$, Eq.(31) becomes as

$$\rho \frac{Dv}{Dt} = -\nabla p + \rho g \tag{32}$$

where this equation is called as **Euler's equation of motion**. If $\eta = 0$, the fluid is called as **perfect fluid**, while if $\eta = 0$ and ρ is constant, the fluid is called as **ideal fluid**.

Examples for the Application of Mass and Momentum Balance

Consider the practical application for the above equations. Let us first find the velocity distribution of the fluid flowing down on the wetted wall as illustrated in Fig. **4**.

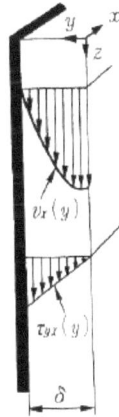

Figure 4: Wetted wall.

Assume Newtonian fluid with constant η, and consider the steady state. The equation of continuities becomes as

$$\nabla \bullet v = 0 \ \text{ or } \ \frac{\partial v_x}{\partial x} + \frac{\partial v_y}{\partial y} + \frac{\partial v_z}{\partial z} = 0 \tag{33}$$

Moreover, assume ρ and η to be constant. Then the Navier-Stokes equation becomes as

$$0 = \eta \nabla^2 v - \nabla p + \rho g \tag{34}$$

As seen in Fig. **4**, the velocity of the fluid is only z-direction. Assume that the velocity is not so high that the **laminar flow** can be considered, and then the following equation holds:

$$v_x = v_y = 0 \tag{35}$$

Substituting this into Eq.(33), we have

$$\frac{\partial v_z}{\partial z} = 0 \tag{36}$$

This equation means that v_z is a function of only x and y such that $v_z(x,y)$. However, the flow pattern is the same in the x-direction, and thus v_z is a function of only y such as $v_z(y)$. Now, Eq.(34) becomes as

$$0 = \eta \left(\frac{\partial^2 v_z}{\partial x^2} + \frac{\partial^2 v_z}{\partial y^2} + \frac{\partial^2 v_z}{\partial z^2} \right) - \frac{\partial p}{\partial x} + \rho g_z \tag{37}$$

Since v_z is a function of only y, the above equation becomes as

$$0 = \eta \frac{d^2 v_z}{dy^2} + \rho g_z \tag{38}$$

Integration of this equation by y gives

$$0 = \eta \frac{dv_z}{dy} + \rho g_z y + C_1 \text{ (integral constant)} \tag{39}$$

Consider the following boundary condition:

$$\tau_{yz} = \frac{dv_z}{dy} = 0 \quad at \quad y = 0, \quad v_z = 0 \quad at \quad y = \delta \quad (at \quad wall) \tag{40}$$

Using the above condition at y=0 for Eq.(39), we have C_1=0. Then by further integration, we have

$$\eta v_z + \rho \frac{g_z y^2}{2} + C_2 = 0 \tag{41a}$$

or

$$v_z = -\rho \frac{g_z y^2}{2\eta} + C_3 \quad (where\ C_3 \equiv -C_2/\eta) \tag{41b}$$

With the boundary condition at y= δ , we have $C_3 = \rho\, g_z\, \sigma^2/2\, \eta$, and the velocity distribution becomes the hyperbolic function as

$$v_z = \rho \frac{g_z \delta^2}{2\eta} [1 - \left(\frac{y}{\delta}\right)^2] \tag{42}$$

Let us consider another example for the flow pattern in the tube. As mentioned before, the Navier-Stokes equation for the non-compressive fluid with ρ and η to be constant is expressed as

$$\rho(\frac{\partial v_r}{\partial t} + v_r \frac{\partial v_r}{\partial r} + \frac{v_\theta}{r}\frac{\partial v_r}{\partial \theta} - \frac{v_\theta^2}{r} + v_z \frac{\partial v_r}{\partial z})$$
$$= \eta[\frac{\partial}{\partial r}(\frac{1}{r}\frac{\partial}{\partial r}(rv_r)) + \frac{1}{r^2}\frac{\partial^2 v_r}{\partial \theta^2} - \frac{2}{r^2}\frac{\partial v_\theta}{\partial \theta} + \frac{\partial^2 v_r}{\partial z^2}] - \frac{\partial p}{\partial r} + \rho g_r \tag{43a}$$

$$\rho(\frac{\partial v_\theta}{\partial t} + v_r \frac{\partial v_\theta}{\partial r} + \frac{v_\theta}{r}\frac{\partial v_\theta}{\partial \theta} + \frac{v_r v_\theta}{r} + v_z \frac{\partial v_\theta}{\partial z})$$
$$= \eta[\frac{\partial}{\partial r}(\frac{1}{r}\frac{\partial}{\partial r}(rv_\theta)) + \frac{1}{r^2}\frac{\partial^2 v_\theta}{\partial \theta^2} + \frac{2}{r^2}\frac{\partial v_r}{\partial \theta} + \frac{\partial^2 v_\theta}{\partial z^2}] - \frac{1}{r}\frac{\partial p}{\partial \theta} + \rho g_\theta \tag{43b}$$

$$\rho(\frac{\partial v_z}{\partial t} + v_r \frac{\partial v_z}{\partial r} + \frac{v_\theta}{r}\frac{\partial v_z}{\partial \theta} + v_z \frac{\partial v_z}{\partial z})$$
$$= \eta[\frac{1}{r}\frac{\partial}{\partial r}(r\frac{\partial v_z}{\partial r}) + \frac{1}{r^2}\frac{\partial^2 v_z}{\partial \theta^2} + \frac{\partial^2 v_z}{\partial z^2}] - \frac{\partial p}{\partial z} + \rho g_z \tag{43c}$$

Consider the velocity distribution of the non-compressive Newtonian fluid flowing in laminar flow in the tube such as blood tube at steady state as shown in Fig. **5**.

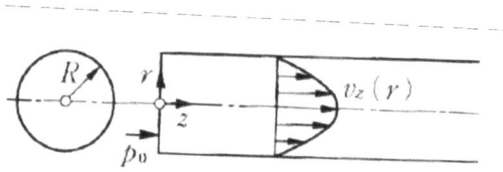

Figure 5: Fluid flowing in the tube.

At steady state, all the terms associated with t becomes 0. If the fluid is assumed to flow at laminar flow, the flow direction is only z-direction, and thus the following equations hold:

$$v_r = v_\theta = 0 \tag{44}$$

Moreover, because of the symmetry of the tube, v is independent of θ , and thus the following equations hold:

$$\frac{\partial v_r}{\partial \theta} = \frac{\partial v_\theta}{\partial \theta} = \frac{\partial v_z}{\partial \theta} = 0 \tag{45}$$

Substituting these equations into (4)', we have

$$\frac{\partial v_z}{\partial z} = 0 \tag{46}$$

which means that v_z is independent of z. Namely, v_z is independent of both θ and z, and thus v_z is only a function of r such that

$$v_z = v_z(r) \tag{47}$$

From the Navier-Stokes equation as given as (43a) and (43b), we have

$$\frac{\partial p}{\partial r} = \frac{\partial p}{\partial \theta} = 0 \tag{48}$$

which means that p is a function of only z. Moreover, from Eq.(43c), we have

$$\eta \frac{1}{r}(r\frac{\partial v_z}{\partial r}) = \frac{\partial p}{\partial z} \tag{49}$$

Note that the LHS of this equation is a function of only r, whereas the RHS is a function of only z. Therefore, in order for Eq.(49) to be satisfied, both sides must be constant. Let this constant be C_1. Then from the LHS, we have

$$\frac{1}{r}\frac{d}{dr}(r\frac{\partial v_z}{\partial r}) = \frac{C_1}{\eta} \tag{50}$$

Multiplying rdr to both sides of this equation, followed by integration, we have

$$\int d(r\frac{dv_z}{dr}) = \frac{C_1}{\eta}\int rdr = \frac{C_1}{\eta}\frac{r^2}{2} + C_2 \text{ (integral constant)}$$

Dividing this equation by r, we have

$$\frac{dv_z}{dr} = C_1\frac{r}{2\eta} + \frac{C_2}{r}$$

Multiplying dr to both sides of this equation, and by integrating, we have

$$\int dv_z = \int (C_1\frac{r}{2\eta} + C_2\frac{1}{r})dr, \rightarrow v_z = C_1\frac{r^2}{4\eta} + C_2 \ln r + C_3 \text{ (integral constant)} \tag{51}$$

Consider then the boundary condition such that

v_z is finite at r=0 (52a) and v_z=0 at r=R   (52b)

Using B.C. at r=0, C_2 must be 0, and using another B.C. at r=R, C_3=-$C_1R^2/4\eta$ is obtained. Therefore, v_z is expressed as

$$v_z = \frac{C_1}{4\eta}(r^2 - R^2) \tag{53}$$

Note that v=v_{max} at r=0, and thus the following equation holds:

$$v_{max} = -\frac{C_1 R^2}{4\eta} \tag{54}$$

where the suffix z was removed for simplicity. In the end, Eq.(53) can be expressed as

$$\frac{v}{v_{max}} = 1 - \left(\frac{r}{R}\right)^2 \tag{55}$$

If the reaction is taking place in the tube, the analysis becomes a little complicated. Let z be the distance from the inlet of the tube and let $0 \leq z \leq L$ as shown in Fig. **6a**, where L denotes the length of the tube. Consider the mass balance for the fluid between z and $z + \Delta z$ at time t as shown in Fig. **6b**.

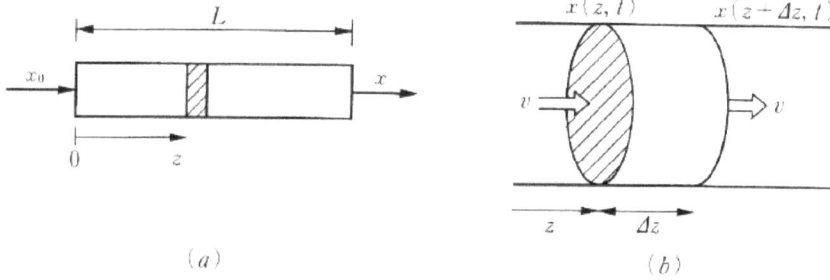

Figure 6: Tubular reactor.

Then the following equation may be derived:

$$\frac{\partial}{\partial t}[x(z,t)A\Delta z] = x(x,t)Av - x(z + \Delta z,t)Av \tag{56}$$

where A is the cross sectional area of the tube, x is the concentration of the component of concern, where x is the function of z and t. v is the velocity of the fluid. If we divide this equation by $A \Delta z$ and let $\Delta z \to 0$. Then we have

$$\frac{\partial}{\partial t}x(z,t) = \lim_{\Delta z \to 0} \frac{x(z,t) - x(z + \Delta z,t)}{\Delta z}v = -v\frac{\partial x(z,t)}{\partial z} \tag{57}$$

Moreover, if we take into account the effect of diffusion as shown in Fig. **7**, we have

$$\frac{\partial}{\partial t}[x(z,t)A\Delta z] = x(x,t)Av - x(z+\Delta z,t)Av - AD_e\frac{\partial x}{\partial z}\bigg|_z + AD_e\frac{\partial x}{\partial z}\bigg|_{z+\Delta z} \tag{58}$$

and finally we have the following equation:

$$\theta\frac{\partial x(\varsigma,t)}{\partial t} = -\frac{\partial x(\varsigma,t)}{\partial\varsigma} + \frac{1}{P_e}\frac{\partial^2 x(\varsigma,t)}{\partial\varsigma^2} \tag{59}$$

where D_e is the effective diffusion coefficient, P_e is called as **Peclet number**, and is defined as $Pe \equiv vL/De$ (dimensionless). θ and ς are defined as $\theta \equiv v/L$, $\varsigma \equiv z/L$.

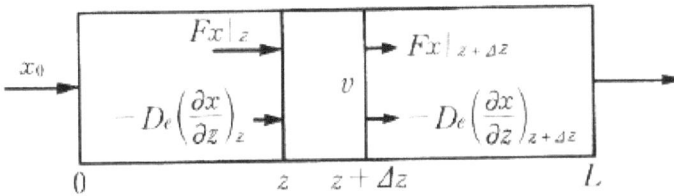

Figure 7: Tubular reactor with diffusion.

Heat Transfer

The phenomenon of heat transport from one place to another is called as **heat transfer**, where the heat transfer occurs from the higher temperature to the lower temperature. Thus the solid or the stationary fluid with higher temperature loses its heat, and the temperature becomes lower, while the reverse occurs for the lower temperature substance. This phenomenon is called as **heat conduction**. When the fluid is moving, the heat is transferred not only by heat conduction but also by the fluid itself, and this phenomenon is called as **heat convection**. There is another heat transfer such as **heat irradiation**, where the typical example is the emission from the sun light.

Consider the substance as shown in Fig. **8**, where the temperature increases with respect to y, and the heat is transferred for the higher temperature to the lower temperature, and the heat transfer rate is proportional to the temperature difference such as dT/dy, and the heat transfer per unit area is called as **heat flux** such that

$$q_y = -k\frac{dT}{dy} \tag{60}$$

where T is the temperature, and q_y is the heat flux. This equation is called as **Fourier's law of heat conduction**, where k is called as **heat conduction coefficient**.

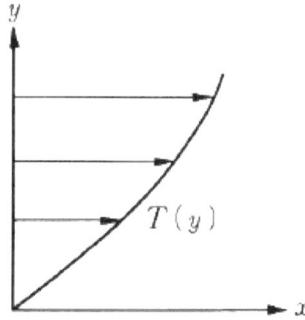

Figure 8: Temperature distribution.

In the similar way as considered for mass transfer and momentum transfer, the heat balance gives the following **equation for heat conduction**:

$$\frac{\partial T}{\partial t} = \alpha\left(\frac{\partial^2 T}{\partial x^2} + \frac{\partial^2 T}{\partial y^2} + \frac{\partial^2 T}{\partial z^2}\right) \tag{61}$$

where α is called as thermal diffusion coefficient.

Consider the heat transfer of flat plate as shown in Fig. **9**, where heat is assumed to be transferred only to the x-direction. From Eq.(57), we have the following equation for this system at steady state:

$$\frac{d^2T}{dx^2} = 0 \tag{62}$$

The boundary conditions are as followed:

T=T_1 at x=0 and T=T_2 at x=L \tag{63}

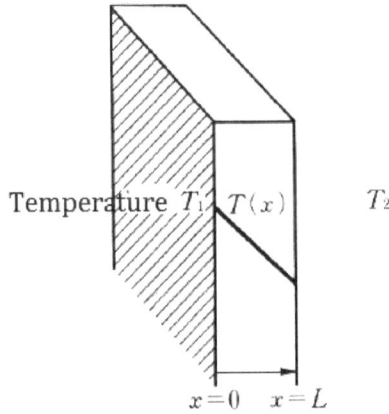

Figure 9: Heat transfer in the flat plate.

Twice integration of Eq.(58) with B.C.s gives the following solution:

$$T(x) = T_1 + \frac{T_2 - T_1}{L} x \qquad (64)$$

which indicates that the temperature distribution is linear with respect to the distance. The amount of heat transferred per time can be expressed from the Fick's law as

$$q_x = -k\frac{dT}{dx} = k\frac{T_2 - T_1}{L} \qquad (65)$$

Therefore, the total amount of heat transfer per time across area A can be expresses as

$$Q = q_x A = kA\frac{T_1 - T_2}{L} = \frac{T_1 - T_2}{L/kA} \qquad (66)$$

This equation resembles the Ohm's law in the electromagnetic system, where T_1-T_2 is the driving force for heat transfer, and L/kA is considered to be **thermal resistance**.

The similar analysis can be made for the cyrindrical coordinate, and the equation of heat conduction is expressed as

$$\frac{\partial T}{\partial t} = \alpha \left[\frac{1}{r}\frac{\partial}{\partial r}\left(r\frac{\partial T}{\partial r} \right) + \frac{1}{r^2}\frac{\partial^2 T}{\partial \theta^2} + \frac{\partial^2 T}{\partial z^2} \right] \tag{67}$$

Consider the case of two conical tube with radius R_1 and R_2, where the heat transfer is assumed to be made only r-direction. Then the above equation becomes at the steady state as

$$\frac{d}{\partial r}\left(r\frac{dT}{dr} \right) = 0 \tag{68}$$

The boundary conditions are

T=T$_1$ at r=R$_1$, and T=T$_2$ at r=R$_2$ \qquad (69)

Integration of Eq.(68) gives rdT/dr + C$_1$ = 0, where C$_1$ is the integral constant. Dividing this equation by r, we have

$$\frac{dT}{\partial r} = -\frac{C_1}{r}$$

Further integration of this equation gives T = - C$_1$ln r + C$_2$, where C$_2$ is the integral constant. Using B.C.s of Eq.(69) we have

$$\frac{T - T_2}{T_1 - T_2} = \frac{\ln(r/R_2)}{\ln(R_1/R_2)} \tag{70}$$

Based on Fourier's law, q$_r$ is expressed as

$$q_r = -k\frac{dT}{dr} = -k\left\{ \frac{T_2 - T_1}{\ln(R_1/R_2)} \right\}\left(\frac{1}{r} \right) \tag{71}$$

which indicates that q$_r$ decreases with the increase in r. The total amount of heat transfer through the internal surface at r=R$_1$ per unit time for the column length L is expressed as

$$Q\big|_{r=R_1} = Aq_r = 2\pi Lk\frac{T_2 - T_1}{\ln(R_2/R_1)} \tag{72}$$

The amount of heat transfer at $r=R_1$, $Q\mid_{r=R1}$ must be equal to $Q\mid_{r=R2}$ at $r=R_2$, and thus

$$Q = 2\pi Lk \frac{T_2 - T_1}{\ln(R_2 / R_1)} = kA_{lm} \frac{T_2 - T_1}{R_2 - R_1} \tag{73}$$

where A_{lm} is the **logarithmic mean** of A_1 ($=2\pi R_1 L$) and A_2 ($=2\pi R_2 L$) such that

$$A_{lm} \equiv 2\pi L \frac{R_2 - R_1}{\ln(R_2 / R_1)} = \frac{A_2 - A_1}{\ln(A_2 / A_1)} \tag{74}$$

Consider the heat transfer in the sphere with radius r next. The amount of heat transfer at r is $Q_r = 4\pi r^2 q \mid_r$ and also at $r+\Delta r$ as $Q_{r+\Delta r} = 4\pi r^2 q \mid_{r+\Delta r}$. Since $Q_r = Q_{r+\Delta r}$ at the steady state, the following equation holds:

$$Q_r - Q_{r+\Delta r} = 4\pi r^2 (q\mid_r - q\mid_{r+\Delta r}) = 0 \tag{75}$$

Divide this equation by $4\pi \Delta r$, and let $\Delta r \to 0$, then the above equation becomes as

$$\lim_{\Delta r \to 0} \frac{r^2 q\mid_r - r^2 q\mid_{r+\Delta r}}{\Delta r} = -\lim_{\Delta r \to 0} \frac{r^2 q\mid_{r+\Delta r} - r^2 q\mid_r}{\Delta r} = -\frac{d}{dr}(r^2 q) = 0 \tag{76}$$

Substtituting the Fourier's law $q = -dT/dr$, we have

$$\frac{d}{dr}\left(r^2 \frac{dT}{dr}\right) = 0 \tag{77}$$

Integration of this equation gives

$$r^2 \frac{dT}{dr} + C_1 = 0 \tag{78}$$

Dividing this equation by r^2 and integrating the resulting equation, we have

$$T - \frac{C_1}{r} + C_2 = 0 \quad (C_2 \text{ is the integral constant}) \tag{79}$$

Consider the boundary conditions such that

$$T=T_1 \text{ at } r=R_1 \text{ and } T=T_2 \text{ at } r=R_2$$

Then C_1 and C_2 are determined as

$$C_1 = \frac{T_1 - T_2}{1/R_1 - 1/R_2} \tag{80a}$$

$$C_2 = -T_2 + \frac{T_1 - T_2}{R_2(1/R_1 - 1/R_2)} \tag{80b}$$

So far, we considered the steady-state heat transfer. Consider next the unsteady heat conduction when the plate with width of $2b$ at the temperature T_1 was heated to T_2 ($T_2 > T_1$) at $t=0$. From the basic equation (Eq.(57)), we have

$$\frac{\partial T}{\partial t} = \alpha \frac{\partial^2 T}{\partial x^2} \tag{81}$$

The boundary conditions are as follows

$$T = T_1 \quad at\, t = 0 \qquad (0 \le x \le 2b) \tag{82a}$$

$$T = T_2 \quad at\, t > 0 \qquad at \quad x = 0 \tag{82b}$$

$$T = T_2 \quad at\, t > 0 \qquad at \quad x = 2b \tag{82c}$$

Introduce the following non-dimensional variables:

$$\theta = \frac{T_2 - T}{T_2 - T_1}, \quad z = \frac{x}{b}, \quad \tau = \frac{at}{b^2}$$

Then Eq.(81) becomes as

$$\frac{\partial \theta}{\partial \tau} = \frac{\partial^2 \theta}{\partial z^2} \tag{83}$$

The B.C.s become as

$$\theta = 1 \quad at\, \tau = 0 \qquad (0 \le z \le 2) \tag{84a}$$

$\theta = 0 \quad at\ \tau > 0 \qquad at \quad z = 0$ (84b)

$\theta = 0 \quad at\ \tau > 0 \qquad at \quad z = 2$ (84c)

The partial differential equation of Eq.(83) can be solved by the **method of separation of variables**. Assume that θ be a function of only z such as f(z) and only τ such as g(τ) such that

$\theta(z,\tau) = f(z)g(\tau)$ (85)

Substituting this into Eq.(83), we have

$$f(z)\frac{dg(\tau)}{d\tau} = g(\tau)\frac{d^2 f(z)}{dz^2}$$ (86a)

or

$$\frac{1}{g(\tau)}\frac{dg(\tau)}{d\tau} = \frac{1}{f(z)}\frac{d^2 f(z)}{dz^2}$$ (86b)

where the LHS of this equation is a function of τ, whereas the RHS is a function of z. For these to be equal, both must be constant. Let this constant be C_1. Then we have

$$\frac{1}{g(\tau)}\frac{dg(\tau)}{d\tau} = C_1 \ (87a) \text{ and } \frac{1}{f(z)}\frac{d^2 f(z)}{dz^2} = C_1$$ (87b)

Solving Eq.(86a), we have $d \ln g = C_1\tau + C_2{}'$, and then we have

$$g(\tau) = C_2 e^{C_1\tau} \quad (C_2 \equiv e^{C_2{}'})$$ (88a)

Since g must be finite at $\tau \to \infty$, C_1 must be negative such that $C_1 = -\beta^2$. Then we have

$$g(\tau) = C_2 e^{-\beta^2}$$ (88b)

Substituting $C_1 = -\beta^2$ into Eq.(87b), we have

$$\frac{1}{f(z)}\frac{d^2 f(z)}{dz^2} = -\beta^2 \tag{89}$$

This may be solved as

$$f(z) = A\sin\beta z + B\cos\beta z \tag{90}$$

By substituting Eq.(88b) and Eq.(90) into Eq.(85), we have

$$\theta(\tau, z) = e^{-\beta^2 \tau}(A'\sin\beta z + B'\cos\beta z) \tag{90'}$$

where A'$\equiv$C$_2$A and B'$\equiv$C$_2$B.

MODELING FOR ENZYME REACTION AND MODEL REDUCTION

For the modeling in practice, consider the simple example to start with. In 1913, Michaelis and Menten proposed the following enzyme reaction scheme:

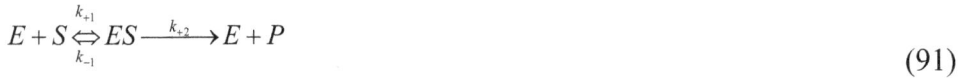

$$E + S \underset{k_{-1}}{\overset{k_{+1}}{\rightleftharpoons}} ES \xrightarrow{k_{+2}} E + P \tag{91}$$

where the substrate S forms the complex with enzyme E, and the product P is formed from the complex ES, while E is reproduced. In the above scheme, k_{+1}, k_{-1}, and k_{+2} are the reaction rate constants. In the above reaction scheme, the first reaction is reversible, while the 2nd reaction is irreversible.

The mass balance for the above system gives the following equations:

$$\frac{d[S]}{dt} = -k_{+1}[S][E] + k_{-1}[ES] \tag{92a}$$

$$\frac{d[E]}{dt} = -k_{+1}[S][E] + (k_{-1} + k_2)[ES] \tag{92b}$$

$$\frac{d[ES]}{dt} = k_{+1}[S][E] - (k_{-1} + k_2)[ES] \tag{92c}$$

$$\frac{d[P]}{dt} = -k_2[ES] \tag{92d}$$

where [·] indicates the concentration. Suppose that the initial condition is given as

$$[S(0)] = S_0 \neq 0, \ [E(0)] = E_0 \neq 0, \ [ES(0)] = [P(0)] = 0$$

If we add Eqs.(92b) and (92c), we have

$$\frac{d([E] + [ES])}{dt} = 0 \tag{93}$$

which indicates that $[E] + [ES] = E_0 =$ constant. Moreover, from Eq.(92d), $[P(t)]$ can be obtained by integration of this equation if $[ES(t)]$ could be obtained. After all, the following equations may be considered

$$\frac{d[S]}{dt} = -k_{+1}E_0[S] + (k_{+1}[S] + k_{-1})[ES] \tag{94a}$$

$$\frac{d[ES]}{dt} = k_{+1}E_0[S] - (k_{+1}[S] + k_{-1} + k_2)[ES] \tag{94b}$$

with initial condition as $[S(0)] = S_0$, $[ES(0)] = 0$, where the following condition was used in deriving the above equations: $[E] = E_0 - [ES]$.

In general, the second reaction step in the reaction scheme as shown in (91) is the limiting (slow as compared to the 1st reaction step), and thus the pseudo-steady state may be assumed to derive the following equation from Eq.(94a):

$$[ES] = \frac{k_{+1}E_0[S]}{k_{+1}[S] + k_{-1} + k_2} \tag{95}$$

By substituting this into Eq.(94b), we have

$$\frac{d[S]}{dt} = -\frac{V_m[S]}{[S] + K} \tag{96}$$

where $V_m \equiv k_2 E_0$, and $K \equiv (k_{-1} + k_2)/k_{+1}$. This is the well-known **Michaelis-Menten equation**.

The Michaelis-Menten equation is often used in practice, where this equation was derived based on the assumption of pseudo-steady state for the fast reaction. This method is a kind of **model simplification** or **model reduction**. Consider the system with different time scales in more detail from the modeling point of view. Eq.(94) can be expressed by the non-dimensional variables as

$$\frac{dx}{dt'} = -x + (x + \kappa - \lambda)y \tag{97a}$$

$$\varepsilon \frac{dy}{dt'} = x - (x + \kappa)y \tag{97b}$$

With initial condition such as x(0)=1, y(0)=0, where the non-dimensional variables and parameters are defined as

$$t' \equiv k_{+1}E_0, \qquad \lambda \equiv \frac{k2}{k_{+1}S_0}, \qquad \kappa \equiv \frac{k_{-1} + k_2}{k_{+1}S_0}, \qquad x \equiv \frac{[S]}{S_0}, \qquad y = \frac{[P]}{E_0}, \qquad \varepsilon \equiv \frac{E_0}{S_0}$$

In general, ε is on the order of 0.01 or 0.001 such that $\varepsilon \ll 1$. Consider expanding x and y around $\varepsilon = 0$ such that

$$x = x_0 + \varepsilon x_1 + \varepsilon^2 x_2 + o(\varepsilon^3) \tag{98a}$$

$$y = y_0 + \varepsilon y_1 + \varepsilon^2 y_2 + o(\varepsilon^3) \tag{98b}$$

where o(ε^3) contains all the terms higher than or equal to ε^3, and o($\cdot$) is called **order**. Substitute Eq.(98) into Eq.(97), and consider the terms with o(1) such that

$$\frac{dx_0}{dt} = -x_0 + (x_0 + \kappa - \lambda)y_0 \tag{99a}$$

$$0 = x_0 - (x_0 + \kappa)y_0 \tag{99b}$$

Moreover, from the initial condition, $x_0(0)=1$, $y_0(0)=0$. This equation gives the Michaelis-Menten equation as mentioned above such that

$$\frac{dx_0}{dt} = -\frac{\lambda x_0}{x_0 + k} \tag{99'}$$

As mentioned before, this equation is useful in many practical applications. Here, let us consider in more detailed modeling for the dynamics around t=0. As shown in Fig. **10**, as compared to x, the dynamic characteristics of y shows abrupt changes on the order of o(ε) around t=0. Consider the change of variables, and expand the time scale around t=0 such that

$$\tau = \frac{t}{\varepsilon} \tag{100}$$

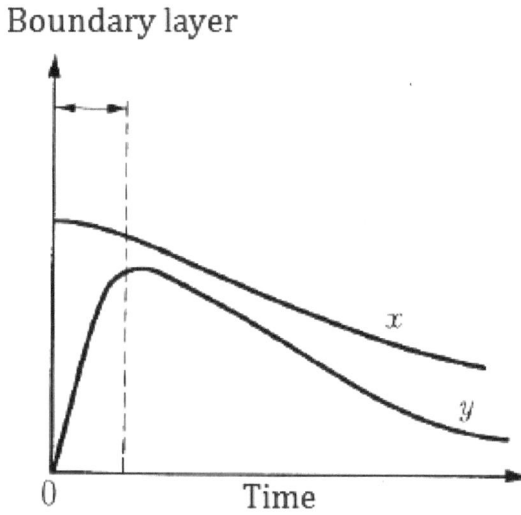

Figure 10: Dynamics with different time scales.

Let X and Y be the variables of x and y, respectively, when the change of variables was made. Then Eq.(97) can be expressed as

$$\frac{dX}{d\tau} = -\varepsilon X + \varepsilon(X + \kappa - \lambda)Y \tag{101a}$$

$$\frac{dY}{d\tau} = X - (X + \tau)Y \tag{101b}$$

Let X and Y be expanded by ε as

$$X(\tau, \varepsilon) = X_0(\tau) + \varepsilon X_1(\tau) + o(\varepsilon^2) \tag{102a}$$

$$Y(\tau, \varepsilon) = Y_0(\tau) + \varepsilon Y_1(\tau) + o(\varepsilon^2) \tag{102b}$$

Substituting this into Eq.(101), and consider the terms of o(1) and o(ε). Then the following equations are obtained:

$$\text{o(1):} \quad \frac{dX_0}{d\tau} = 0 \tag{103a}$$

$$\frac{dY_0}{d\tau} = X_0 - (X_0 + \kappa)Y_0 \quad (Y_0(0)=0) \tag{103b}$$

$$\text{o(ε):} \quad \frac{dX_1}{d\tau} = -X_0 + (X_0 + \kappa - \lambda)Y_0 \quad (X_1(0)=0) \tag{104a}$$

$$\frac{dY_1}{d\tau} = (1 - Y_0)X_1 - (X_0 + \kappa)Y_1 \quad (Y_1(0)=0) \tag{104b}$$

The initial condition can be met by X_0 and Y_0, and thus the initial conditions for higher terms such as X_1, X_2,... and Y_1, Y_2,... can be all set to 0. The solution to Eq.(103) becomes as

$$X_0(\tau) = 1 \tag{105a}$$

$$Y_0(\tau) = \frac{1}{1+\kappa}(1 - e^{-(1+\kappa)\tau}) \tag{105b}$$

Recall that X(τ) and Y(τ) show the dynamics around t=0, while x(t) and y(t) represent the dynamics a little far away from t=0 (Fig. **10**). These variables must be connected (**asymptotic matching**) at the boundary. Namely, the followings must be satisfied at the boundary layer.

$$\lim_{\tau \to \infty}[X(\tau), Y(\tau)] = \lim_{t \to 0}[x(t), y(t)] \tag{106}$$

The solution to Eq.(99) becomes as

$$x_0(t) + \kappa \ln x_0(t) = c - \lambda t \quad \text{(c: integral constant)} \tag{107a}$$

$$y_0(t) = \frac{x_0(t)}{x_0(t) + \kappa} \tag{107b}$$

Consider the boundary condition of Eq.(106) such that

$$X_0(\tau) = 1 \qquad \Rightarrow \qquad X_0(\infty) = 1 \qquad \Rightarrow \qquad x_0(0) = 1$$
$$\Rightarrow \qquad c = x_0(0) + \kappa \ln x_0(0) = 1 \tag{108a}$$

$$Y_0(\infty) = \frac{1}{1+\kappa}, \qquad y_0(0) = \frac{x_0(0)}{x_0(0)+\kappa} = \frac{1}{1+\kappa} \tag{108b}$$

The higher order terms can be also obtained in the similar way, and the overall solution becomes as

$$x(t,\varepsilon) = x_0(t) + o(\varepsilon), \qquad x_0(t) + \kappa \ln x_0(t) = 1 - \lambda t \tag{109a}$$

$$y(t,\varepsilon) = \begin{cases} Y_0(\tau) + o(\varepsilon) = \left(\dfrac{1}{1+\kappa}\right)[1 - e^{-(1+\kappa)t/\varepsilon}] + o(\varepsilon) & (0 \le t \ll 1) \qquad (109b) \\[2em] y_0(t) + o(\varepsilon) = \dfrac{x_0(t)}{x_0(t)+\kappa} + o(\varepsilon) & (\varepsilon \ll t) \qquad (109c) \end{cases}$$

This analysis method is called as **singular perturbation method** sometimes used for the model reduction of the system with different time scales.

VARIOUS MODELING FOR BIOSYSTEMS

For the modeling of the metabolism in practice, the network structure which defines the network of interconnected elements such as concentrations of metabolites *etc.* together with stoichiometric interaction must be defined. Once the model structure was defined, the next step is to determine the expression that define the interactions between the different components. Kinetic rate equations may be derived from the actual reaction mechanisms with different degrees of details depending on the purpose of modeling, or in certain case, may be represented by the appropriate or simplified expression which captures the essential feature of the reaction. The former approach is explained in Chapter 4 in more detail, and the latter approach is briefly explained in the present chapter.

When the network structure and the interactions between different components were determined by kinetic rate expression or simplified expression, the model can be described as a set of balance equations such as mass balance equations which determines the dynamics of the system showing the time trajectories of the concentrations of concern.

Consider the following general state equations derived by mass balances, which describe the dynamics of the system of concern:

$$\frac{dx(t)}{dt} = f(x(t), v(x(t), u(t), \theta), \theta) \tag{110a}$$

$$with \quad x(0) = x_0$$
$$y(t) = h(x(t), u(t), \theta) \tag{110b}$$

where $\mathbf{x}(t)$ is the m-dimensional state vector ($x(t) \in R^m$) which includes time-dependent state variables, $\mathbf{v}(x(t), u(t), \theta)$ is an n-dimensional vector of reaction rates or fluxes ($v(t) \in R^n$), $\mathbf{u}(t)$ is a vector of input variables to be used for the manipulation or control of the system, and θ is a set of model parameters. The $\mathbf{h}(x(t), u(t), \theta)$ is the output or measurement vector, typically a subset of state variables. Although the time trajectories of the state variables can be obtained by numerically integrating Eq.(110a) with initial condition, computational burden is demanding in the case of using non-linear kinetic rate expression, and thus the application is limited to the central metabolic pathways as explained in Chapter 8. The state equations may be simplified with stoichiometric information for the flux balance analysis such that

$$\frac{dx(t)}{dt} = S \bullet v(x(t), u(t), \theta) \tag{111a}$$

$$with \quad x(0) = x_0$$
$$y(t) = c^T v \tag{111b}$$

where S is the stoichiometric coefficient matrix with $m \times n$ dimension ($S \in R^m \times R^n$), and $\mathbf{c}$ is the measurement vector. This mass balance represents the principal constraint in FBA, and defines a feasible solution space for the set of fluxes. At (quasi-) steady state, Eq.(111) can be simplified as

$$S \bullet v = 0 \tag{112a}$$

$$y = c^T v \tag{112b}$$

These mass balances represent the principal constraints in FBA, and defines a feasible solution space for the set of fluxes.

The degree of freedom of the system is the number of fluxes minus the number of reactions, where the system is uniquely determined if m and n are equal with S to be non-singular. The system becomes over-determined if m is larger than n, while it becomes under-determined if m is smaller than n. In the case where the system is under-determined, additional constraints such as cofactor balance equations or some appropriate objective functions must be introduced such that

$$J = \max_v c^T v \tag{113}$$

under the constraint of Eq.(112) where c is the weight coefficient vector. The typical objective functions may be the yield, titer, and productivity of the desired product such as [7]

$$J_1 = \frac{\int_0^T X(t)\rho(t)dt}{\int_0^T X(t)v(t)dt} \tag{114a}$$

$$J_2 = \int_0^T X(t)\rho(t)dt \tag{114b}$$

$$J_3 = \frac{1}{T}\int_0^T X(t)\rho(t)dt \tag{114c}$$

where X(t) is the biomass concentration at time t, ρ (t) and v (t) are the specific product production rate, and the specific substrate consumption rate, respectively. T is the cultivation time, where T itself may be a parameter for optimization.

As mentioned in introduction, in the typical FBA and its extension to genome-scale, a vector valued objective function may be introduced (see Chapter 5).

Consider the simple example as illustrated in Fig. **11a** to understand the network structure, where S is the substrate, A and B are the intracellular metabolites, and C,D, and E are the extracellular metabolites. Let r_i (i=1,2,...,5) be the flux or the reaction rate at i-th pathway. Then the mass balance for the intracellular metabolites gives the following expressions:

$$\frac{dA}{dt} = r_1 - r_2 - r_3 \tag{115a}$$

$$\frac{dB}{dt} = r_2 - r_4 - r_5 \tag{115b}$$

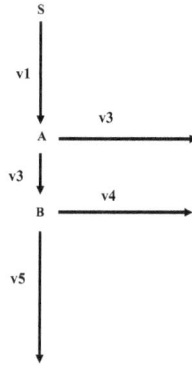

(a)

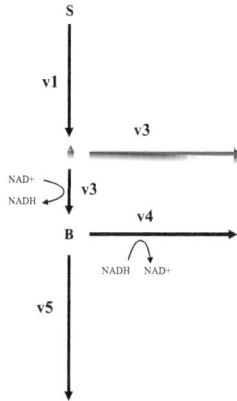

(b)

Figure 11: Simple example for the network analysis.

At steady state, these equations can be expressed as follows:

$$\begin{bmatrix} 1 & -1 & -1 & 0 & 0 \\ 0 & 1 & 0 & -1 & -1 \end{bmatrix} \begin{bmatrix} r_1 \\ r_2 \\ r_3 \\ r_4 \\ r_5 \end{bmatrix} = 0 \qquad (116)$$

The degree of freedom of the system is 5-2=3. Therefore, three fluxes among five fluxes must be measured to uniquely determine the system. Namely, if r_1, r_3, and r_5 could be measure, Eq.(116) becomes as

$$\begin{bmatrix} 1 & -1 & 0 \\ 0 & 0 & -1 \end{bmatrix} \begin{bmatrix} r_1 \\ r_3 \\ r_5 \end{bmatrix} + \begin{bmatrix} -1 & 0 \\ 1 & -1 \end{bmatrix} \begin{bmatrix} r_2 \\ r_4 \end{bmatrix} = 0 \qquad (117a)$$

from which, r_2 and r_4 can be determined by the following equation:

$$\begin{bmatrix} r_2 \\ r_4 \end{bmatrix} = -\begin{bmatrix} -1 & 0 \\ 1 & -1 \end{bmatrix}^{-1} \begin{bmatrix} 1 & -1 & 0 \\ 0 & 0 & -1 \end{bmatrix} \begin{bmatrix} r_1 \\ r_3 \\ r_5 \end{bmatrix} = \begin{bmatrix} 1 & -1 & 0 \\ 1 & -1 & -1 \end{bmatrix} \begin{bmatrix} r_1 \\ r_3 \\ r_5 \end{bmatrix} \qquad (117b)$$

If co-factor balance such as NADH balance can be used as shown in Fig. **11b**, another constraint such that $r_2=r_4$ may be used, and the resulting degrees of freedom becomes 5-3=2.

Metabolic flux analysis is made based on the stoichiometric equations and mass balances as mentioned above.

UNSTRUCTURE MODELS AND VARIOUS CULTIVATION METHODS

So far, many models which can describe the cell growth rate have been proposed, where those may be classifies as **structure model** and **unstructured model**, **lumped-parameter system** and **distributed-parameter system**, **deterministic model** and **stochastic model**.

Consider the exponentially growing cells as expressed as

$$\frac{dX}{dt} = \mu X \tag{118}$$

Where X is the cell concentration [g/L], and μ is the specific growth rate [h^{-1}]. This equation can be also used in the economic analysis, where μ is called Marsus coefficient. Now, if μ is assumed to be constant, Eq.(118) can be solves as

$$X(t) = X(0)e^{\mu t} \tag{119}$$

where the **doubling time** t_D is obtained as

$$\frac{X(t)}{X(0)} = 2 \rightarrow t_D = \ln 2/\mu \tag{120}$$

As mentioned above, Eq.(119) may be applied only during the period of exponentially growing cells without any nutrient limitation. In fact, the cell growth declines as illustrated by the following **logistic model**:

$$\frac{dX}{dt} = kX(1 - \frac{X}{X_m}) \tag{121}$$

Where k and X_m are the model parameters, and X_m represents the maximum cell concentration. This equation can be explicitly solved as

$$X(t) = \frac{X(0)X_m e^{kt}}{[X_m - X(0) + X(0)e^{kt}]} \tag{122}$$

And the doubling time is obtained as

$$t_D = \frac{1}{k} \ln \left[2(\frac{X_m - X(0)}{X_m - 2X(0)}) \right] \tag{123}$$

Eq.(122) indicates that the cell's state is determined if the initial condition X(0) is given. This is not true in practice, and may be extended as will be explained next. Although the application of the logistic model is limited, this may be used for the analysis of ecosystem with very little information about the organisms in detail as will be mentined later.

Unstructure Models

As mentioned above, the application of the above model is difficult in practice, since the specific growth rate μ is not constant, but a function of substrate S, metabolite P, as well as culture condition such as pH and temperature T *etc.*, such that

$$\mu = f(S, P, pH, T, ...) \tag{124}$$

As for the effect of substrate, the following **Monod model** is often used:

$$\mu(S) = \frac{\mu_m S}{K_s + S} \tag{125}$$

where μ_m and K_s are the model parameters, and μ_m is the maximum specific growth rate $[h^{-1}]$, and K_s is the saturation constant $[g/L]$. The saturation constant is related to the affinity of the reaction molecule with the enzyme. Namely, K_s is small if the affinity is high, while it is large if the affinity is low. If the substrate inhibition term was considered, the following equation may be considered:

$$\mu(S) = \frac{\mu_m S}{K_s + S + S^2 / K_i} \tag{126}$$

Several other equations may be considered for the practical application such that

$$\mu(S) = \mu_m (1 - e^{-S/K_s}) \text{ (Teissier model)} \tag{127a}$$

$$\mu(S) = \frac{\mu_m S^n}{K_s + S^n} \text{ (Moser model)} \tag{127a}$$

$$\mu(S) = \frac{\mu_m S}{K_s X + S} \text{ (Contois model)} \tag{127c}$$

These equations are rather empirical models, which may fit well depending upon the growth condition for a variety of fermentations.

The product inhibition may be also considered as

$$\mu(S,P) = \mu(S)(1-\frac{P}{P_m})^n \tag{128a}$$

or

$$\mu(S,P) = \mu(S)e^{-P} \tag{128b}$$

where the above models are also empirical, and sometimes used for alcohol fermentation.

In the case of recombinant protein production, the following equations may be considered for the plasmid-harboring cells and plasmid-free cells:

$$\frac{dX_+}{dt} = (1-\alpha)\mu_+ X_+ \tag{129a}$$

$$\frac{dX_-}{dt} = \mu_- X_- + \alpha\mu_+ X_+ \tag{129b}$$

where the suffix "+" indicates plasmid-bearing cell, while "-" indicates plasmid-free cell. The above equations imply that the plamid bearing cells lose plasmid at the rate α ($0 \leqq \alpha < 1$).

In relation to the cell growth rate, the substrate consumption rate may be expressed as

$$-\frac{dS}{dt} = \frac{1}{Y_{X/S}}\mu X + mX \tag{130}$$

where $Y_{X/S}$ is the yield coefficient ($\Delta X/-\Delta S$), and m is the **maintenance coefficient**.

Let the specific substrate consumption rate be defined as

$$v \equiv -\frac{1}{X}\frac{dS}{dt} \tag{131}$$

Then Eq.(130) becomes as

$$-v = -\frac{\mu}{Y_{X/S}} + m \tag{132}$$

If m can be neglected such as during the exponential growth phase, the cell concentration and the substrate concentration is related as

$$S(0) - S(t) = \frac{X(t) - X(0)}{Y_{X/S}} \tag{133}$$

As for metabolite formation, the metabolite formation patterns in relation to the cell growth rate may be classified as the growth-associated type such as alcohol fermentation *etc.* and the growth-non associated type such as antibiotic fermentation *etc.* There is yet another pattern showing the pattern in-between the above two types such as citric acid fermentation and glutamic acid fermentation, where the product formation is somewhat growth-associated, but it is not so clear as compared to the case of growth associated pattern.

Modeling for the Fermentation Process

Consider the bioreactor system as shown in Fig. **12**. The mass balance gives the following basic equations for this system:

$$\frac{d(XV)}{dt} = F_{in}X_F - F_{out}X + \mu XV \tag{134a}$$

$$\frac{d(SV)}{dt} = F_{in}S_F - F_{out}S - vXV \tag{134b}$$

$$\frac{d(PV)}{dt} = F_{in}P_F - F_{out}P + \rho XV \tag{134c}$$

$$\frac{dV}{dt} = F_{in} - F_{out} \tag{134d}$$

where X,S,P are the cell, substrate, and product concentrations [g/L], respectively, and X_F, S_F, and P_F are the respective concentrations of the feed or input flow. F_{in} and F_{out} are the flow rate [L/h] of the input and output flows, respectively, and V is the volume [L] of the culture broth. μ, v, ρ are the specific growth rate [h^{-1}], the specific substrate consumption rate [h^{-1}], and the specific product formation rate $[h^{-1}]$, respectively. If the multiple products or metabolites are formed, Eq.(134c) may be replaced by

$$\frac{d(P_i V)}{dt} = F_{in} P_{iF} - F_{out} P_i + \rho_i XV \quad (i=1,2,\ldots,N) \tag{134c'}$$

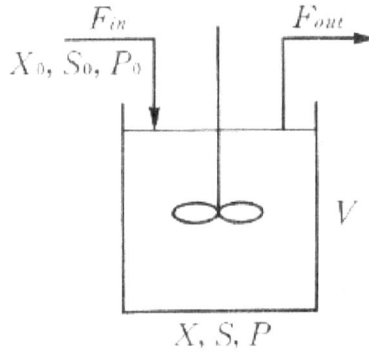

Figure 12: Bioreactor system.

In the similar way, Eq.(134b) may be modified for the case of multiple substrates, and also Eq.(134a) may be modified for the case of the mixed culture or co-culture of different types of strains.

a) Batch Culture

In the case of batch culture, $F_{in}=F_{out}=0$, and V is constant, and thus the above equations can be simplified as

$$\frac{dX}{dt} = \mu X \tag{135a}$$

$$\frac{dS}{dt} = -\nu X \tag{135b}$$

$$\frac{dP}{dt} = \rho X \tag{135c}$$

b) Fed-Batch Culture

In the case where the substrate concentration be kept low at constant level to avoid by-product formation such as acetate formation (overflow metabolism) in

E.coli, and ethanol formation in baker's yeast cultivation (Crabtree effect) *etc.* Since $F_{out}=0$ but $F_{in} \neq 0$, the basic equations become as

$$\frac{d(XV)}{dt} = \mu XV \tag{136a}$$

$$\frac{d(SV)}{dt} = F_{in}S_F - vXV \tag{136b}$$

$$\frac{d(PV)}{dt} = \rho XV \tag{136c}$$

$$\frac{dV}{dt} = F_{in} \tag{136d}$$

where $X_F=P_F=0$ was assumed without loss of generality in practice.

Consider the case where the amount of substrate is kept constant throughout the fermentation such that $d(SV)/dt = 0$. From Eq.(136b), $XV=F_{in}S_F/ v$ holds. Substituting this relationship into Eq.(136a), the following equation is obtained:

$$\frac{d(XV)}{dt} = Y_{X/S}F_{in}S_F \tag{137}$$

where $Y_{X/S} \equiv \mu / v$. If $Y_{X/S}$ is constant, all the terms in the RHS of the above equation becomes constant, and thus XV changes linearly.

In the case when the substrate concentration was kept low at constant level such that $dS/dt=0$. Then from Eq.(136b), we have

$$F_{in}(t) = \frac{vXV}{S_F - S} \tag{138}$$

where S is the constant substrate concentration level. If μ is assumed to be constant, Eq.(136a) can be solved as $X(t)V(t)=X(0)V(0)\exp(\mu t)$. Then Eq.(138) becomes as

$$F_{in}(t) = \frac{vX(0)V(0)}{S_F - S}\exp(\mu t) \tag{138'}$$

This feeding policy is called **exponentially fed-batch** culture. This is the feed-forward method, and this must be compensated by the feedback control such as DO-stat or pH stat.

c) Continuous Culture

Continuous culture is sometimes called as **chemostat**, and is useful in practice for the laboratory scale investigation such as to study the effect of the cell growth rate on the fermentation characteristics. However, it must be careful for the long-term operation due to the possibility of contamination or mutation. In the continuous culture, $F_{in} = F_{out}$ and $dV/dt = 0$ in Eq.(134). Without loss of generality, let $X_F = 0$ and $P_F = 0$ be satisfied as assumed above. Then Eq.(134) becomes as

$$(\mu - D)X = 0 \tag{139a}$$

$$D(S_F - S) - vX = 0 \tag{139b}$$

$$\rho X - DP = 0 \tag{139c}$$

where D is the dilution rate defined as F/V [h^{-1}]. The solution to Eq.(139a) becomes either X=0 or μ =D. The case where X=0 implies the trivial solution, and corresponds to washout in practice. Consider the non-trivial case where μ =D, and μ is assumed to be expressed as Monod model as mentioned before. Then the solution gives the following equation:

$$S = \frac{K_s D}{\mu_m - D} \tag{140a}$$

Substituting this into Eq.(139b) and (139c), we have

$$X = Y_{X/S}(S_F - \frac{K_s D}{\mu_m - D}) \tag{140b}$$

and

$$P = \frac{\rho}{D} X = \frac{\rho}{D} Y_{X/S}(S_F - \frac{K_s D}{\mu_m - D}) \tag{140c}$$

Consider the productivity of the cell defined as

$$J_X = DX = DY_{X/S}(S_F - \frac{K_s D}{\mu_m - D})$$ (141)

The optimal dilution rate that gives the maximum value of J_X can be derived from $dJ_X/dD=0$ as

$$D^* = \mu_m \left(1 - \sqrt{\frac{K_s}{K_s + S_F}}\right)$$ (142)

As the dilution rate (=the cell growth rate) was increased, the cell concentration decreases, while the substrate concentration increases as shown in Fig. **13**. This characteristics will be also discussed in Chapter 8 for the experimental data with simulation result. The importance of the continuous culture in practice is that the cell growth rate can be controlled by manipulating the dilution rate because $\mu = D$ is satisfied at steady state.

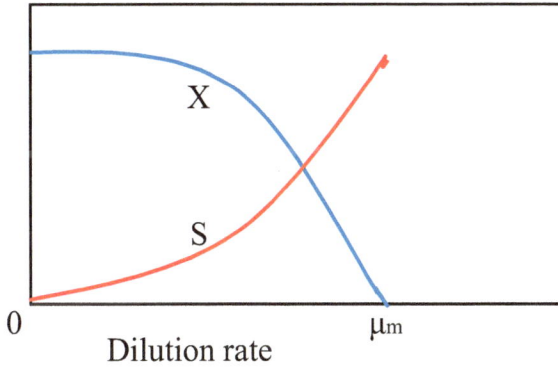

Figure 13: Effect of dilution rate on the steady states.

d) *CO_2 Evolution Rate and O_2 Consumption Rate*

In the bioreactor operation, modeling, and control, **CO_2 evolution rate (CER)** and O_2 consumption rate or **O_2 uptake rate (OUR)** become important. Those can be obtained using CO_2/ O_2 analyzer and let F_{in} and F_{out} be the inlet and outlet flow rate of air [L/h], respectively.

From the mass balance for CO_2

$$CER = F_{out}C_{CO2out} - F_{in}C_{CO2in}$$ (143)

From the nitrogen balance

$$F_{out}C_{N2out} = F_{in}C_{N2in} \tag{144}$$

From Eq.(144)

$$F_{out} = F_{in}C_{N2in} / C_{N2out}$$

Substituting this into Eq.(143), we have

$$CER = F_{in}(\frac{C_{N2in}}{C_{N2out}}C_{CO2out} - C_{CO2in}) \tag{145}$$

Since $C_{N2out} = 1 - C_{O2out} - C_{CO2out}$, the specific CO_2 evolution rate (CER) and the specific Oxygen uptake rate (OUR) can be expressed as follows:

$$CER = \frac{F_{in}}{V}(\frac{C_{N2in}C_{CO2out}}{1-C_{O2out}-C_{CO2out}} - C_{CO2in})$$

$$OUR = \frac{F_{in}}{V}(C_{O2in} - \frac{C_{N2in}C_{O2out}}{1-C_{O2out}-C_{CO2out}}) \tag{146}$$

where C_{in} and C_{out} are the concentrations in the input and output gas. V is the volume of the fermentor. C_{N2in} and C_{O2in} are the concentrations of the N_2 and O_2 in the air supplied by the compressor to the fermentor, such that C_{N2in}=79.4 %. C_{O2in}=20.6%, and C_{CO2in}=0.

F_{in} can be corrected by assuming the ideal gas as

$$F_{in} = \frac{PF_{in}*}{RT} \tag{147}$$

P=1 atm, T=20°C→T=20+273 K, R=0.082, and F_{in} *=measured flow rate

POPULATION DYNAMICS IN THE ECOSYSTEM

Let the population dynamics in nature be expressed as only one organism's population such as

$$\frac{dn}{dt} = f(n) \tag{148}$$

where n is the number of the living species of concern. In nature, multiple species are present, and interact among them. Thus the following equation may be considered for such a case:

$$\frac{dn_i}{dt} = \{a_i - b_i n_i + \sum_{\substack{k=1 \\ k \neq i}} c_{ik} n_k\} n_i \qquad (i = 1, 2, ..., n) \tag{149}$$

where n_i is the number of i-th specie, and a_i, b_i, and c_{ik} are the model parameters. The above equations are called **Lotka-Volterra equations**, where logistic equation is the special case of such equations.

Lotka-Volterra Model

a) Lotka-Volterra Two Competing Model

Consider the case of two species competing for the same prey or the living space, where the following equations may express such situation:

$$\frac{dn_1}{dt} = a_1 \{1 - \frac{(n_1 + s_2 n_2)}{K_1}\} n_1 \tag{150a}$$

$$\frac{dn_2}{dt} = a_2 \{1 - \frac{(s_1 n_1 + n_2)}{K_2}\} n_2 \tag{150b}$$

where a_1, a_2, K_1, K_2, s_1, and s_2 are the positive model parameters.

b) Lotka-Volterra Prey-Predator Model

Let n_1 be the number of prey, and n_2 be the number of predator, and consider the following prey-predator model:

$$\frac{dn_1}{dt} = a(1 - \frac{n_1}{K} - bn_2)n_1 \tag{151a}$$

$$\frac{dn_2}{dt} = (-c + dn_1)n_2 \tag{151b}$$

where a, b, c, d, and K are the model parameters.

c) Prey-Predator Model for Three Species

Consider the case of adding one more specie which prey the other two competing species, where the dynamics may be expressed as

$$\frac{dn_1}{dt} = a_1\{1 - \frac{(n_1 + s_2 n_2)}{K_1} - k_1 n_3\}n_1 \tag{152a}$$

$$\frac{dn_2}{dt} = a_2\{1 - \frac{(s_1 n_1 + n_2)}{K_2} - k_2 n_3\}n_2 \tag{152b}$$

$$\frac{dn_3}{dt} = (-r + \alpha_1 n_1 + \alpha_2 n_2)n_3 \tag{152c}$$

where a_1, a_2, K_1, K_2, s_1, s_2, k_1, k_2, α_1, α_2, and r are the model parameters.

Age Distribution Model

Let the characteristic value such as age or cell size be x, and let $n(x,t)$ be the density distribution of x. Consider how n changes with respect to x and t assuming no growth or death. This may be derived from the equation of continuity such that

$$\frac{\partial n}{\partial t} = -\frac{\partial J}{\partial x} \tag{153}$$

where J is the flux of x.

Consider the case where x of every individual changes with the constant speed v, then $J(x,t)=vn(x,t)$, and Eq.(153) becomes as

$$\frac{\partial n}{\partial t} = -\frac{\partial(vn)}{\partial x} \tag{154}$$

Let x be the age, let v=1 in Eq.(154) and by introducing the death rate as d(x). Then the balance equation becomes as

$$\frac{\partial n}{\partial t} = -\frac{\partial n}{\partial x} - d(x)n \tag{155}$$

If the growth rate of the species is expressed as a function of x such that dx/dt=g(x), then the above equation becomes as

$$\frac{\partial n}{\partial t} = -\frac{\partial \{g(x)n\}}{\partial x} - d(x)n \tag{156}$$

Assume that the size increment of individual depends on the size at time t, and independent of the past history. Then X(t) changes to X(t+Δt) after time Δt, and the average and the variance of ΔX=X(t+Δt)-X(t) are defined as

$$E[\Delta x] = g(x)\Delta t + o(\Delta t) \tag{157a}$$

$$Var[\Delta x] = h(x)\Delta t + o(\Delta t) \tag{157b}$$

where o(Δt) denotes the higher order terms than Δt. With these relationships, Eq.(156) becomes as

$$\frac{\partial n}{\partial t} = -\frac{\partial \{g(x)n\}}{\partial x} + \frac{1}{2}\frac{\partial^2 \{h(x)n\}}{\partial x^2} - d(x)n \tag{158}$$

where this equation is known as **Fokker-Plank equation** except the last term.

So far, we assumed that the system is described by mathematical expressions derived by applying (mass) balances. In certain cases, it may be difficult to obtain such information, but the experimental data are available. This is the next topic.

MODELING BY ARTIFICIAL NEURAL NETWORKS

In the case where the mathematical description is difficult, but many data are available, the data driven approach may be considered. One approach is to utilize **artificial intelligence (AI)**, and among them the **Artificial Neural Networks (ANNs)** have been recognized to be useful for the data-driven modeling approach. The ANNs have been developed based on the knowledge on brain information processing with parallel distributed processing (PDP). The main features of ANNs are the self-adaptability or homeostasis to the environment, self-organization, self-adjustability, and learning ability.

Consider the neuron model as depicted in Fig. **14**, where the multi-inputs and single output were considered by McCulloch and Pitts in 1943. Let z be the membrane potential expressed as the summation of weighted inputs such that

$$z = \sum_{i=1}^{n} w_i x_i \tag{159}$$

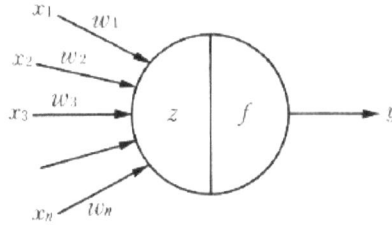

Figure 14: Neuron model proposed by McCulloch-Pitt.

Then the output y is a non-linear function of z as expressed as the **sigmoid function** such as

$$f(z) = \frac{1}{1 + e^{-z}} \tag{160}$$

There are several types of ANNs, where those may be roughly classified as either multi-layered NN model or fully connected NN model. The basic idea came from the learning law proposed by Hebb in 1949 in "The Organization of Behavior" [8]. The earlier NN is called **Perceptron** proposed by Rosenblatt in 1958 [9], where it consists of sensory layer (S), association layer (A), and reaction layer (R), and the information is processed in the direction: S→A→R.

However, Minky and Papert pointed out that perceptron has the limitation in pattern recognition in 1967 [10]. After this, the research on ANNs became slow down, but some important notion on self-organization has been considered [11]. In 1980 or later, the learning of ANNs by error back-propagation method has been proposed, and many active researches have been initiated since then [12-16].

The typical ANN is expressed as the multiple layers such as **input layer**, **middle layer** or **hidden layer**, and **output layer** as shown in Fig. **15**a for the three layers. Let N_A, N_B, and N_C be the number of neurons in the respective layer, and let a_k, b_k, and c_k be the output of the k-th neuron at each layer. Let I_k be the k-th input, and let v_{km} and w_{ki} be the connection weights (or connectivity strength or connectivity coefficient), where the former is for the connection between the m-th output in the input layer and the k-th neuron in the hidden layer, while the latter is for the connection between the i-th neuron of the hidden layer and the k-th neuron in the output layer.

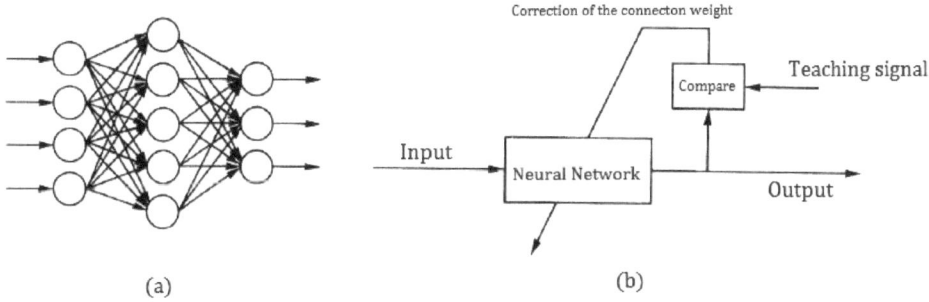

Figure 15: Neural network (a) and its learning (b)

One of the important features of ANN is its learning ability, where this is made by adjusting the connection weights such that the difference between the output of ANN and the desired signal (or teaching signal) is made small as much as possible as shown in Fig. **15b**. The typical learning is made by repetitively adjusting the connection weights to minimize the difference between the output of ANN c_k and the teaching data d_k (k=1,2,...,N_c) by the gradient method. Let this mean square error be expressed as E_1, then

$$E_1 = \frac{1}{2}\sum_{k=1}^{N_c}(c_k - d_k)^2 \tag{161}$$

where 1/2 is just for convenience. The following equations hold at the k-th neuron

$$c_k = f(z_k) \tag{162a}$$

$$z_k = \sum_{i=1}^{N_B} w_{ki} b_i \tag{162b}$$

where f($\cdot$) is often used as sigmoidal function as given in Eq.(160).

Consider next how much E_1 changes with respect to the perturbation of w_{ki} such that

$$\frac{\partial E_1}{\partial w_{ki}} = \left(\frac{\partial E_1}{\partial c_k}\right)\left(\frac{\partial c_k}{\partial z_k}\right)\left(\frac{\partial z_k}{\partial w_{ki}}\right) \tag{163}$$

where

$$\frac{\partial E_1}{\partial c_k} = c_k - d_k$$

$$\frac{\partial c_k}{\partial z_k} = f'(z_k)$$

$$\frac{\partial z_k}{\partial w_{ki}} = b_i$$

and thus the following equation is obtained:

$$\frac{\partial E_1}{\partial w_{ki}} = \delta_k^{(3)} b_i \tag{164a}$$

where

$$\delta_k^{(3)} = (c_k - d_k) f'(z_k) \tag{164b}$$

In the case where f(·) is expressed as the sigmoid function, the following equation is satisfied:

$$f'(z_k) = f(z_k)(1 - f(z_k)) \tag{165}$$

Then consider how much E_1 changes with respect to the perturbation of v_{km} such that

$$\frac{\partial E_1}{\partial v_{km}} = \sum_{i=1}^{N_c} \left[\left(\frac{\partial E_1}{\partial c_i} \right) \left(\frac{\partial c_i}{\partial z_i} \right) \left(\frac{\partial z_i}{\partial b_k} \right) \right] \left(\frac{\partial b_k}{\partial y_k} \right) \left(\frac{\partial y_k}{\partial v_{km}} \right)$$

$$= \sum_{i=1}^{N_c} \left[(c_i - d_i) f'(z_i) w_{ki} \right] f'(y_k) a_m = \delta_k^{(2)} a_m \tag{166a}$$

where

$$\delta_k^{(2)} \equiv f'(y_k) \sum_{i=1}^{N_c} \delta_i^{(3)} w_{ki} \tag{166b}$$

In the end, we may be able to minimize the value of E_1 by changing the values of w_{ki} and v_{km} using Eqs.(163) and (166a) such that

$$w_{ki}(n+1) = w_{ki}(n) - \eta \delta_k^{(3)} b_i \qquad \qquad (167a)$$

$$v_{km}(n+1) = w_{km}(n) - \eta \delta_k^{(2)} a_m \qquad \qquad (167b)$$

where $(\,\cdot\,)$ indicates the iteration number, and η is a parameter called as **learning rate**, where $0 < \eta < 1$. Since $\delta_k^{(3)}$ is required for the computation of $\delta_k^{(2)}$ as seen above Eq.(166b), and thus the correction of the connection weight is made recursively from the output layer toward the input layer, This algorithm is called as **generalized delta rule** or **error back-propagation method**.

Since the local optimum may be obtained inherent in the gradient method, one approach to overcome this problem, or move out of the local minimum is to introduce the momentum term, thus proceeding over the local optimum without reducing the step size when approaching to the optimal point. The correction of the weights may be made by introducing **momentum coefficient** α such that

$$\Delta w_{ki}(n+1) = -\eta \delta_k^{(3)} b_i + \alpha \Delta w_{ki}(n) \qquad \qquad (168a)$$

$$\Delta v_{km}(n+1) = -\eta \delta_k^{(2)} a_m + \alpha \Delta v_{km}(n) \qquad \qquad (168b)$$

where

$$\Delta w_{ki}(n+1) \equiv w_{ki}(n+1) - \Delta w_{ki}(n) \qquad \qquad (169a)$$

$$\Delta v_{km}(n+1) = v_{km}(n+1) - v_{km}(n) \qquad \qquad (169b)$$

The overall algorithm for the error back-propagation method is as follows, where it may be convenient to normalize all the data in the range [0,1] or [-1,1]:

1) Assign arbitrary values between 0 and 1 or -1 to 1 for all v_{km} and w_{ki}.

2) Compute the output of the neurons in the input layer for the normalized input values I_i ($i=1,2,\ldots,N_A$) by the following equation:

$$x_i = I_i \qquad (i=1,2,\ldots,N_A)$$

$$a_i = f(xi) = \frac{1}{1+e^{-x_i}} \qquad (i = 1,2,...,N_A)$$

Compute then the output of each neuron in the hidden layer by the following equation

$$b_i = f\left(\sum_{m=1}^{N_A}(v_{im}a_m) + \theta_B\right) \qquad (i = 1,2,...,N_B)$$

where f is the sigmoid function, and θ_B is the threshold value. Moreover, compute the outputs of the neurons in the output layer for the outputs of the hidden layer by

$$c_i = f\left(\sum_{k=1}^{N_B}(w_{ik}b_k) + \theta_C\right) \qquad (i = 1,2,...,N_C)$$

3) Repeat the above procedure for the M sets of learning data. Then compute the square error between the outputs of ANN and the teaching data by

$$E = \sum_{i=1}^{M}\sum_{k=1}^{N_C}(c_k^{(i)} - d_k^{(i)})^2$$

where $c_k^{(i)}$ is the k-th output of ANN and $d_k^{(i)}$ is the corresponding teaching data for the i-th learning or training data.

4) Compute $\delta_k^{(3)}$ for the k-th neuron (k=1,2,…,N_C) for the 1st set of input-output data by

$$\delta_k^{(3)} = (c_k - d_k)f(z_k)(1 - f(z_k))$$

where

$$z_k = \sum_{i=1}^{N_B}w_{ki}b_i$$

Then compute $\delta_k^{(2)}$ by

$$\delta_k^{(2)} = f'(y_k)\sum_{i=1}^{N_C}\delta_i^{(3)}w_{ki}$$

where

$$y_k = \sum_{i=1}^{N_A} v_{ki} a_i + \theta_B$$

5) Compute the correction for the weights by

$$\Delta w_{ki}(n+1) = -\eta \delta_k^{(3)} b_i + \alpha \Delta w_{ki}(n)$$

$$\Delta v_{km}(n+1) = -\eta \delta_k^{(2)} a_m + \alpha \Delta v_{km}(n)$$

Then correct w and v by the following equations:

$$w_{ki}(n+1) = w_{ki}(n) + \Delta w_{ki}(n+1)$$

$$v_{km}(n+1) = v_{km}(n) + \Delta v_{km}(n+1)$$

6) Repeat the above procedure from (2) to (5) until the squared error becomes within the pre-specified tolerance. The same procedure is made for the other sets of data.

CONCLUDING REMARKS

In the present Chapter, the basis for the modeling is given based on the transport phenomena such as mass, momentum, and heat transport. Then a variety of modeling approach was explained for several bioprocess systems.

As mentioned later, models may be classified as structure or unstructured models, lumped-parameter or distributed-parameter models, and deterministic or stochastic models. The distributed models were not mentioned in the present book, where those may appear for the single cell analysis with the change of components of interest in time and space in the cytosol *etc*. This type of modeling require the distributed sensing system with simulation by super-computer. Moreover, stochastic modeling approach was not considered in the present book, where the uncertainty must be analyzed based on the statistical consideration.

DISCLOSURE

Part of this chapter has been previously published in [17].

REFERENCES

[1]   van Riel NAW. Dynamic modeling and analysis of biochemical networks: mechanism-based models and model-based experiments. Br Bioinf 2006; 7, 364-374

[2]   Kumar A, Chrisofides P, Daoutidis P. Singular perturbation modeling of nonlinear processes with nonexplicit time-scale multiplicity. Chem Eng Sci 1998; 53, 1491-1504.

[3]   Gerdtzen ZP, Daoutidis P, Hu WS. Non-linear reduction for kinetic models of metabolic reaction networks. Metab Eng 2004; 6, 140-154.

[4]   Bird RB, Stewart WE, Lightfoot EN. Transport phenomena, John Wiley & Sons, NY 1960.

[5]   Slattery JC. Momentum, energy, and mass transport in continua, McGraw-Hill 1972

[6]   Fahien RW. Fundamentals of transport phenomena, McGraw-Hill 1973

[7]   Almquist J, Cvijovic M, Hatzimanikatis V, Nielsen J, Jirstrand M (2014) kinetic models in industrial biotechnology- improving cell factory performance. Metab Eng 2014; 24, 38-60

[8]   Hebb DO. The first stage of perceptron: Growth of the assembly, in "The organization of behavior", John-Wiley 1949: 60-78

[9]   Rosenblatt. The perceptron: A probabilistic model for information storage and organization in the brain. Psychological Review 1958; 68, 386-408.

[10]   Minsky M, Papert S. Perceptrons: An introduction to computational geometry, MIT press, Cambridge, MA 1969

[11]   Fukushima K. Cognitron: A self-organizing multilayered neural network. Biol Cybern 1975; 20, 121-136.

[12]   Rumelhart DE, McLelland JL (eds.) in Parallel distribute processing, MIT 1988

[13]   Baughman DR, Liu YA. Neural networks in bioprocess and chemical engineering, Acad Press 1995

[14]   Zupan J, Gasteiger. Neural networks for chemist, VCH Weinheim, Germany 1993.

[15]   Kohonen T. Self-organized formation of topologically correct feature maps. Biol Cybern 1982; 43, 59-69.

[16]   Hopfield JJ. Neural networks and physical systems with emergent collective computational abilities. PNAS USA 1982; 79, 2554-2558.

[17]   Shimizu K, Mathematical approaches to the analysis of biosystems, Corona Pub Co., Japan, 1999 (in Japanese).

CHAPTER 4

Kinetic Modeling for the Main Metabolic Pathways

Abstract: Kinetic models for the main metabolic pathways such as glycolysis, TCA cycle, pentose phosphate pathway together with anaplerotic pathways and gluconeogenic pathways are explained. Kinetic models for substrate uptake pathways such as glucose PTS, glycerol uptake pathways, and xylose assimilating pathways are then explained. Kinetic models for the fermentation pathways under anaerobic condition are also explained. Moreover, kinetic models for amino acid synthetic pathways such as glutamic acid/ glutamine synthetic pathways and lysine synthetic pathways are explained. Unlike flux balance analysis based on the stoichiometric constraints, kinetic modeling approach based on enzymatic reactions for the metabolic pathways can be easily extended for the inclusion of enzyme level and transcriptional regulations, and the kinetic models can reasonably express the dynamics.

Keywords: Kinetic modeling, Embden-Meyerhoff-Parnas pathway, glycolysis, pentose phosphate pathway, TCA cycle, gluconeogenic pathway, amino acid synthesis, carbohydrate metabolism, nitrogen metabolism, anaerobic fermentation, lysine synthetic pathway.

INTRODUCTION

Metabolism is defined as the overall chemical or enzymatic reactions that occur in the living organisms, where the living organisms assimilate nutrient with high enthalpy and low entropy and get free energy during the process of its breakdown to low enthalpy and high entropy substances, and thus keep the cells alive. Let us define the **metabolite** as either substrate, intermediate, or product of each metabolic reaction in the cell, and let the **metabolic pathways** be a series of metabolic or enzymatic reactions. The typical central metabolic pathways of *Escherichia coli* are shown in Fig. **1**, where the central metabolism is common to many organisms, since it is the important source for energy generation and for the production of precursor metabolites for biosynthesis. **Catabolism** is defined as all chemical or enzymatic reactions that are involved in the breakdown of organic materials such as protein, sugar, fatty acid *etc.* in order to get energy, while **anabolism** is defined as the biosynthetic reactions that lead to the building of cell materials such as proteins, DNA, RNA, lipids *etc.* using the energy obtained by the process of catabolism.

The cell generates energy as ATP (adenosine triphosphate) typically at the glycolysis pathway by breaking down carbohydrates to low molecules such as

Kazuyuki Shimizu and Yu Matsuoka

pyruvate *via* **substrate level phosphorylation** as shown in Fig. **1**. Moreover, the reducing equivalents such as NADH and FADH₂ produced at the glycolysis and TCA (tri carboxylic acid) cycle are oxidized in the respiratory chain, where ATP is produced *via* **oxidative phosphorylation** process. Namely, those energy generating processes are termed as **catabolism**. The cell constituents such as proteins, cell membranes *etc.* are formed from their precursor metabolites [1, 2].

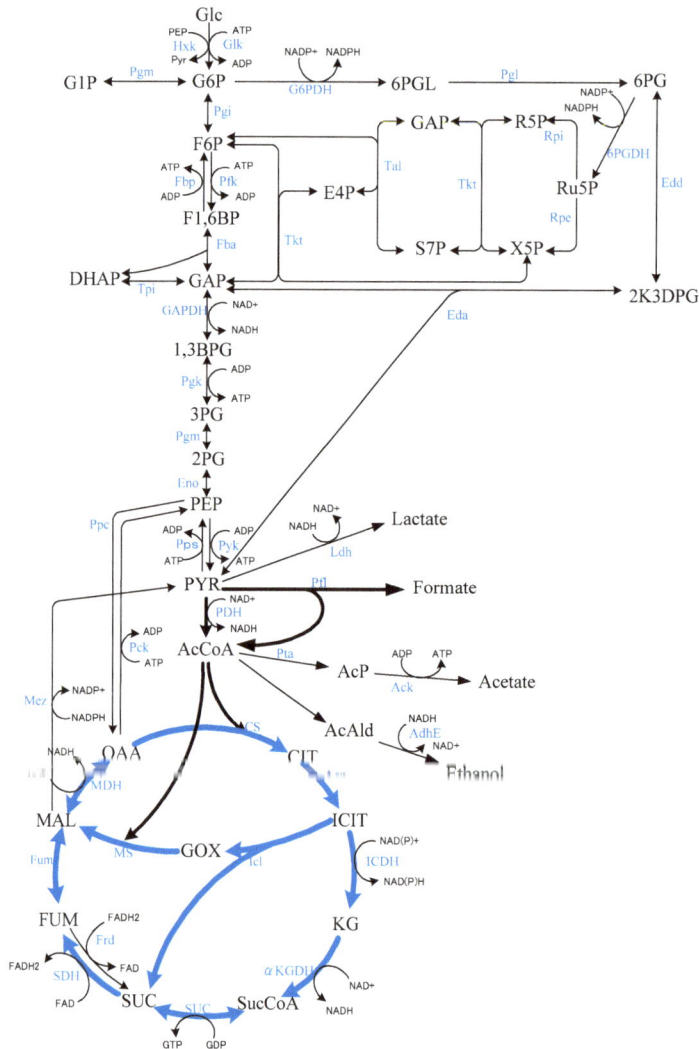

Figure 1: Main metabolic pathways of *E.coli*.

The typical carbohydrate to be used as a carbon source is glucose (Glc), and it is converted to pyruvate (PYR) mainly *via* three different pathways such as

Embden-Meyerhof-Parnas (EMP) pathway, pentose phosphate (PP) (or hexose monophosphate: HMP) pathway, and Entner-Doudoroff (ED) pathway (Fig. **1**). The EMP pathway is often called as glycolysis, but PP pathway and ED pathway are also sometimes called glycolysis. The pyruvate thus formed goes down to the TCA cycle to get energy and precursor metabolites under aerobic condition.

Here, the modeling for the kinetics of the main metabolic pathways are explained together with different carbohydrate uptake pathways. The fermentation pathways under anaerobic condition is also explained. Then amino acid synthetic pathways such as glutamate/ glutamine synthetic pathways are considered in relation to nitrogen regulation. Moreover, the modeling for lysine synthetic pathways is considered. The enzyme which catalyzes the metabolic reaction can be identified by EC number (**Appendix A**), and the reaction mechanism may be classified as given in **Appendix B**. The rate equations are considered mainly for bacteria such as *E.coli* and *Zymomonas mobilis*, but some equations are also considered for yeast metabolism. As for lysine synthetic pathways, the rate equations for *Corynebacterium glutamicum* are mainly considered.

PTS AND EMP PATHWAYS

The first step in the glucose breakdown is the transport and phosphorylation of glucose, where this is typically made by phosphotransferase system (PTS) in bacteria, where the reaction is catalyzed by a sequence of enzymes such as EI, HPr, $EIIA^{Glc}$ and $EIICB^{Glc}$ as will be also explained in Chapter 7. The kinetic rate equation for PTS may be derived for the glucose conversion through PTS as the following equation, where Glc and PEP are the substrates for the reaction, and G6P and PYR are the products, which inhibit the reaction rate as those are accumulated [3, 4]:

$$v_{PTS} = \frac{v_{PTS}^{max}[Glc^{ex}]\dfrac{[PEP]}{[PYR]}}{\left(K_{a1} + K_{a2}\dfrac{[PEP]}{[PYR]} + K_{a3}[Glc^{ex}] + [Glc^{ex}]\dfrac{[PEP]}{[PYR]}\right)\left(1 + \dfrac{[G6P]^{n_{G6P}}}{K_{G6P}}\right)} \quad (1a)$$

where [·] denotes to the concentration.

Glucose can be also transported by other transporters such as GalP, MglCAB, and Man in the case of *pts* mutation in *E.coli* (see Chapter 7), and the phosphorylation of the transported glucose is made by glucokinase (Glk) (ATP: D-glucose 6-

phophotransferase, EC 2.7.1.2), where this reaction requires one mole of ATP, and the rate equation may be expressed as [5]

$$v_{Glk} = \frac{v_{Glk}^{max}[Glc][ATP]}{\left(K_m^{Glc} + [Glc] + \dfrac{K_m^{Glc}[ATP]}{K_{m,ATP}\left(1 + \dfrac{[G6P]}{K_{i,G6P}}\right)} + \dfrac{[Glc][ATP]}{K_{m,ATP}\left(1 + \dfrac{[G6P]}{K_{i,G6P}}\right)} + K_m^{Glc}\dfrac{[G6P]}{K_{i,G6P}} \right) \left(K_{m,ATP}\left(1 + \dfrac{[G6P]}{K_{i,G6P}}\right) \right)}$$

(1b)

More generally, the rate equation for hexokinase (Hxk) (ATP: D-hexose phosphotransferase, EC 2.7.1.1) may be expressed as follows [6]:

$$v_{Hxk} = \frac{v_{Hxk}^{max}[Glc][ATP]}{\left(K_m^{Glc} + [Glc] \right)\left(K_m^{ATP} + [ATP] \right)}$$

(1c)

The next step is the isomerization of G6P to form fructose 6-phosphate (F6P) by glucose phosphate isomerase, Pgi (D-glucose-6-phosphate-isomerase, EC 5.3.1.9), where this pathway is reversible. Pgi is the first enzyme in the EMP pathway (Fig. **2**), where two Pgi isoenzymes may be lumped into one rate equation as follows [3, 7]:

$$v_{Pgi} = \frac{v_{Pgi}^{max}\left([G6P] - \dfrac{[F6P]}{K_{eq}}\right)}{K_{G6P}\left(1 + \dfrac{[F6P]}{K_{F6P}\left(1 + \dfrac{[F6P]}{K_{inh}^{F6P}}\right)} + \dfrac{[6PG]}{K_{inh}^{6PG}}\right) + [G6P]}$$

(2)

where the minus term in the numerator is for the backward reaction.

The next step requires additional ATP for the phosphorylation to produce fructose 1,6-bisphosphate (F16BP or FBP) from F6P. This reaction is catalyzed by the important enzyme, phosphofruct kinase (Pfk) (ATP: D-fructose-6-phosohate 1 phosphotransferase, EC 2.7.1.11), that regulates the flow through EMP pathway, and sometimes becomes rate-limiting for the EMP pathway. Pfk is encoded by *pfkA* and *pfkB* in *E.coli*, where Pfk-1 encoded by *pfkA* is dominant [8]. Here, we consider only Pfk-1 encoded by *pfkA*, where it is allosterically inhibited by PEP

[9], and its reaction rate may be expressed as the next equation, where F6P is the substrate for this reaction, and PEP inhibits v_{Pfk} as it accumulates [3].

$$v_{Pfk} = \frac{v_{Pfk}^{max}[F6P][ATP]}{\left\{[ATP]+K_{ATP,s}\left(1+\frac{[ADP]}{K_{ADP,c}}\right)\right\}\left\{[F6P]+K_{F6P,s}\frac{1+\frac{[PEP]}{K_{PEP}}+\frac{[ADP]}{K_{ADP,b}}+\frac{[AMP]}{K_{AMP,b}}}{1+\frac{[ADP]}{K_{ADP,a}}+\frac{[AMP]}{K_{AMP,a}}}\right\}\times\left(1+\frac{L_{Pfk}}{\left(1+[F6P]\frac{1+\frac{[ADP]}{K_{ADP,a}}+\frac{[AMP]}{K_{AMP,a}}}{1+\frac{[PEP]}{K_{PEP}}+\frac{[ADP]}{K_{ADP,b}}+\frac{[AMP]}{K_{AMP,b}}}\right)^{n_{Pfk}}}\right)} \quad (3a)$$

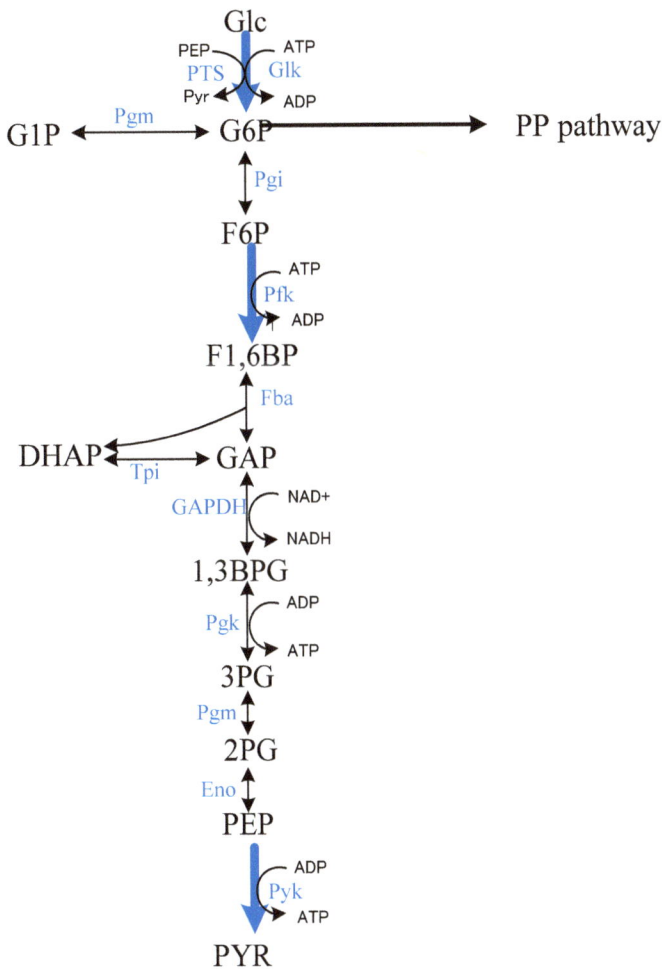

Figure 2: EMP pathway.

Since the concentrations of ATP, ADP, AMP may be considered to be constant at certain culture condition, this equation may be simplified as [10].

$$v_{Pfk} = \frac{v_{Pfk}^{max} K_{ATP}[F6P]}{K_{(ATP,ADP)}\left([F6P]+K_s^{F6P}\dfrac{K_{b(ADP,AMP)}+\dfrac{[PEP]}{K_{PEP}}}{K_{a(ADP,AMP)}}\right)\times\left(1+\dfrac{L_{Pfk}}{\left(1+[F6P]\dfrac{K_a}{K_s^{F6P}\left(K_b+\dfrac{[PEP]}{K_{PEP}}\right)}\right)^{n_{Pfk}}}\right)} \tag{3b}$$

The next step is the cleavage of FBP (F16BP) to 2 moles of triose phosphates. The enzyme which catalyzes this reaction is fructose bisphosphate aldolase (Fba) (fructose-1,6-bisphosphate: D-glyceraldehyde-3-phosphate-lyase, EC 4.1.2.13). For Fba reaction, the following equation may be considered [3], where FBP is the substrate and GAP and DHAP are the products.

$$v_{Fba} = \frac{v_{Ald}^{max}\left([FDP]-\dfrac{[GAP][DHAP]}{K_{eq}}\right)}{\left(K_{FDP}+[FDP]+\dfrac{K_{GAP}[GAP]}{K_{eq}K_{blf}}+\dfrac{K_{DHAP}[DHAP]}{K_{eq}K_{blf}}+\dfrac{[FDP][GAP]}{K_{GAP,inh}^{PEP}}+\dfrac{[DHAP][GAP]}{K_{eq}K_{blf}}\right)} \tag{4}$$

The cleavage of FBP to 2 moles of triose phosphates such as GAP and DHAP can be interchangeable by the reversible triose phosphate isomerase (Tpi) (D-glyceraldehyde-3-phosphate ketol-isomerase, EC 5.3.1.1). The equilibrium lies from DHAP toward GAP as long as the EMP pathway functions in the proper way. The corresponding equation may be expressed as follows [3]:

$$v_{Tpi} = \frac{v_{Tpi}^{max}\left([DHAP]-\dfrac{[GAP]}{K_{eq}}\right)}{\left(K_{DHAP}\left(1+\dfrac{[GAP]}{K_{DGAP}}\right)+[DHAP]\right)} \tag{5}$$

Next is the oxidation step catalyzed by glyceraldehyde 3-phosphate dehydrogenase GAPDH (D-glyceraldehyde-3-phosphate: NAD oxidoreductase, EC 1.2.1.12), and produce nicotinamide adenine dinucleotide (NADH) and high energy component 1,3-bisphospho- D-glycerate (1,3BPG). For GAPDH reaction, the following equation may be considered [3], where GAP is the substrate, and 1,3BPG is the product for this reaction. Moreover, this reaction is inhibited by NADH as [NADH]/[NAD] ratio increases.

$$v_{GAPDH} = \frac{v_{GAPDH}^{max}\left([GAP] - \frac{[13BPG][NADH]}{K_{eq}[NAD]}\right)}{\left(K_{GAP}\left(1 + \frac{[13BPG]}{K_{13BPG}}\right) + [GAP]\right)\left(\frac{K_{NAD}}{[NAD]}\left(1 + \frac{[NADH]}{K_{NADH}}\right) + 1\right)} \quad (6)$$

In the next step, high-energy compound 1,3 BPG releases one phosphate group as 1 mole of ATP by the reaction catalyzed by phosphoglycerate kinase (Pgk) (ATP: 3-phospho-D-glycerate 1-phospho-transferase, EC 2.7.2.3) to produce 3PG. The rate equation for this reaction may be expressed as [3]:

$$v_{Pgk} = \frac{v_{Pgk}^{max}\left([3PG]] - \frac{[ATP][3PG]}{K_{eq}}\right)}{\left(K_{ADP}\left(1 + \frac{[ATP]}{K_{ATP}}\right) + [ADP]\right)\left(K_{13BPG}\left(1 + \frac{[3PG]}{K_{3PG}}\right) + [13BPG]\right)} \quad (7)$$

The conversion of 3PG to 2-phosphoglycerate (2PG) is catalyzed by phosphoglycerate mutase (Pgm) (2,3-diphospho-D-glycerate: D-phosphoglycerate 2,3-phosphomutase, EC 5.4.2.11). In this reaction, the phosphate group is transferred from the 3rd position to the 2nd position, and the rate equation may be expressed as [3].

$$v_{Pgm} = \frac{v_{Pgm}^{max}\left([3PG] - \frac{[2PG]}{Keq}\right)}{[3PG] + K_{3PG}\left(1 + \frac{[2PG]}{K_{2PG}}\right)} \quad (8)$$

The next reaction step is catalyzed by enolase (Eno) (2-phospho-D-glycerate hydro-lyase, EC 4.2.1.11) to form PEP, and the rate equation may be expressed as [3]:

$$v_{Eno} = \frac{v_{Eno}^{max}\left([2PG] - \frac{[PEP]}{K_{eq}}\right)}{[2PG] + K_{2PG}\left(1 + \frac{[PEP]}{K_{PEP}}\right)} \quad (9)$$

This reaction is connected to the intracellular electron shift often referred to the intra-molecular oxidation-reduction reaction. By this reaction, PEP becomes more energy-rich as compared to 2PG.

From GAP/ DHAP to PEP, the reactions are reversible, and the pathways may be lumped together into one such as EMP, and the rate equation may be expressed as follows, where the products are the PEP and NADH by ignoring the effect of ATP at Pgk [10]:

$$v_{EMP} = \frac{v_{EMP}^{max}\left([GAP] - \frac{[PEP][NADH]}{K_{eq}[NAD]}\right)}{\left(K_{GAP}\left(1 + \frac{[PEP]}{K_{PEP}}\right) + [GAP]\right)\left(\frac{K_{NAD}}{[NAD]}\left(1 + \frac{[NADH]}{K_{NADH}}\right) + 1\right)} \quad (6)'$$

In the final step of EMP pathway, the reaction is catalyzed by pyruvate kinase (Pyk) (ATP: pyruvate phosphotransferase, EC 2.7.1.40) to produce PYR, and the phosphate group of PEP is transferred to adenosine diphosphate (ADP) to produce ATP. Pyk is the 2nd allosteric enzyme in the EMP pathway. It is catalyzed by two isoenzymes, PykI and PykII, encoded by *pykF* and *pykA*, respectively in *E.coli*. PykI is activated by FBP and inhibited by ATP, whereas PykII is activated by AMP. Here, let us lump these together, and the rate equation may be expressed as the next equation, where PEP and ADP are the substrates, and it is inhibited by ATP [3]. Moreover, this equation indicates that FBP and AMP are the activators.

$$v_{Pyk} = \frac{v_{Pyk}^{max}[PEP]\left(\frac{[PEP]}{K_{PEP}} + 1\right)^{n_{Pyk}-1}[ADP]}{K_{PEP}\left(L_{Pyk}\left(\frac{1 + \frac{[ATP]}{K_{ATP}}}{\frac{[FBP]}{K_{FDP}} + \frac{[AMP]}{K_{AMP}} + 1}\right)^{n_{Pyk}} + \left(\frac{[PEP]}{K_{PEP}} + 1\right)^{n_{Pyk}}\right)([ADP] + K_{ADP})} \quad (10)$$

PENTOSE PHOSPHATE PATHWAY

Fig. **3** shows the overall pentose phosphate (PP) pathway, where this pathway connects to the EMP pathway at G6P, F6P, and GAP. In the PP pathway, G6P is oxidized by the NADP$^+$-linked G6P dehydrogenase (G6PDH) (D-glucose-6-phosphate: NADP oxidoreductase, EC 1.1.1.49) to produce D-glucono-δ-lactone 6-phosphate (PGL). G6PDH is an important regulatory enzyme, where it is inhibited by nicotinamide adenine dinucleotide phosphate (NADPH), and the rate equation may be expressed as [3]:

$$v_{G6PDH} = \frac{v_{G6PDH}^{max}[G6P]}{\left([G6P]+K_{G6P}\right)\left(1+\frac{[NADPH]}{K_{NADPH}^{G6P}}\right)\left(\frac{K_{NADP}}{[NADP]}\left(1+\frac{[NADPH]}{K_{NADPH}^{NADP}}\right)+1\right)} \quad (11)$$

The product of the G6PDH reaction is almost immediately hydrolyzed to form 6-phosphogluconate (6PG) by gluconolactonase (Pgl) (D-glucono-δ-lactone hydrolase, EC 3.1.1.17). The rate equation may be expressed as [5].

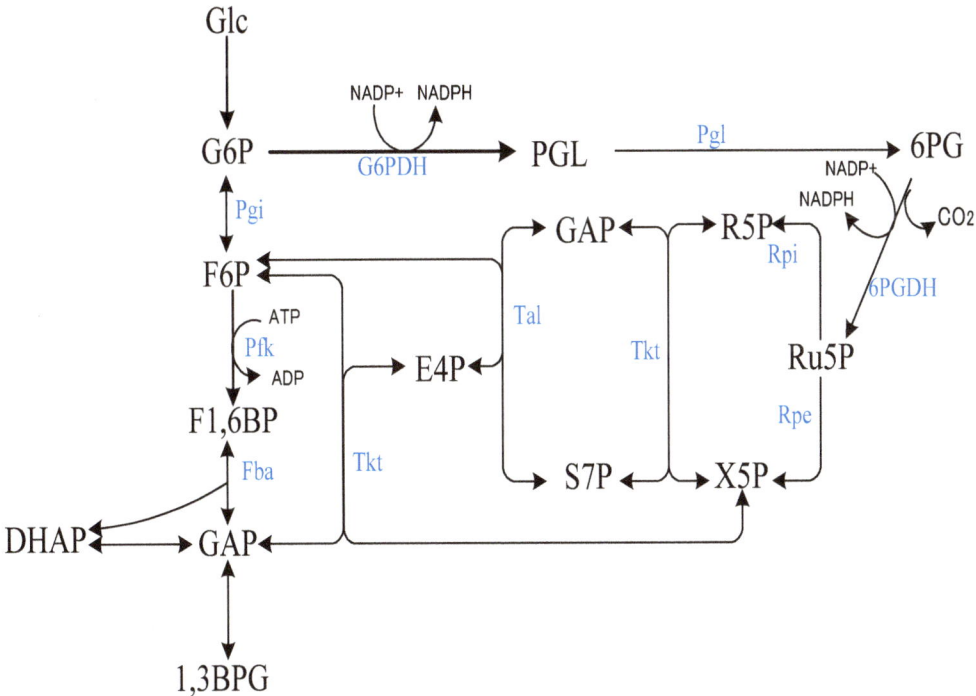

Figure 3: Pentose phosphate pathway.

$$v_{Pgl} = \frac{v_{Pgl}^{max} \dfrac{[PGL]}{K_{m,PGL}\left(1+\dfrac{[G6P]}{K_{i,G6P}}\right)}}{1 + \dfrac{[PGL]}{K_{m,PGL}\left(1+\dfrac{[G6P]}{K_{i,G6P}}\right)}} \qquad (12)$$

where this reaction is fast, and this may be combined with G6PDH reaction.

The second NADP$^+$-linked oxidation occurs to produce D-ribulose 5-phosphate (Rib5P) from 6PG. This reaction is catalyzed by 6-phosphogluconate dehydrogenase (6PGDH) (6-phospho-D-gluconate: NADP$^+$ 2-oxidoreductase, EC 1.1.1.44), where C-1 atom of 6PG is released as CO_2 to form ribulose 5-phosphate (Ru5P). 6PGDH is also an important regulatory enzyme, where it is inhibited by NADPH. For 6PGDH reaction, the following equation may be considered [3].

$$v_{6PGDH} = \frac{v_{6PGDH}^{max}[6PG]}{\left([6PG]+K_{6PG}\right)\left(1+\dfrac{K_{NADP}}{[NADP]}\left(1+\dfrac{[NADPH]}{K_{NADPH}}\right)\left(1+\dfrac{[ATP]}{K_{ATP}}\right)\right)} \qquad (13)$$

where the pathway reactions from G6P to Ru5P is uni-direction, and these pathways are called as the **oxidative PP pathway** due to the production of reducing equivalent NADPH, while the other reactions in the PP pathway are reversible and called as the **nonoxidative PP pathway**.

Ru5P is attacked partly by two different enzymes such as ribulose phosphate 3-epimerase (Rpe) (D-ribulose-5-phosphate 3-epimerase, EC 5.1.3.1) which converts Ru5P to xylulose 5-phosphate (X5P), and ribose 5-phosphate isomerase (Rpi) (D-ribose-5-phosphate ketol-isomerase, EC 5.3.1.6), which converts Ru5P to R5P. The rate equations may be simply expressed as follows [3]:

$$v_{Rpe} = v_{Rpe}^{max}\left([Ru5P] - \frac{[X5P]}{K_{eq}^{Rpe}}\right) \qquad (14)$$

$$v_{Rpi} = v_{Rpi}^{max}\left([Ru5P] - \frac{[R5P]}{K_{eq}^{Rpi}}\right) \qquad (15)$$

Both intermediates such as X5P and R5P are required for the cleavage reaction catalyzed by transketolase (Tkt) (sedoheptulose-7-phosphate: D-glyceraldehyde-3-phosphate glycolaldehyde transferase, EC 2.2.1.1), which yields GAP and sedoheptulose 7-phosphate (S7P). The second reaction cleaves both intermediates from transketolase reaction to produce F6P and erythrose 4-phosphate (E4P), where this reaction is catalyzed by transaldorase (Tal) (D-glyceraldehyde-3-phosphatedehydroxyacetonetransferase,EC2.2.1.2). The third cleavage reaction is carried out by the same transketolase as the first stage reaction, and it cleaves E4P and X5P to form GAP and F6P (Fig. **3**). The rate equations may be expressed as [3]

$$v_{TktA} = v_{TktA}^{max}\left([R5P][X5P] - \frac{[S7P][GAP]}{K_{eq}^{TktA}}\right) \tag{16a}$$

$$v_{TktB} = v_{TktB}^{max}\left([X5P][E4P] - \frac{[F6P][GAP]}{K_{eq}^{TktB}}\right) \tag{16b}$$

$$v_{Tal} = v_{Tal}^{max}\left([GAP][S7P] - \frac{[E4P][F6P]}{K_{eq}^{TktB}}\right) \tag{17}$$

R5P is the important precursor for purine and pyrimidine production, and that E4P is the important precursor for aromatic amino acid production for the cell synthesis.

The major roles of the PP pathway is to produce NADPH, important reducing equivalent for the cell synthesis, and to produce precursor metabolites such as R5P, E4P, and possibly S7P for the cell synthesis. Namely, the main role for the PP pathway is to produce necessary compounds for anabolism.

ENTNER DOUDOROFF PATHWAY

As shown in Fig. **4**, Entner-Doudoroff (ED) pathway connects to PP pathway at 6PG, where 6PG is converted to 2-keto-3-deoxy-6-phosphogluconate (KDPG) by dehydration reaction catalyzed by phosphogluconate dehydratase (Edd) (6-phosphogluconate hydro-lyase, EC 4.2.1.12). The rate equation may be expressed as [5]

$$v_{Edd} = \frac{v_{Edd}^{max} \dfrac{[6PG]}{K_{m,6PG}\left(1+\dfrac{[GAP]}{K_{i,GAP}}\right)}}{1+\dfrac{[6PG]}{K_{m,6PG}\left(1+\dfrac{[GAP]}{K_{i,GAP}}\right)}} \tag{18}$$

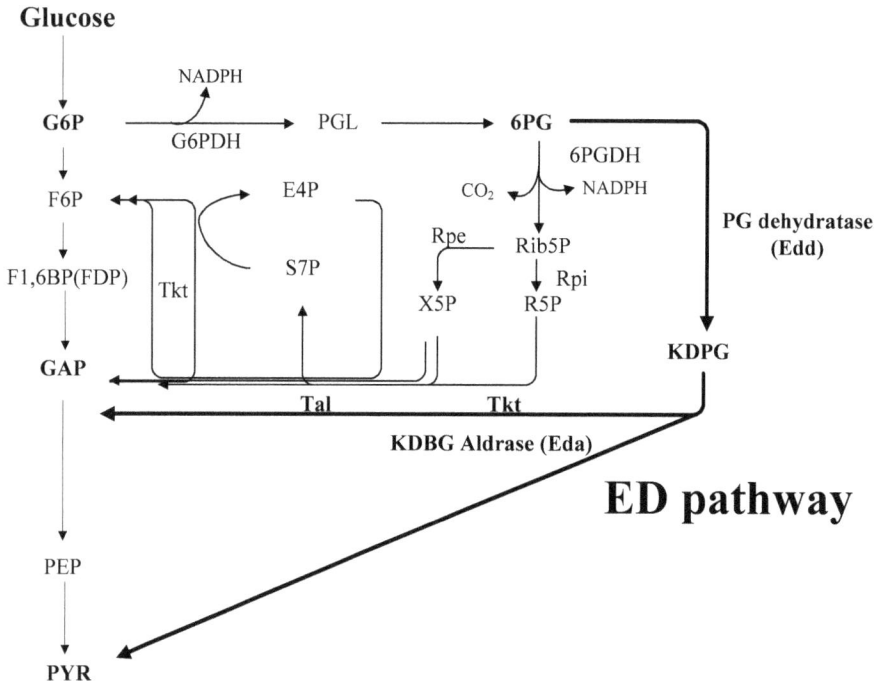

Figure 4: ED pathway.

The next step is the cleavage of KDPG to GAP and PYR by phospho-2-keto-3-deoxy-gluconate aldorase (Eda) (6-phospho-2-keto-3-deoxy- D-gluconate D-glyceraldehyde-3-phosphate lyase, EC 4.1.2.14). The rate equation for Eda pathway may be expressed as [5]

$$v_{Eda} = \frac{v_{Eda}^{max} \dfrac{[KDPG]}{K_{m,KDPG}}}{1+\dfrac{[KDPG]}{K_{m,KDPG}}} \tag{19}$$

Although ED pathway is active in such microorganism as *Zymomonas mobilis*, this may be induced in other bacteria such as *E.coli* depending on the genetic and culture conditions.

PDH AND TCA CYCLE

As shown in Fig. **1**, the terminal product of EMP and ED pathway is PYR, where it is converted to a two carbon acid derivative such as AcCoA by releasing CO_2 from the 1st carbon of PYR by pyruvate dehydrogenase complex (PDHc) reaction. This enzyme is a multi-enzyme complex consisting of three different enzymes, and requires such cofactors as thiamine pyrophosphate (TPP), lipoic acid, and NAD^+. The three enzymes are pyruvate dehydrogenase (pyruvate: lipoate oxidoreductase, EC 1.2.4.1), lipoate acetyltransferase (acetyl-CoA: dihydrolipoate S-acetyltransferase, EC 2.3.1.12), and lipoamid dehydrogenase (reduced-NAD: lipoamid oxidoreductase, EC 1.6.4.3).

The rate equation of PDHc may be considered with respect to PYR, and PDHc is inhibited by NADH as [NADH]/[NAD] ratio increases as given below [11].

$$v_{PDH} = \frac{\dfrac{v_{PDH}^{max}}{[NAD]}\left(\dfrac{1}{1+K_i\dfrac{[NADH]}{[NAD]}}\right)\left(\dfrac{[PYR]}{K_m^{PYR}}\right)\left(\dfrac{1}{K_m^{NAD}}\right)\left(\dfrac{[COA]}{K_m^{COA}}\right)}{\left(1+\dfrac{[PYR]}{K_m^{PYR}}\right)\left(\dfrac{1}{[NAD]}+\dfrac{1}{K_m^{NAD}}+\dfrac{[NADH]}{K_m^{NADH}[NAD]}\right)\left(1+\dfrac{[COA]}{K_m^{COA}}+\dfrac{[AcCoA]}{K_m^{AcCoA}}\right)} \tag{20}$$

The AcCoA thus produced from PYR goes into a series of reactions called as either the tricarboxylic acid (TCA) cycle, Krebs cycle, or citric acid cycle (Fig. **5**). In the first step of the TCA cycle, AcCoA gives the acetyl group to the four carbon dicarboxylic acid such as oxaloacetate (OAA) to form a six-carbon tricarboxylic acid such as citric acid (CIT). This reaction is catalyzed by citrate synthase (CS) (citrate oxaloacetate-lyase, EC 4.1.3.7), where free CoA is generated in this reaction, and this can be reutilized in the formation of AcCoA. Note that the end products of the TCA cycle is NADH and CO_2, where NADH allosterically inhibits the activity of CS. For this reaction, the following equation may be considered [12], where AcCoA and OAA are the substrates, and the reaction is inhibited by NADH.

$$v_{CS} = \frac{v_{CS}^{\max}[AcCoA][OAA]}{\left(K_d^{AcCoA}K_m^{OAA} + K_m^{AcCoA}[OAA]\right) + \left([AcCoA]K_m^{OAA}\left(1 + \frac{[NADH]}{K_{i1}^{NADH}}\right)\right) + [AcCoA][OAA]\left(1 + \frac{[NADH]}{K_{i2}^{NADH}}\right)} \quad (21)$$

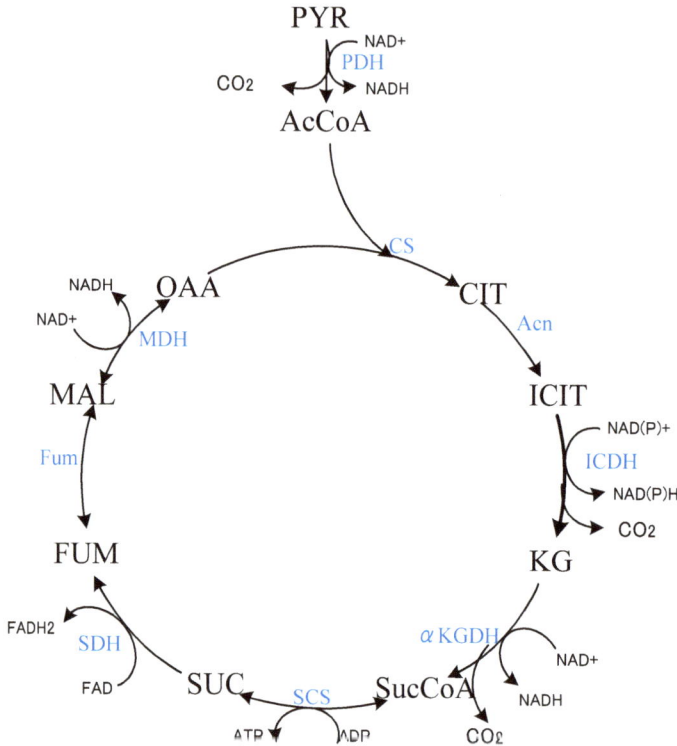

Figure 5: TCA cycle.

The next step is the formation of *cis*-aconitate and then isocitrate (ICIT) by the enzyme aconitate hydratase or aconitase (Acn) (citrate (iso-citrate) hydro-lyase, EC 4.2.1.3). This reaction is fast, and therefore the reaction may be combined with CS reaction. The rate equation for this reaction may be expressed as [13]

$$v_{Acn} = \frac{v_{Acn}^{\max}\left(k_{Acn,f}\frac{[CIT]}{K_m^{CIT}} - k_{Acn,r}\frac{[ICIT]}{K_m^{ICIT}}\right)}{1 + \frac{[CIT]}{K_m^{CIT}} + \frac{[ICIT]}{K_m^{ICIT}}} \quad (22)$$

ICIT is then converted to α-ketoglutaric acid (αKG) or 2-oxoglutarate (2KG) by the reaction catalyzed by isocitrate dehydrogenase (ICDH) (threo-Ds-isocitrate: NADP oxidoreductase, EC 1.1.1.42). NAD(P)H and CO_2 are formed through this reaction, where the microorganisms possess predominantly the $NADP^+$-specific ICDH, whereas fungi and yeast possess the NAD^+-specific ICDH [14].

ICIT is the important branch point in the TCA cycle, where ICDH (high affinity to ICIT, K_m = 8 mM) and isocitrate lyase (Icl) (low affinity to ICIT, K_m=406 mM) compete for ICIT, where Icl forms part of the glyoxylate pathway as will be explained later in the anaplerotic reaction section. This branch point is under both gene level regulation by *aceBAK* operon under the control of isocitrate lyase regulator (IclR) *etc.*, and enzyme level regulation by the phosphorylation/ dephosphorylation of ICDH. Here, we consider the rate equation for ICDH as follows [15], where ICIT is the substrate, and the reaction is inhibited by NAD(P)H as [NAD(P)H]/[NAD(P)] ratio increases.

$$v_{ICDH} = \cfrac{v_{ICDH}^{max}\dfrac{K_f}{K_m^{ICIT}}K_d^{NADP}\left([ICIT]-\dfrac{[NADPH][\alpha KG]}{K_{eq}^{ICDH}[NADP]}\right)}{\left(\begin{array}{l}\dfrac{1}{[NADP]}+\dfrac{[ICIT]K_m^{NADP}}{K_m^{iICIT}K_d^{NADP}[NADP]}+\dfrac{1}{K_d^{NADP}}+\dfrac{[ICIT]}{K_m^{ICIT}K_d^{NADP}}+ \\[3mm] \dfrac{[ICIT]}{K_d^{ICIT}[NADP]}\dfrac{[NADPH]K_m^{NADP}}{K_m^{ICIT}K_d^{NADP}K_{einh}^{NADPH}}+\dfrac{[NADPH]K_{eknh}^{\alpha KG}}{K_m^{\alpha KG}K_{enhe}^{NADPH}[NADP]}+\dfrac{[\alpha KG]K_m^{NADPH}}{K_m^{\alpha KG}K_{enhe}^{NADPH}[NADP]}+ \\[3mm] \dfrac{[\alpha KG]}{K_m^{2KG}}\dfrac{[NADPH]}{K_{enhe}^{NADPH}[NADP]}+\dfrac{[\alpha KG]K_m^{NADPH}}{K_m^{2KG}K_{enhe}^{NADPH}}\dfrac{[NADPH]}{K_{ekn}^{NADPH}[NADP]}\end{array}\right)} \tag{23}$$

where v_{ICDH}^{max} is a function of isocitrate dehydrogenase kinase (ICDK) (ATP: isocitrate dehydrogenase phosphotransferase, EC 2.7.11.5) and isocitrate dehydrogenase phosphatase (ICDP) (EC 3.1.3.-) such as

$$v_{ICDK} = \cfrac{k_f[ICDK]\dfrac{[ICDH]}{K_m^{ICDH}}}{\left(1+\dfrac{[ICDH]}{K_m^{ICDH}}\right)\left(1+\dfrac{[3PG]}{K_i^{3PG}}\right)} \tag{24a}$$

$$v_{ICDP} = \cfrac{k_f[ICDP][ICDH]\dfrac{K_A^{3PG}+\beta[3PG]}{K_a^{3PG}+\beta[3PG]}}{[ICDH]+K_m^{ICDH}} \tag{24b}$$

The next reaction step is the conversion of αKG to succinyl CoA (SucCoA) by 2-oxoglutarate dehydrogenase complex (KGDH) (2-oxoglutarate lipoate oxidoreductatse, EC 1.2.4.2), a multi-enzyme complex system, similar to PDHc. The KGDH complex requires the participation of thiamine pyrophosphate (TPP), α-lipoic acid, CoA, NAD^+, Mg^{2+}, and produce CO_2 and NADH. For KGDH reaction, the following equation may be considered, where αKG and CoA are the substrates, and the reaction is assumed to be inhibited by NADH as [NADH]/[NAD] ratio increases [16].

$$v_{KGDH} = \frac{v_{KGDH}^{max}[\alpha KG][CoA]}{\left(\begin{array}{l} \dfrac{K_m^{NAD}[\alpha KG][CoA]}{[NAD]} + K_m^{CoA}[\alpha KG] + K_m^{\alpha KG}[CoA] + [\alpha KG][CoA] + \\[2mm] \dfrac{K_m^{\alpha KG}K_z[SUC][NADH]}{K_I^{SUC}[NAD]} + \dfrac{K_m^{NAD}[\alpha KG][CoA][NADH]}{K_I^{NADH}[NAD]} + \\[2mm] \dfrac{K_m^{CoA}[\alpha KG][SUC]}{K_I^{SUC}} + \dfrac{K_m^{\alpha KG}K_z[\alpha KG][SUC][NADH]}{K_I^{\alpha KG}K_I^{SUC}[NAD]} \end{array}\right)} \tag{25}$$

SucCoA is then converted to succinate (SUC) by succinyl-CoA synthetase (SCS) (succinate: CoA ligase, EC 6.2.1.5). Through this reaction step, CoA is released and ATP (GTP) is formed. The rate equation for SCS reaction may be expressed as [17]

$$v_{SCS} = \frac{v_{SCS}^{max}\left(k_f \dfrac{[SucCoA][ADP][P_i]}{K_m^{SUSCoA}K_m^{ADP}K_m^{Pi}} - k_r \dfrac{[SUC][ATP][CoA]}{K_m^{SUC}K_m^{ATP}K_m^{CoA}}\right)}{\left\{\left(1+\dfrac{[SucCoA]}{K_m^{SUCCoA}}\right)\left(1+\dfrac{[P_i]}{K_m^{Pi}}\right)+\dfrac{[SUC]}{K_m^{SUC}}\right\}\left\{\left(1+\dfrac{[ATP]}{K_m^{ATP}}\right)\left(1+\dfrac{[CoA]}{K_m^{CoA}}\right)+\dfrac{[ADP]}{K_m^{ADP}}\right\}} \tag{26}$$

The next step is the dehydrogenation of SUC, where SUC is oxidized to form fumarate (FUM) by succinate dehydrogenase (SDH) (succinate: oxidoreductase, EC 1.3.99.1). SDH is closely linked to the electron transport chain, and enters this system at the flavoprotein level, where $FADH_2$ is released by this reaction step. For SDH reaction, the next equation may be considered [16].

$$v_{SDH} = \frac{v_{SDH}^{max}\left([SUC]-\dfrac{[FUM]}{K_{eq}}\right)}{K_m^{SUC}v_{SDH2} + v_{SDH2}[SUC] + \dfrac{v_{SDH1}[FUM]}{K_{eq}}} \tag{27}$$

Fumarate thus formed is then hydrated at the double bond to form malic acid (MAL) by fumarase or fumarate hydratase (Fum) (L-malate hydro-lyase, EC 4.2.1.2). For fumarase reaction, the next equation may be considered, where FUM is the substrate, and the reaction is inhibited by MAL [16].

$$v_{Fum} = \frac{v_{Fum}^{max}\left([FUM] - \dfrac{[MAL]}{K_{eq}}\right)}{K_m^{Fum}v_{Fum1} + v_{Fum2}[FUM] + \dfrac{v_{Fum1}[MAL]}{K_{eq}}} \tag{28}$$

The reactions from αKG to MAL may be lumped together into one equation such as [18]

$$v_{KGMAL} = \frac{v_{KGMAL}^{max}[\alpha KG]}{K_{m,\alpha KG} + [\alpha KG]} \tag{25'}$$

where one enzyme KGMAL may be considered for these pathways.

The final reaction in the TCA cycle is the dehydrogenation of MAL to oxaloacetate (OAA) catalyzed by malate dehydrogenase (MDH) (L-malate: NAD oxidoreductase, EC 1.1.1.37), where NADH is produced in this reaction. For MDH reaction, the following equation may be considered [16], where MAL is the substrate, and the reaction is inhibited by OAA and NADH as these concentrations increase [16].

$$v_{MDH} = \frac{v_{MDH}^{max}\left([MAL] - \dfrac{[OAA]}{K_{eq}}\right)}{\left(\begin{array}{l}\dfrac{K_I^{NAD}K_m^{MAL}v_{MDH2}}{[NAD]} + K_m^{MAL}v_{MDH2} + \dfrac{K_m^{NAD}V_{MDH2}[MAL]}{[NAD]} + v_{MDH2}[MAL] + \\[3mm] \dfrac{K_m^{OAA}v_{MDH1}[NADH]}{K_{eq}[NAD]} + \dfrac{K_m^{NADH}v_{MDH1}[OAA]}{K_{eq}[NAD]} + \dfrac{v_{MDH1}[NADH][OAA]}{K_{eq}[NAD]} + \\[3mm] \dfrac{v_{MDH1}K_m^{OAA}[NADH]}{K_{eq}K_I^{NAD}} + \dfrac{v_{MDH2}K_m^{NAD}[MAL][OAA]}{K_I^{OAA}[NAD]} + \dfrac{v_{MDH2}[MAL][NADH]}{K_I^{NADH}} + \\[3mm] \dfrac{v_{MDH1}[MAL][NADH][OAA]}{K_{eq}K_I^{MAL}[NAD]} + \dfrac{v_{MDH2}[MAL][OAA]}{K_{II}^{OAA}} + \dfrac{v_{MDH1}[NADH][OAA]}{K_{II}^{NAD}K_{eq}} + \\[3mm] \dfrac{K_I^{NAD}v_{MDH2}[MAL][NADH][OAA]}{K_{II}^{NAD}K_m^{OAA}K_I^{NADH}}\end{array}\right)} \tag{29}$$

ACETATE METABOLISM

The microorganism such as *E.coli* produces acetate (ACE) from AcCoA by the so-called overflow metabolism, while ethanol (ETOH) is formed from PYR in yeast by the similar mechanism. There are two major acetate producing pathways such as phosphoacetyl transferase/acetate kinase (Pta-Ack) and pyruvate oxidase (Pox) pathways as shown in Fig. **6**. The phosphate acetyltransferase (Pta) (acetyl-CoA: phosphate acetyltransferase, EC 2.3.1.8) reversibly converts AcCoA and inorganic phosphate to acetyl phosphate (AcP) and CoA, while acetate kinase (Ack) (ATP: acetate phosphotransferase, EC 2.7.2.1) reversibly converts AcP and ADP to acetate and ATP. The other acetate producing pathway is through the pyruvate oxidase (Pox) (pyruvate: ubiquinone oxidoreductase, EC 1.2.5.1), a peripheral membrane protein that converts PYR, ubiquinone, and H_2O to acetate, ubiquinol and CO_2. This pathway is usually induced at the early stationary phase, where acetyl-CoA synthetase (ACS) (acetate: CoA ligase, EC 6.2.1.1) is also activated to convert acetate to AcCoA at the stationary phase.

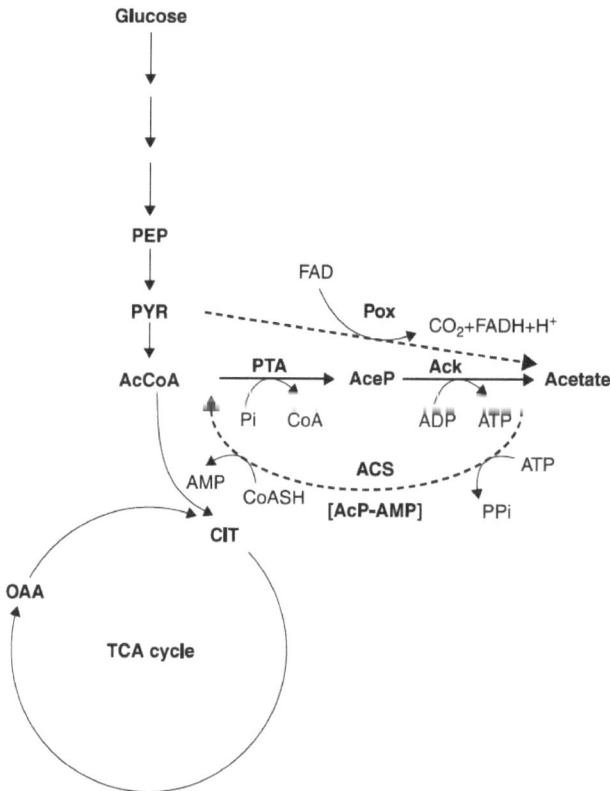

Figure 6: Acetate metabolism.

The acetate formation is important in practice for the cell growth and the specific metabolite production in *E. coli*. When enough glucose is present around the cell, the cell produces acetate and excretes it. For the Pta reaction, the following equation may be considered [11, 19]:

$$v_{Pta} = \frac{v_{Pta}^{max}\left(\dfrac{1}{K_i^{AcCoA}K_m^P}\right)\left([AcCoA][P_i]-\dfrac{[AcP][CoA]}{K_{eq}}\right)}{\left(1+\dfrac{[AcCoA]}{K_i^{AcCoA}}+\dfrac{[P_i]}{K_i^{P_i}}+\dfrac{[AcP]}{K_i^{AcP}}+\dfrac{[CoA]}{K_i^{CoA}}+\left(\dfrac{[AcCoA][P_i]}{K_i^{AcCoA}K_m^{P_i}}\right)+\left(\dfrac{[AcP][CoA]}{K_m^{AcP}K_i^{CoA}}\right)\right)}$$

$$(30)$$

The rate equation for Ack may be expressed as follows [9]:

$$v_{Ack} = \frac{v_{Ack}^{max}\left(\dfrac{1}{K_m^{ADP}K_m^{AcP}}\right)\left([AcP][ADP]-\dfrac{[ACE][ATP]}{K_{eq}}\right)}{\left(1+\dfrac{[AcP]}{K_m^{AcP}}+\dfrac{ACE}{K_m^{ACE}}\right)\left(1+\dfrac{[ADP]}{K_m^{ADP}}+\dfrac{[ATP]}{K_m^{ATP}}\right)}$$

$$(31)$$

Once glucose was consumed or became low level, the cells begin to utilize organic acid such as acetate instead of excreting it. This acetate-associated metabolic switch occurs just as the cells begin to decelerate growth, or go into stationary phase in the batch culture. During the stationary phase when glucose was depleted, the cells scavenge the acetate by using Acs. This reaction catalyzed by Acs proceeds through an enzyme-bound acetyl-adenylate (Acetyl-AMP) intermediate (Fig. **6**). Acs has high affinity to acetate while Ack-Pta has low affinity. The rate equation for Acs may be expressed as [20].

$$v_{Acs} = \frac{v_{Acs}^{max}[ACE][NADP]}{\left(K_m+[ACE]\right)\left(K_{eq}+[NADP]\right)}$$

$$(32)$$

In the case of yeast, ethanol is formed from PYR *via* pyruvate decarboxylase (PDC) (2-oxo-acid carboxy-lyase, EC 4.1.1.1) and alcohol dehydrogenase (ADH) (alcohol: NAD^+ oxidoreductase, EC 1.1.1.1) reaction by the overflow metabolism. The rate equations may be expressed as [5].

$$v_{PDC} = \frac{v_{PDC}^{max}\left(\dfrac{[PYR]}{K_{m,PYR}}\right)}{1 + \dfrac{[PYR]}{K_{m,PYR}}} \qquad (33)$$

$$v_{ADH} = \frac{v_{ADH}^{max}\left(\dfrac{[AcAld][NADH]}{K_{m,AcAld}K_{NADH}}\right)}{1 + \dfrac{[AcAld]}{K_{m,AcAld}} + \dfrac{[NADH]}{K_{m,NADH}} + \dfrac{[AcAld][NADH]}{K_{m,AcAld}K_{m,NADH}}} \qquad (34)$$

where CO_2 is released at PDC reaction, and acetoaldehyde (AcAld) is the substrate for ADH reaction, where ADH requires NADH.

ANAPLEROTIC PATHWAYS

Catabolism for energy generation and anabolism for biosynthesis are closely related. Some of the intermediates in the main metabolic pathways are necessary as precursors for biosynthesis. The cell can no longer survive without these precursor metabolites. Thus the organisms must have an ancillary system that take care of replenishment of these intermediates. The routes required for this replenishment is called as **anaplerotic** pathways [14]. Among the intermediates, OAA is often critical, since its concentration is relatively low due to its fast utilization at CS reaction, while it is also an important precursor for many amino acids such as aspartate (Asp), lysine (Lys) *etc*. There are, therefore, several anaplerotic pathways to prevent OAA shortage (Fig. **7**).

The typical anaplerotic pathway is phosphoenol pyruvate carboxylase (Ppc) (orthophosphate: oxaloacetate carboxy-lyase, EC 4.1.1.31) that catalyzes the reaction of replenishing OAA from PEP. The reaction catalyzed by Ppc is usually low without any activator, where AcCoA is a potent activator, and FBP alone exhibits modest activation, but it produces a strong synergistic activation with AcCoA [21]. Based on these observations, the following equation may be considered [22]:

$$v_{Ppc} = \frac{K_1 + K_2[AcCoA] + K_3[FDP] + K_4[AcCoA][FDP]}{1 + K_5[AcCoA] + K_6[FDP]}\left(\frac{[PEP]}{K_m + [PEP]}\right) \qquad (35)$$

There is another anaplerotic pathway from PYR to OAA catalyzed by a biotin-dependent pyruvate carboxylase (Pyc) (pyruvate:carbon-dioxide ligase, EC 6.4.1.1) where *E.coli* does not have this pathway, while *Corynebacteria, Bucilli etc.* posseses this pathway.

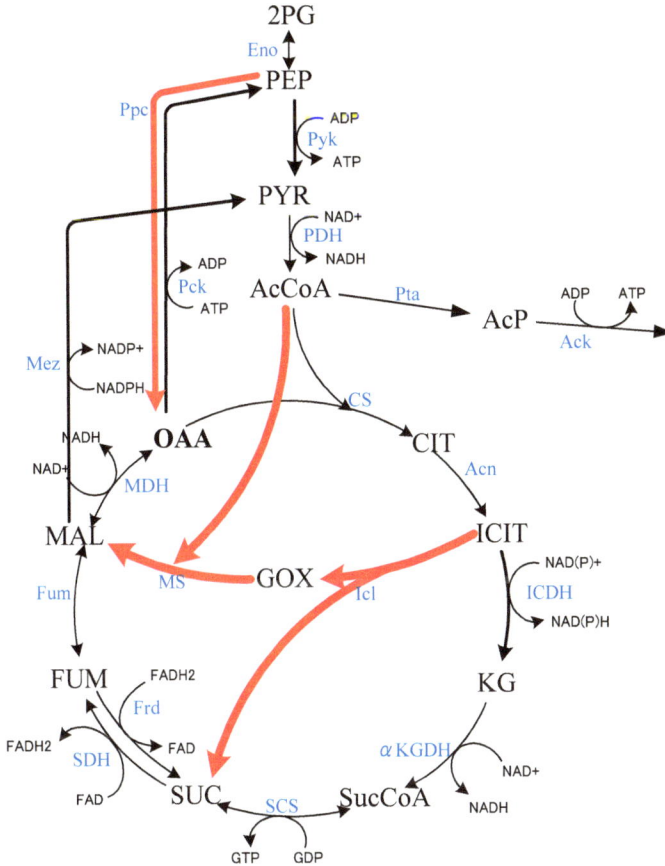

Figure 7: Anaplerotic pathways.

In relation to Ppc, consider the reverse reaction, where this is the gluconeogenic pathway such as phosphoenol pyruvate carboxykinase (Pck) (ATP: oxaloacetate carboxy-lyase, EC 4.1.1.49) which converts OAA to PEP, and this can be also utilized as the anaplerotic pathway by reversing the reaction. The kinetics of the Pck in *E.coli* may be assumed to follow the rapid equilibrium mechanism [23]. Thus, the reaction rate may be expressed as the following equation, where OAA and ATP are the substrates for this reaction, and it is inhibited by PEP and ADP [24]:

$$v_{Pck} = v_{Pck}^{max} \left(\cfrac{[OAA][ATP]}{\begin{array}{l} K_{i,ATP}K_{OAA}[ATP] + K_{ATP}[OAA] + [OAA][ATP] \\[2mm] + \cfrac{K_{i,ATP}K_{OAA}[PEP]}{K_{i,PEP}} + \cfrac{K_{i,ATP}K_{OAA}[ADP]}{K_{iADP}} + \cfrac{K_{i,ATP}K_{OAA}[PEP][ADP]}{K_{PEP}K_{i,ADP}} + \cfrac{K_{i,ATP}K_{OAA}[ATP][PEP]}{K_{i,PEP}K_{i,ATP}} + \cfrac{K_{i,ATP}K_{OAA}[OAA][ADP]}{K_{i,ADP}K_{i,OAA}} \end{array}} \right)$$

$$(36a)$$

Since intracellular ATP and ADP in *E.coli* and others are much higher than their respective K_s and K_i, this equation may be simplified as follows:

$$v_{Pck} = v_{Pck}^{max} \left(\cfrac{[OAA]\cfrac{[ATP]}{[ADP]}}{\begin{array}{l} K_m^{OAA}\cfrac{[ATP]}{[ADP]} + [OAA]\cfrac{[ATP]}{[ADP]} + \cfrac{K_i^{ATP}K_m^{OAA}}{K_i^{ADP}} + \cfrac{K_i^{ATP}K_m^{OAA}}{K_m^{PEP}K_i^{ADP}}[PEP] + \\[4mm] \cfrac{K_i^{ATP}K_m^{OAA}}{K_i^{PEP}K_I^{ATP}}\cfrac{[ATP][PEP]}{[ADP]} + \cfrac{K_i^{ATP}K_m^{OAA}}{K_i^{ADP}K_I^{OAA}}[OAA] \end{array}} \right)$$

$$(36b)$$

Fig. **8** shows that the flux is insensitive to the change in [ATP]/[ADP] ratio under typical growth condition [22]. On the other hand, [PEP] and [OAA] are less than 1 mM and 2 mM, respectively, and thus the reaction rate is sensitive to these pool sizes under typical growth condition [22].

Another important anaplerotic pathway is the **glyoxylate pathway**, which is consisted of isocitrate lyase (Icl) (threo-D₅-isocitrate-glyoxylate-lyase, EC 4.1.3.1) and malate synthase (MS) (L-malate glyoxylate-lyase, EC 4.1.3.2). The ICIT undergoes an aldo cleavage to SUC and glyoxylate (GOX) by Icl. As for the glyoxylate pathway, the following equation may be considered for Icl, where ICIT is the substrate, and it is inhibited by SUC and GOX [25].

$$v_{Icl} = \cfrac{v_{Icl}^{max}\cfrac{[ICIT]}{K_m^{ICIT}}}{\left(1 + \cfrac{[ICIT]}{K_m^{ICIT}} + \cfrac{[SUC]}{K_m^{SUC}} + \cfrac{[GOX]}{K_m^{GOX}} + \cfrac{[ICIT]}{K_m^{ICIT}}\cfrac{[SUC]}{K_m^{SUC}} + \cfrac{[SUC]}{K_m^{SUC}}\cfrac{[GOX]}{K_m^{GOX}} + \cfrac{I}{K_I} \right)} \quad (37)$$

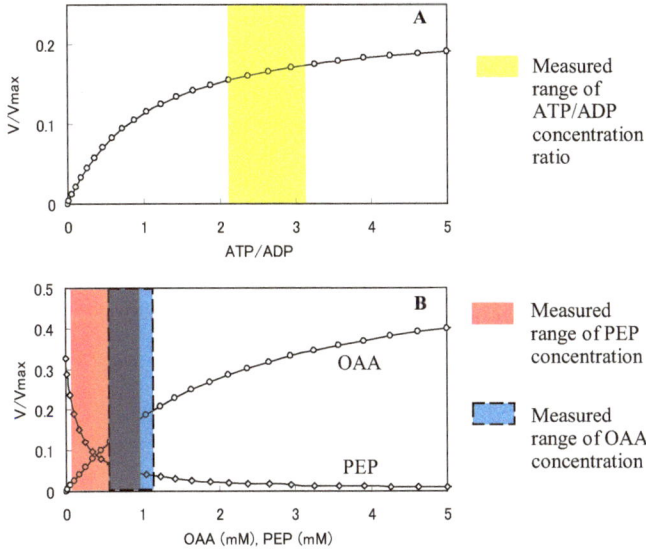

Figure 8: Effects of intracellular ATP/ADP ratio, and PEP and OAA concentrations on the Pck flux.

For MS reaction, the following equation may be considered, where GOX and AcCoA are the substrates, and the reaction is inhibited by MAL [25].

$$v_{MS} = \frac{v_{MS\_f}^{\max} \dfrac{[GOX]}{K_m^{GOX}} \dfrac{[AcCoA]}{K_m^{AcCoA}} - v_{MS\_r}^{\max} \dfrac{[MAL]}{K_m^{MAL}}}{\left(1 + \dfrac{[GOX]}{K_m^{GOX}} + \dfrac{[MAL]}{K_m^{MAL}} + \left(1 + \dfrac{[AcCoA]}{K_m^{AcCoA}}\right)\right)} \qquad (38)$$

This glyoxylate pathway leads to the proposition of the existence of a cyclic mechanism for replenishing C_4 acids from the TCA cycle for biosynthesis, and this forms the bypass of the TCA cycle forming the upper TCA cycle, and plays an important role in the metabolism of short-chain fatty acids [14] as well as part of gluconeogenic pathway, and is sometimes called as the **glyoxylate cycle** or **glyoxylate shunt**.

GLUCONEOGENIC PATHWAYS

Without sugars available, the organic acids such as acetate, alcohol such as ethanol, or fatty acid *etc.* can be assimilated and can be metabolized *via* TCA cycle and glyoxylate pathway, and the biosynthesis can be made *via* **gluconeogenesis**, where acetate is first converted to AcCoA by ACS (Fig. **9**).

AcCoA goes into TCA cyclcle and the glyoxylate pathways as stated above. Then MAL thus formed *via* either TCA cycle or glyoxylate pathway is converted to PYR by malic enzyme (Mez) (L-malate: $NADP^+$ oxidoreductase, EC1.1.1.40), which catalyzes the oxidative carboxylation of MAL forming $NADPH + CO_2$. For malic enzyme, the following equation may be considered [16].

$$v_{Mez} = \frac{v_{Mez}^{max}[MAL][NADP]}{\left(K_{MAL} + [MAL]\right)\left(K_{eq} + [NADP]\right)} \tag{39}$$

The pyruvate thus formed is phosphorylated to form PEP by PEP synthase (Pps) (ATP: pyruvate, water phosphotransferase, EC 2.7.9.2), where the reaction rate may be expressed as

$$v_{Pps} = \frac{v_{Pps}^{max}\left(k_f \dfrac{[PYR][ATP]}{K_m^{PYR}K_m^{ATP}} - k_r \dfrac{[PEP][AMP][P_i]}{K_m^{PEP}K_m^{AMP}K_m^{P_i}}\right)}{\left\{\left(1 + \dfrac{[PEP]}{K_m^{PEP}}\right)\left(1 + \dfrac{[P_i]}{K_m^{P_i}}\right) + 1 + \dfrac{[PYR]}{K_m^{PYR}}\right\}\left(1 + \dfrac{[AMP]}{K_m^{AMP}} + \dfrac{[ATP]}{K_m^{ATP}}\right)} \tag{40}$$

Most of the EMP pathway reactions are reversible, but Pfk must be replaced by other enzyme such as fructose bisphosphatase (Fbp) (D-fructose-1,6-bisphosphate 1-phosphohydrolase, EC, 3.1.3.11) for converting FBP to F6P in the gluconeogenesis, where the reaction rate may be expressed as

$$v_{Fbp} = \frac{v_{Fbp}^{max}\dfrac{[FBP]}{K_m^{FDP}}}{\left(1 + \dfrac{[FBP]}{K_m^{FDP}}\right)\left(1 + \dfrac{[AMP]}{K_i^{AMP}}\right)} \tag{41}$$

In this way, the EMP pathway is reversed and goes up to G6P, where it enters into the PP pathway, but its flux will be quite small [26].

CARBOHYDRATE UPTAKE PATHWAYS

So far, glucose was considered as a main carbon source. Other than glucose, various carbohydrates can be utilized as carbon sources. Here, we consider the typical other carbon sources such as xylose (XYL) and glycerol (Gly), which are typically utilized in the biofuels production.

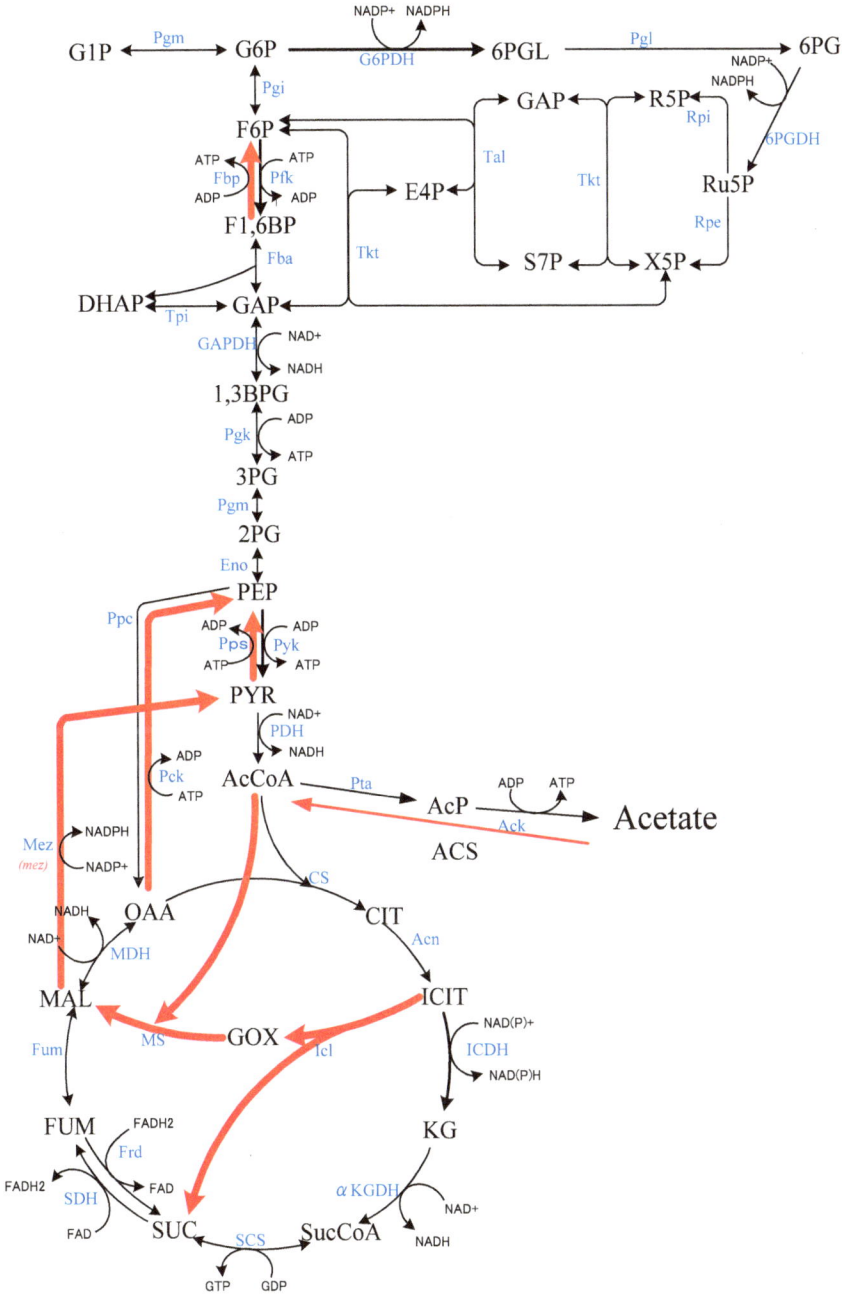

Figure 9: Gluconeogenic pathways.

Xylose Assimilating Pathway

D-xylose transported *via* xylose transporter into cytosol is converted to D-xylulose (XYLU) by xylose isomerase (Xyi) (D-xylose keto-isomerase, EC 5.3.1.5) (Fig. **10**). Xylose transporter is subject to catabolite repression by glucose, and this will be discussed in Chapter 7. The reaction rate by Xyi may be expressed as [5]

$$v_{Xyi} = \frac{v_{Xyi}^{max}[XYL]}{K_{m,XYL} + [XYL]} \tag{42}$$

D-xylulose is subsequently phosphorylated by xylulokinase (Xyk) (ATP: D-xylulose 5-phosphotransferase, EC 2.7.1.17) to form D-xylulose 5-phosphate (X5P) [5].

$$v_{Xyk} = \frac{v_{Xyik}^{max}\dfrac{[XYLU]}{K_{m,XYLU}}\dfrac{[ATP]}{K_{m,ATP}}}{1 + \dfrac{[XYLU]}{K_{m,XYLU}} + \dfrac{[ATP]}{K_{m,ATP}} + \dfrac{[XYLU][ATP]}{K_{m,XYLU}K_{m,ATP}}} \tag{43}$$

Under anaerobic condition, xylulose reductase (XR) is induced, and xylitol and xylitol 5-phosphate are produced, where they may inhibit the cell growth.

Glycerol Assimilating Pathway

Glycerol is transported by glyceroporin, GlpF encoded by *glpF*, where this may be modeled by assuming unidirectional transport [27] A simple diffusion model may be used for GlpF [28, 29].

Glycerol transported into cytosol is oxidized to dihydroxyacetone (DHA) by a glycerol dehydrogenase (GLYDH) (glycerol: NAD oxidoreductase, EC 1.1.1.6) encoded by *glyDH*. DHA is then phosphorylated by a dihydroxyacetone kinase (DHAK) (phosphoenolpyruvate: glycerone phosphotransferase, EC 2.7.1.121) using ATP (Fig. **10**). The two enzymes involved in the fermentative glycerol dissimilation pathway [30] may be modeled with GLYDH encoded by *gldA* assuming reversible [31] and DHAK encoded by *dhaKLM* assuming irreversible [29, 32].

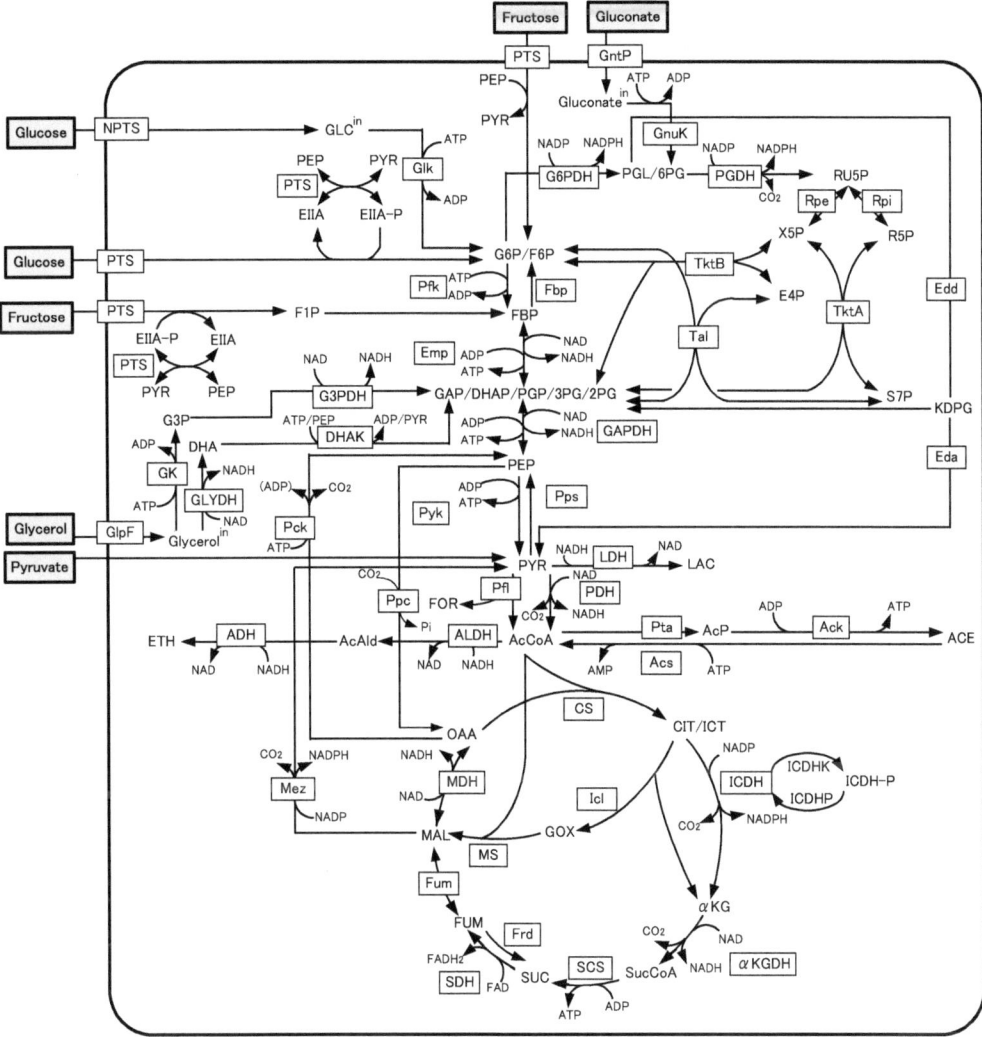

Figure 10: Various carbohydrates uptake pathways.

$$v_{GLYDH} = \frac{\dfrac{v_{GLYDH}^{max}}{K_{m,GLYC}}\left([Glyc] - \dfrac{[DHA]}{K_{eq,DHA}}\right)}{1 + \dfrac{[Glyc]}{K_{m,GLYC}} + \dfrac{[DHA]}{K_{m,DHA}}} \tag{44}$$

Another pathway for glycerol utilization is that glycerol is phosphorylated by glycerol kinase (Glyk) (ATP: glycerol phosphotransferase, EC 2.7.1.30) to form L-glycerol 3-phosphate (GLY3P), which is then converted to DHAP by glycerol-

3-phosphate dehydrogenase (G3PDH) (sn-glycerol-3-phosphate:NAD(P)+ 2-oxidoreductase, EC 1.1.1.94)

The rate equation for DHAK may be expressed as [29]

$$v_{DHAK} = \frac{v_{DHAK}^{max}[DHA]}{K_{m,DHA}+[DHA]} \tag{45}$$

ANAEROBIC FERMENTATION PATHWAYS

In the absence of oxygen or other electron accepters, respiratory chain cannot be utilized, and thus ATP is generated *via* substrate level phosphorylation through the process of degradation of carbon source in the metabolic pathways. Under such fermentation condition, the cells such as *E.coli* excretes such metabolites as lactate, ethanol, succinate, formate (CO_2 and H_2 as well) as well as acetate, where the relative production rates for these metabolites are governed by the demand for redox neutrality (Fig. **11**). Succinate is formed from PEP *via* Ppc by the reversed pathway from OAA to SUC, where SDH pathway is reversed by fumarate reductase (Frd) (succinate: quinone oxidoreductase, EC 1.3.5.4). The rate equation for Frd may be expressed as

$$v_{Frd} = \frac{v_{Frd}^{max}\dfrac{[SUC]}{K_m^{SUC}} - v_{Frd,r}^{max}\dfrac{[FUM\}}{K_m^{FUM}}}{1+\dfrac{[SUC]}{K_m^{SUC}}+\dfrac{[FUM]}{K_m^{FUM}}} \tag{46}$$

Pyruvate serves as a common substrate for pyruvate formate-lyase (Pfl) (acetyl-CoA: formate acetyltransferase, EC 2.3.1.54) and PDHc, and this branch point involves the cleavage of PYR. AcCoA is the product of both Pfl and PDHc reactions, and it is converted to either acetate and ethanol or subsequently undergo further oxidation in the TCA cycle down to αKG (Fig. **11**).

Under anaerobic condition, the energy generation is made only by substrate level phosphorylation and/ or anaerobic respiration, and thus the specific glucose consumption rate is increased to enhance energy production through EMP pathway [33]. The possible fermentation pathways are given in Fig. **11**, where NADH reoxidation is critical for regeneration of NAD^+ for the metabolism to be continued.

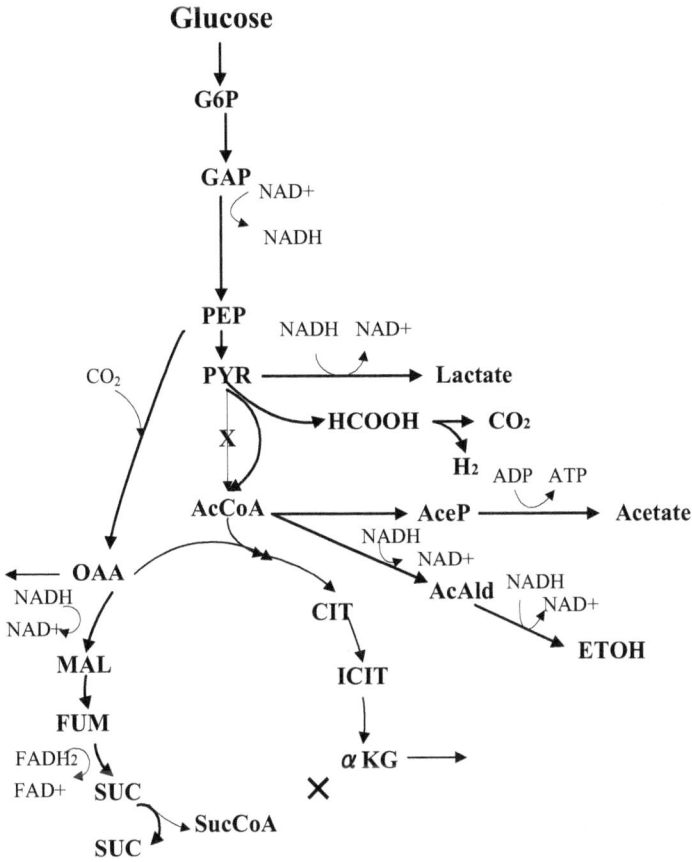

Figure 11: Fermentation pathways under anaerobic condition.

The lactate is formed from PYR by lactate dehydrogenase LDH (*(R)*-lactate: NAD^+ oxidoreductase, EC 1.1.1.28), and the rate equation may be expressed as [11]

$$v_{LDH} = \frac{v_{LDH}^{max} \dfrac{1}{K_{m,PYR} K_{mNADH}} \left([PYR][NADH] - \dfrac{[Lac][NAD]}{K_{eq}} \right)}{\left(1 + \left(\dfrac{[PYR]}{K_{m,PYR}} + \dfrac{[Lac]}{K_{m,Lac}} \right) \right) \left(1 + \dfrac{[NADH]}{K_{m,NADH}} + \dfrac{[NAD]}{K_{m,NAD}} \right)} \tag{47}$$

As stated above, AcCoA is formed from PYR *via* PDHc or Pfl, where the kinetics of Pfl reaction encoded by *pflB* may be modeled by assuming the ping-pong mechanism as [29, 34]

$$v_{Pfl} = \frac{v_{Pfl,f}^{\max}[PYR][CoA]}{[PYR][CoA]+K_{m,PYR}[CoA]+K_{m.CoA}[PYR]} - \frac{v_{Pfl,r}^{\max}[AcCoA][FORM]}{[AcCoA][FORM]+K_{m,AcCoA}[FORM]+K_{m,Pf}l[AcCoA]}$$

$$(48)$$

The acetaldehyde dehydrogenase (ALDH) (acetaldehyde: NAD$^+$ oxidoreductase, EC 1.2.1.10) is used to convert AcCoA into acetaldehyde (AcAld), and the subsequent reduction of AcAld to ethanol, where ALDH reaction may be modeled by assuming bi-uni-uni ping-pong mechanism [35], and ADH reaction may be modeled by assuming reversible Michaelis –Menten type as follows [29]:

$$v_{ALDH} = \frac{v_{ALDH}^{\max}[AcCoA]}{Km,AcCoA+[AcCoA]\left(1+\dfrac{K_{m,NADH}}{[NADH]}+\dfrac{K_{m,NADH}[[AcAld]}{K_{ia}[NADH]}\right)} - \frac{v_{ALDH}^{\max}[AcAld][[NAD][CoA]}{K_{eq,ALDH}A}$$

where

$$A \equiv K_{ia}K_{m,NAD}[CoA]+K_{m,CoA}[AcAld][NAD]+K_{m,NAD}[[AcAld][CoA]+K_{m,AcAld}[NAD][CoA]+[AcAld][NAD][CoA]$$

$$(49)$$

$$v_{ADH} = \frac{v_{ADH}^{\max}\left([AcAld]-\dfrac{[ETOH]}{K_{eq,ADH}}\right)}{Km,AcAld\left(1+\dfrac{[AcAld]}{K_{m,AcAld}}+\dfrac{[ETOH]}{K_{m,ETOH}}\right)}$$

$$(50)$$

In the case of yeast, ethanol is formed from PYR *via* PDC and ADH as stated before as (33) and (34).

AMINO ACID SYNTHETIC PATHWAYS

Amino acids synthesis is important for biosynthesis or anabolism point of view, where the precursors are formed in the main metabolic pathways as shown in Fig. **12**. Here, the important amino acids synthetic pathways such as glutamate/glutamine synthetic pathways and lysine synthetic pathways are considered in the followings.

Glutamic Acid and Glutamine Synthetic Pathways

Glutamic acid (Glu) is synthesized from αKG in the TCA cycle by glutamate dehydrogenase (GDH) (L-glutamate: NADP$^+$ oxidoreductase, EC 1.4.1.4) with NH$_3$ and NADPH (Fig. **13**). The produced glutamic acid inhibits GDH and controls the enzyme synthesis as well. Note that glutamic acid regulates the synthesis of Ppc and CS. Gutamine (Gln) is synthesized only from glutamate by glutamine synthetase (GS) (L-glutamate: ammonia ligase, EC 6.3.1.2) with NH_3 and ATP (Fig. **13**). On the other hand, glutamic acid can be formed from α KG by GDH and from Gln by glutamate synthase (GOGAT) (L-glutamate: $NADP^+$ oxidoreductase, EC 1.4.1.13) with NADPH (under N-limitation).

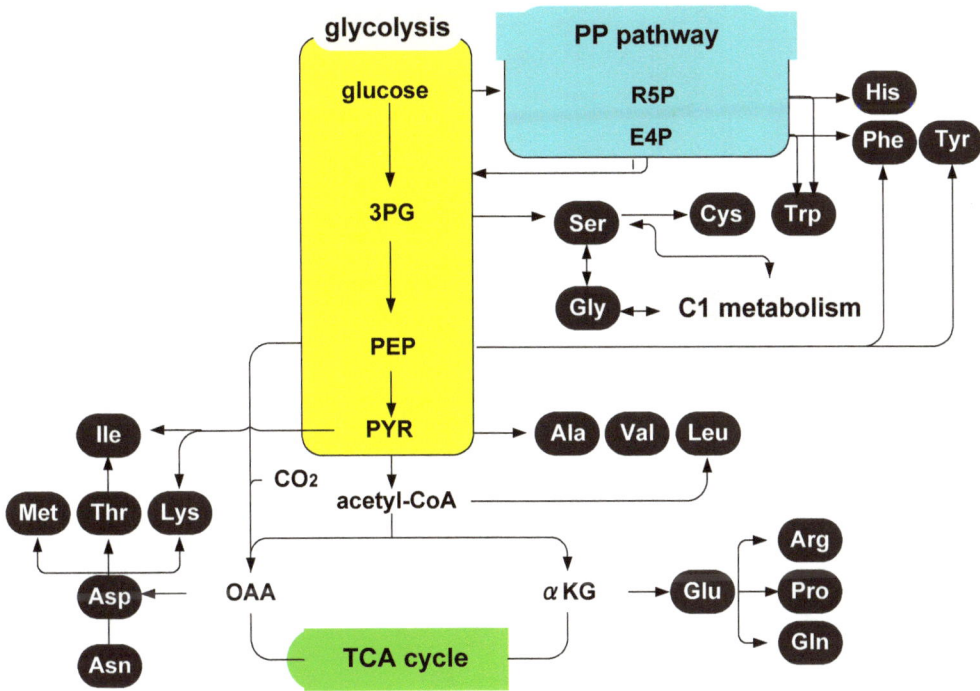

Figure 12: Amino acid synthesis from the precursors in the main metabolic pathways.

Glutamate and glutamine play key roles in the cellular metabolism and serve as precursors for protein synthesis. GS is active during low ammonium concentration, while GDH is active at higher ammonium concentration, since GS has higher affinity than GDH for ammonia (Km= 0.1mM and 1.1mM, respectively) [36, 37].The activity of GS is controlled by cyclic interrelated protein, P_{II} which acts in response to the concentration of glutamine and αKG [38, 39], as will be explained in Chapter 7.

At present, several kinetic models have been proposed for ammonium assimilation [36, 40, 41]. GDH reaction plays a central role in the overall nitrogen homeostasis. It requires αKG, NH_4^+ and NADPH in order to be activated. The activity of GDH becomes most important at higher ammonium concentration [37]. The reaction rate of GDH may be expressed as [36].

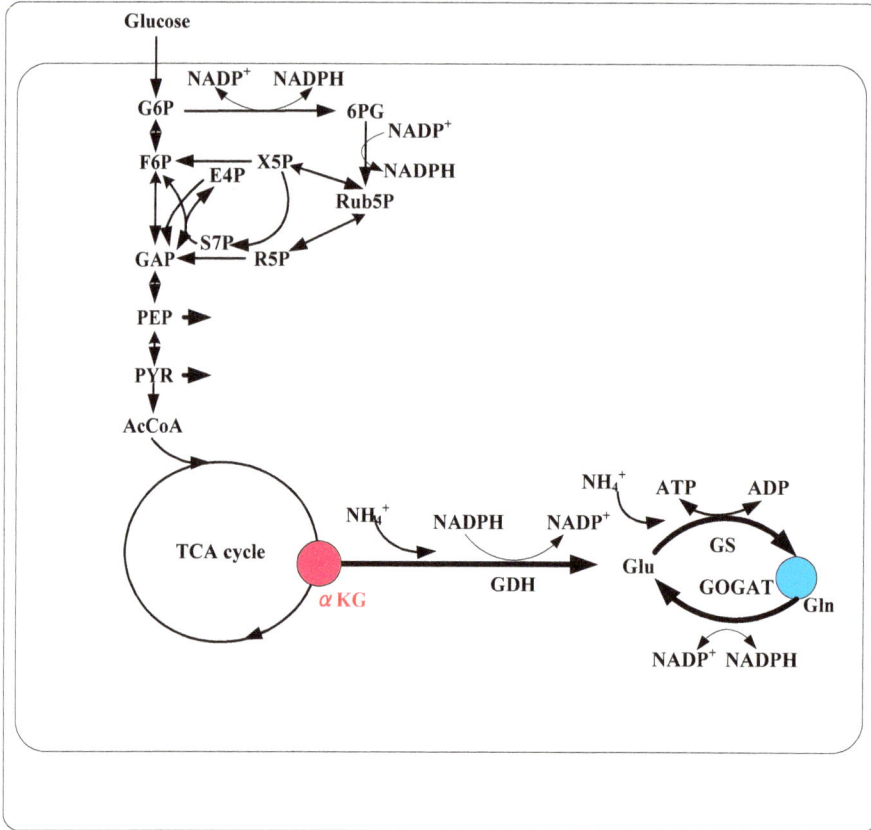

Figure 13: Glutamate/ glutamine synthetic pathways.

$$v_{GDH} = \frac{\dfrac{v_{GDH}^{max}}{K_{2KG}K_{NH_4^+}K_{NADPH}}[\alpha KG][NH_4^+][NADPH] - \dfrac{[Glu][NADP]}{K_{eq}}}{\left(1 + \dfrac{[NH_4^+]}{K_{NH_4^+}}\right)\left(1 + \dfrac{[\alpha KG]}{K_{2KG}} + \dfrac{[Glu]}{K_{GLU}}\right)\left(1 + \dfrac{NADPH}{K_{NADPH}} + \dfrac{[NADP]}{K_{NADP}}\right)} \qquad (51)$$

The activation of GS requires UTP, ATP and αKG, and is inhibited by glutamin. The rate equation for GS may be expressed as [36]

$$v_{GS} = \frac{\dfrac{v_{GS}^{max}}{K_{ATP}K_{NH_4^+}K_{GLU}}[ATP][NH_4^+][Glu] - \dfrac{[ADP][G\ln][P]}{K_{eq}}}{\left(1 + \dfrac{[ATP]}{K_{ATP}} + \dfrac{[ADP]}{K_{ADP}} + \dfrac{[P]}{K_P} + \dfrac{[ADP][P]}{K_{ADP}K_P}\right)\left(1 + \dfrac{[NH_4^+]}{K_{NH_4^+}} + \dfrac{[G\ln]}{K_{GLN}} + \dfrac{[Glu]}{K_{GLU}} + \dfrac{[G\ln][NH_4^+]}{K_{GLN}K_{NH_4^+}} + \dfrac{[Glu][NH_4^+]}{K_{GLU}K_{NH_4^+}}\right)}$$

(52)

GOGAT reaction may be expressed as [36]

$$v_{GOGAT} = \frac{v_{GOGAT}^{max}\dfrac{[Glu][\alpha KG][NADPH]}{K_{GLN}K_{\alpha KG}K_{NADPH}}}{\left(1 + \dfrac{[MET_{GLU}]}{K_{METGLU}}\right)\left(1 + \dfrac{[G\ln]}{K_{GLN}} + \dfrac{[Glu]}{K_{GLU}}\right)\left(1 + \dfrac{[\alpha KG]}{K_{\alpha KG}} + \dfrac{[Glu]}{K_{GLU}}\right)\left(1 + \dfrac{[NADPH]}{K_{NADPH}} + \dfrac{[NADP]}{K_{NADP}}\right)}$$

(53)

Lysine Synthetic Pathways

Fig. **14** shows the lysine (Lys) synthetic pathways together with related amino acids synthetic pathways. L-aspartate (Asp) is formed from OAA by transamination by aspartate transaminase (ART) (L-aspartate: 2-oxoglutarate aminotransferase, EC 2.6.1.1). Asp is then activated *via* phosphorylation by aspartokinase (Ask) (ATP: L-aspartate 4-phosphotransferase, EC 2.7.2.4) and reduced to give L-aspartate semialdehyde (ASA) in the first two steps. ASA is the branch point to enter into either threonine, methionine, and isoleusine synthesis or lysine synthesis. Dihydrodipicolinate synthase (DHPS) (L-aspartate-4-semialdehyde hydro-lyase, EC 4.3.3.7) and dihydrodipicolinate reductase (DHPR) ((S)-2,3,4,5-tetrahydropyridine- 2,6-dicarboxylate: NAD(P)$^+$ 4-oxidoreductase, EC 1.17.1.8) catalyze the third and fourth steps in the lysine synthetic pathway, respectively, and are the enzymes which commit flux to the biosynthesis of meso-diaminopimelate (DAP) and lysine. The synthesis of DAP from L-tetrahydrodipicolinate (THDP) is accomplished by three separate routes: the succinylase and acetylase pathways, in which N-succinylated or N-acetylated intermediates are generated, and the infrequently encountered dehydrogenase pathway. The synthesized meso-DAP can be either used for cell wall synthesis or decarboxylated by diaminopimelate decarboxylase (DAPDC) (meso-2,6-diaminoheptanedioate carboxy-lyase, EC 4.1.1.20) to produce lysine.

The first step in the bacterial synthesis of lysine is the phosphorylation of L-aspartate to L-aspartyl phosphate (BAP) by the action of Ask (Fig. **15**). Only one kind of Ask has been observed in *C. glutamicum* so far. This is in contrast to

E. coli and *Bacillus* sp. which have three distinct isoenzymes [42, 43]. The presence of both lysine and threonine (Thr) cause a significant inhibition of Ask activity, even at low concentrations, whereas lysine or threonine alone does not show the inhibitory effect [44].

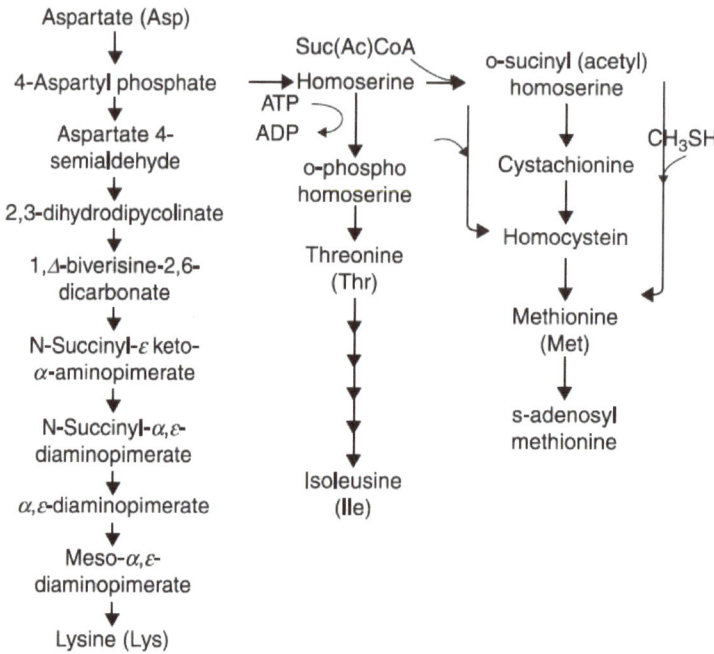

Figure 14: Lysine, threonine, and isoleusine synthetic pathways.

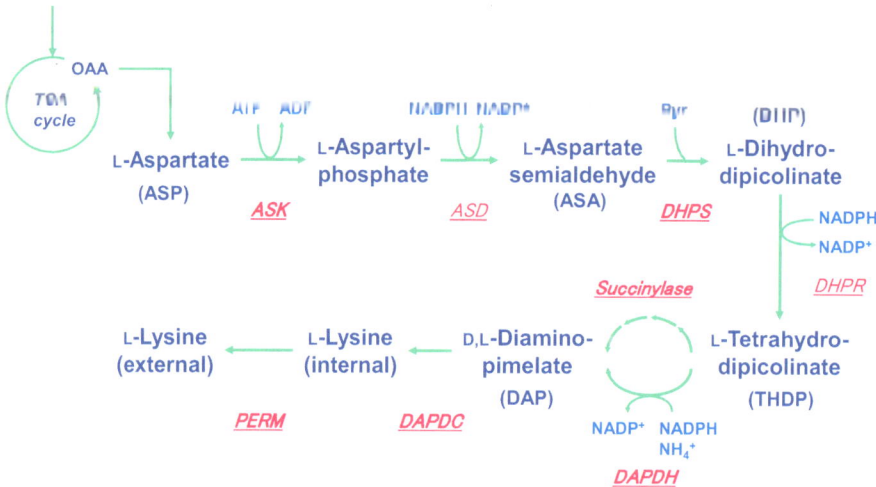

Figure 15: Lysine synthetic pathways.

Consider the rate equations for lysine synthetic pathways [45]. The Ask reaction is a rapid equilibrium system with random addition of the two substrates, L-Asp and ATP [46]. Thus, the simplest expression may be considered by assuming the rapid equilibrium random bi-bi mechanism [47]. If we assume that binding of the first pair of substrates does not affect the binding of the second substrate, then the rate expression may be expressed as

$$
v_{Ask} = \frac{v_{Ask}^{max}([ASP][ATP] - \dfrac{[ADP][BAP]}{K_{eq,Ask}})}{K_{ASK}K_{ATP} + K_{ATP}[ASP] + K_{ASP}[ATP] + [ASP][ATP] + \dfrac{K_{ASP}K_{ATP}[ADP]}{K_{ADP}} + \dfrac{K_{ASP}K_{ATP}[BAP]}{K_{BAP}} + \dfrac{K_{ASP}K_{ATP}[ADP][BAP]}{K_{ADP}K_{BAP}}}
$$

$$
\cdot \frac{1}{1 + L_{Lys-Thr}(1 + \dfrac{[Lys][Thr]}{K_{Lys-Thr}})^8} \tag{54}
$$

where [BAP] is given as

$$
[BAP] = \frac{[ASA][NADP^+][P_i]}{K_{eq,Ask}[NADPH][H^+]} \tag{55}
$$

where $K_{eq,Ask}$ is the equilibrium constant for Ask. The last term of the expression represents the concerted inhibition by lysine and threonine. In order to determine the reaction rate, the intracellular levels of Asp, ATP, BAP, and ADP must be known. The reversible condensation reaction catalyzed by DHPS [48] may be represented as PYR + L-ASA $\Leftrightarrow$ DHP + $2H_2O$. This reaction proceeds *via* a ping-pong mechanism [49], and a general rate expression may be derived [46]. Since the condensation occurs in the direction of product synthesis [50], and the kinetic expression for DHPS may be described as

$$
v_{DHPS} = \frac{v_{DHPS}^{max}[PYR][ASA]}{K_{m,ASA}[PYR] + K_{m,PYR}[ASA] + \dfrac{K_{m,PYR}K_{i,ASA}}{K_{i,DHP}}[DHP] + [PYR][ASA] + \dfrac{K_{m,PYR}}{K_{i,DHP}}[ASA][DHP]} \tag{56}
$$

DHP concentration cannot be measured because of its instability, and may be estimated using the following equation by assuming DHPR to be in equilibrium:

$$
[DHP] = \frac{[THDP][NADP^+]}{K_{eq,DHPR}[NADPH][H^+]} \tag{57}
$$

where $K_{eq,DHPR}$ is the equilibrium constant for DHPR. Substitution of Eq. (57) into Eq. (56) yields

$$v_{DHPS} = \frac{v_{DHPS}^{max}[PYR][ASA]}{K_{m,ASA}[PYR]+K_{m,PYR}[ASA]+\frac{K_{m,PYR}K_{i,ASA}}{K'_{i,DHP}}\frac{[NADP^+]}{[NADPH]}[THDP]+[PYR][ASA]+\frac{K_{m,PYR}[NADP^+]}{K'_{i,DHP}[NADPH]}[ASA][THDP]}$$

(58)

where $K'_{i,DHP} \equiv K_{i,DHP}K_{eq,DHPR}[H^+]$.

In the synthesis of meso-diaminopimelate, succinylase and dehydrogenase variants can be used in parallel in *C. glutamicum*, and 30% of the lysine produced by the bacterium is generated *via* the dehydrogenase, with the remainder being generated using the succinylase pathway enzymes [51, 52].

The NADPH-dependent DAPDH catalyzes the direct conversion of THDP into meso-DAP [53] such that

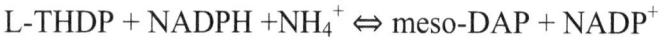

L-THDP + NADPH +NH$_4^+$ $\Leftrightarrow$ meso-DAP + NADP$^+$

The DAPDH reaction is modeled by non-mechanistic three-substrate kinetics such as

$$v_{DAPDH} = \frac{v_{DAPDH}^{max}}{\left(1+\frac{K_{NADPH}}{[NADPH]}\right)\left(1+\frac{K_{NH_4^+}}{[NH_4^+]}\right)\left(1+\frac{K_{THDP}}{[THDP]}\right)}$$

(59)

The reaction symbolized by succlnylarse presents a connection of the enzymes such as tetrahydrodipicolinate succinylase, succinyl-amino-ketopimelate transaminase, succinyldiaminopimelate desuccinylase, and diaminopimelate epimerase [54]. The resulting rate expression is empirically modeled by the Michaelis-Menten kinetics such that

$$v_{succinylase} = \frac{v_{succinylase}^{max}[THDP]}{K_{m,THDP}+[THDP]}$$

(60)

The rate of synthesis of meso-DAP is, therefore, the sum of the fluxes as through both DAPDH and succinylase pathways.

The last conversion to lysine involves decarboxylation of DAP by DAPDC such that meso-DAP →L-Lys + CO_2. The rate equation for DAPDC may be expressed as Michaelis-Menten kinetics as

$$v_{DAPDC} = \frac{v_{DAPDC}^{max}[DAP]}{K_{m,DAP} + [DAP]} \tag{61}$$

As for lysine permease (PERM), the export of positively charged L-lysine takes place together with two hydroxide ions in a symport mechanism, and the reorientation of the positively charged carrier protein is driven by the negative membrane potential [55]. The quantitative description of this mechanism may be expressed elsewhere [56].

CONCLUDING REMARKS

Here, we considered only enzyme reactions, which can be used for the simulation of enzyme level regulation. The transcriptional regulation by transcription factors will be mentioned in Chapters 7 and 8.

Although most of the rate equations are considered for such bacteria as *E.coli*, *Zymomonas mobilis*, and *Corynebacterium* sp., those can be also used for other organisms with some modification, since the enzymatic reaction mechanisms are quite common to the variety of organisms. Some of the rate equations introduced in the present Chapter are used for the simulation as explained in Chapter 8.

Some of the kinetic model equations are functions of cofactors such as ATP, NAD(P)H *etc*. However, those concentrations are in general difficult to predict by simulation, and those may be assumed to be constant. Moreover, the model parameter identification and its statistical reliability will be important in practice, and those are also mentioned in Chapters 5 and 8.

NOMENCLATURE

Metabolites

2KG	=	2-Keto-D-gluconate
6PG	=	6-Phosphogluconolactone
ACE	=	Acetate
AcP	=	Acetyl phosphate

AcCoA	=	Acetyl-CoA
ASP	=	Aspartate
ADP	=	Adenosine diphosphate
ATP	=	Adenosine-5'-triphosphate
AMP	=	Adenosine monophosphate
DHAP	=	Dihydroxyacetone phosphate
E4P	=	Erythrose 4-phosphate
F6P	=	Fructose 6-phosphate
FDP	=	Fructose 1,6-bisphosphate
FUM	=	Fumarate
G6P	=	Glucose-6-phosphate
GAP	=	Glyceraldehyde 3-phosphate
ICIT	=	Isocitrate
NAD/ NADH	=	Nicotinamide adenine dinucleotide
NADP/ NADPH	=	Nicotinamide adenine dinucleotide phosphate
MAL	=	Malate
OAA	=	Oxaloacetate
P	=	Phosphate
PEP	=	Phosphoenolpyruvate
PYR	=	Pyruvate
R5P	=	Ribulose 5-phosphate
Ru5P	=	Ribose 5-phosphate
S7P	=	Sedoheptulose 7-phosphate
SUC	=	Succinate
X5P	=	Xylulose 5-phosphate

Protien (enzyme)

2KGDH	=	2-Keto-D-gluconate Dehydrogenase
Ack	=	Acetate kinase
Acs	=	Acetyl coenzyme A synthetase
AcP	=	Acetyl phosphate
Aldo	=	Aldolase
CS	=	Citrate synthase
EI	=	Cytoplasmatic protein (enzyme I)
EII, EIIB, EIIC,	=	Carbohydrate specific (enzyme II)
EIIA^{Glc,} / EIICB^{Glc}	=	Enzymes for glucose transport
Eno	=	Enolase
Fum	=	Fumarase
G6PDH	=	Glucose-6-phosphate dehydrogenase
GAPDH	=	Glyceraldehyde 3-phosphate dehydrogenase
ICDH	=	Isocitrate dehydrogenase
HPr	=	Histidine containing protien
Icl	=	Isocitrate lyase
MDH	=	Malate dehydrogenase
Mez	=	Malic enzyme
Ms	=	Malate synthase
Pck	=	Phosphoenolpyruvate carboxykinase
PDH	=	Pyruvate dehydrogenase
Pfk/Pfk-1	=	Phosphofructokinase-1
Pgi	=	Phosphoglucose isomerase / Glucosephosphate isomerase
Pgk	=	Phosphoglycerate kinase
Pgm	=	Phosphoglycerate mutase

Ppc	= PEP carboxylase
Pps	= Phosphoenolpyruvate synthase
Pta	= Phosphotransacetylase
PTS	= Phosphotransferase system
Pyk	= Pyruvate kinase
Rpe	= Ribulose phosphate 3-epimerase
Rpi	= Ribulose 5-phosphate 3-isomerase
SDH	= Succinate dehydrogenase
Tal	= Transaldolase
TktA	= TransketolaseI
TktB	= TransketolaseII

APPENDIX A: EC NUMBER FOR ENZYME IDENTIFICATION

Enzymes can be identified by the 4 digits of number with EC (Enzyme Commission) such as EC 1. 2. 3. 4., where the 1st number indicates the type of enzyme such that

1. Oxide-reductases such as dehydrogenase, oxidases, and oxygenases,

2. Transferases such as aminotransferases, transaminases, and mutases,

3. Hydrolases such as esterases, peptidases, and glycosidases,

4. Lyases such as decarboxylases and dehydratases,

5. Isomerases such as racemases, epimerases, and mutases,

6. Ligases or Synthases such as aminoacyl-tRNA synthetases and peptide synthetases.

For example, alcohol dehydrogenase is expresses as EC 1. 1. 1. 1., where the 1st digit indicates the oxidoreductase as mentioned above, 2nd digit indicates CH-OH is the donor, the 3rd digit indicates that NAD^+ or NADH is the acceptor, and the last digit indicates the detailed classification.

APPENDIX B: MECHANISM OF ENZYMATIC REACTIONS

In relation to the modeling of the kinetic reaction mechanism, the reaction types may be classified. Consider, for example, the enzyme reaction with one substrate and two products such as phosphatase such that [58]

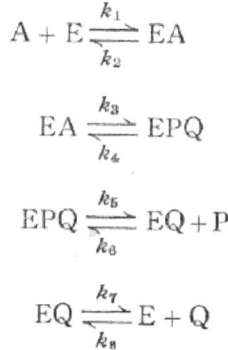

$$A + E \underset{k_2}{\overset{k_1}{\rightleftharpoons}} EA$$

$$EA \underset{k_4}{\overset{k_3}{\rightleftharpoons}} EPQ$$

$$EPQ \underset{k_6}{\overset{k_5}{\rightleftharpoons}} EQ + P$$

$$EQ \underset{k_8}{\overset{k_7}{\rightleftharpoons}} E + Q$$

where A denotes organic phosphoric compound such as glycerol phosphate *etc.*, P denotes the alcohol part of A such as glycerol *etc.*, Q denotes inorganic phosphate, and E denotes enzyme. The above reaction scheme may be expressed as

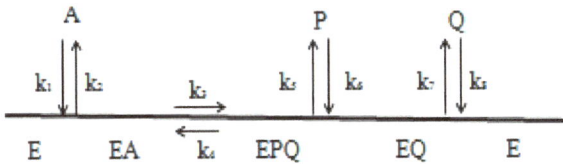

This may be expressed using positive direction of the reactions such as

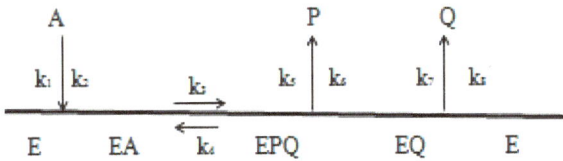

This may be further simplified as

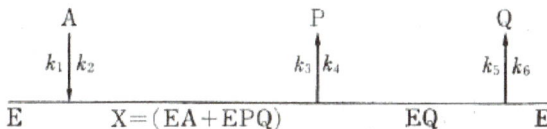

Let the number of substrates and products be expressed as Uni, Bi, Ter, Quad *etc.* Then the reaction types may be classified as

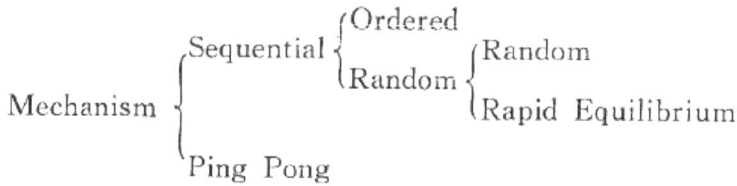

$$
\text{Mechanism}
\begin{cases}
\text{Sequential}
\begin{cases}
\text{Ordered} \\
\text{Random}
\begin{cases}
\text{Random} \\
\text{Rapid Equilibrium}
\end{cases}
\end{cases} \\
\text{Ping Pong}
\end{cases}
$$

where sequential mechanism is referred to the case where the reaction product is released after all the substrates finish binding, while Ping Pong mechanism is referred to the case where the product is releases before all the substrates bind. Some of the typical reaction mechanisms are shown below:

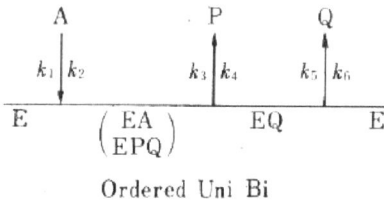

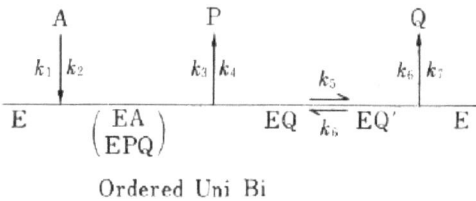

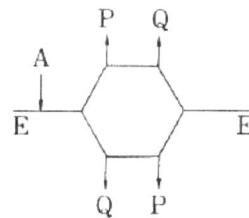

Ordered Uni Bi

Ordered Uni Bi

Random Uni Bi

Iso Ordered Uni Bi

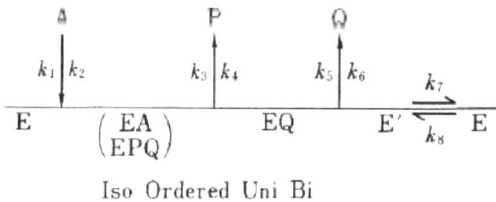

The sequential mechanism includes the two such as the ordered mechanism where the substrate binding sequence is determined, or the random mechanism where its order is not determined. Moreover, if the enzyme state becomes isomerization, it is called as Iso.

Ordered Bi Bi

Iso Ordered Bi Bi

Random Bi Bi

Ping Pong Bi Bi

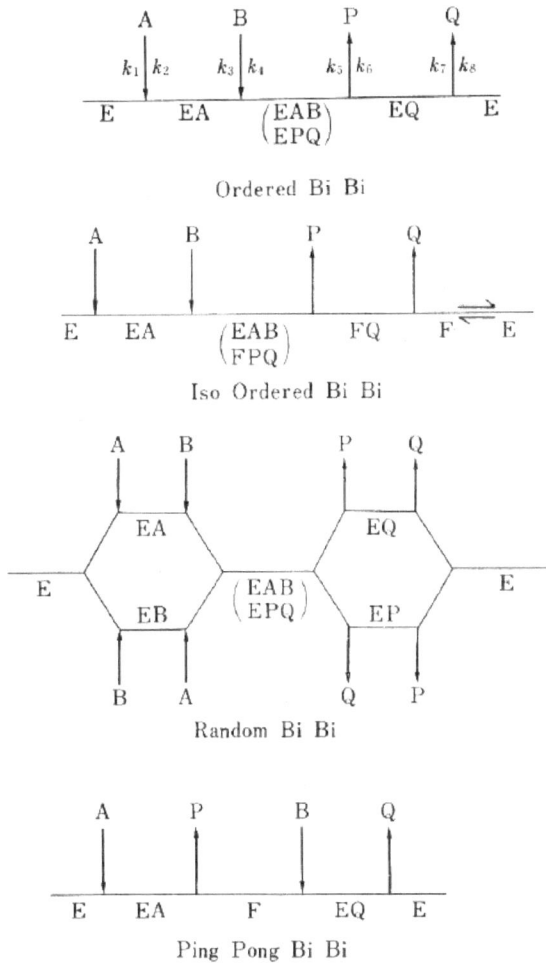

APPENDIX C KINETIC PARAMETER VALUES

Enzyme	Parameter	Value	Unit	References
Cell growth	μ_m	0.6	h^{-1}	estimated
	K_s	0.1	M	
Glk	v_{Glk}^{max}	4.503	mmol/gDCWh	
	K_m^{GLC}	0.12	mM	[6]
	K_m^{AATP}	0.5	mM	
PTS	v_{PTS}^{max}	25.739	mmol/gDCWh	[3,4]

Enzyme	Parameter	Value	Unit	References
	K_{a1}	1.0	mM	
	K_{a2}	0.01	mM	
	K_{a3}	1.0		
	n_{G6P}	4		
	K_{G6p}	0.5	mM	
Pgi	v_{PGI}^{max}	26.3711	mmol/gDCWh	
	K_{G6P}	2.46	mM	
	K_{F6P}	0.2	mM	
	K_{eq}	0.43	mM	
	K_{6pginh}^{G6P}	0.2	mM	
	K_{6pginh}^{F6P}	0.2	mM	[3]
Pfk	v_{Pfk}^{max}	24.613	mmol/gDCWh	
	$K_{(ATP)}$	4.27	mM	
	$K_{(ATP,ADP)}$	4.6944	mM	
	$K_{a(ADP,AMP)}$	1.1118	mM	
	$K_{b(ADP,AMP)}$	98.88	mM	
	n_{Pfk}	4		
	L_{Pfk}	4000000		
	K_s^{F6P}	0.14	mM	
	K_{PEP}	3.26	mM	[3]
Fbp	k_f	8.76E+02	min^{-1}	
	K_m^{FDP}	0.0154	mM	
	K_i^{AMP}	0.0027	mM	
Aldo	v_{Aldo}^{max}	2.8337	mmol/gDCWh	
	K_{FDP}	0.133	mM	
	K_{GAP}	0.088	mM	
	K_{DHAP}	0.088	mM	[3]

Enzyme	Parameter	Value	Unit	References
	K_{inh}^{GAP}	0.6	mM	
	K_{eq}	0.14	mM	
	V_{blf}	2		
GAPDH	v_{GAPDH}^{max}	121.29	mmol/gDCWh	
	K_{GAP}	0.15	mM	
	K_{PGP}	0.1	mM	
	K_{NAD}	0.45	mM	
	K_{NADH}	0.02	mM	
	K_{eq}	0.63		[3]
Ppc	v_{Ppc}^{max}	0.1885	mmol/gDCWh	
	K_m^{PEP}	0.323	mM	
	k_1	0.03176		
	k_2	1.2878	mM	
	k_3	0.05425	mM	
	k_4	0.8139	mM	
	k_5	0.0939	mM	
	k_6	0.2693	mM	[22]
Pck	v_{Pck}^{max}	4.5116	mmol/gDCWh	
	K_m^{ATP}	0.06	mM	
	K_I^{ATP}	0.04	mM	
	K_i^{ATP}	0.04	mM	
	K_m^{OAA}	0.67	mM	
	K_i^{PEP}	0.06	mM	
	K_I^{OAA}	0.67	mM	
	K_m^{PEP}	0.07	mM	
	K_i^{ADP}	0.04	mM	[24]
Pyk	v_{Pyk}^{max}	1.0849	mmol/gDCWh	
	K_{PEP}	0.31	mM	[3]

Enzyme	Parameter	Value	Unit	References
	K_{FDP}	0.19	mM	
	K_{AMP}	0.2	mM	
	n_{Pyk}	4		
	K_{ATP}	22.5	mM	
	K_{ADP}	0.26	mM	
	L_{Pyk}	1000		
Pps	k_f	1.75E+06	min^{-1}	
	k_r	1.33E+05	min^{-1}	
	K_m^{PYR}	0.083	mM	
	K_m^{ATP}	0.028	mM	
	K_m^{PEP}	0.037	mM	
	K_m^{AMP}	0.11	mM	
	K_m^{Pi}	38	mM	
Mez	v_{Mez}^{max}	0.069945	mmol/gDCWh	
	K_m^{MAL}	0.37	mM	[16]
	K_{eq}	0.10		
PDH	v_{PDH}^{max}	27171	mmol/gDCWh	
	K_m^{PYR}	1	mM	
	K_m^{NAD}	0.4	mM	
	K_m^{AcCoA}	0.008	mM	
	K_m^{NADH}	0.1	mM	
	K_m^{COA}	0.014	mM	
	K_i^{PDH}	46.4	mM	[11]
Pta	v_{Pta}^{max}	12.585	mmol/gDCWh	
	K_i^{AcCoA}	0.2	mM	
	K_m^P	2.6	mM	
	K_i^P	2.6	mM	[11]

Enzyme	Parameter	Value	Unit	References
	K_m^{ACP}	0.7	mM	
	K_i^{ACP}	0.2	mM	
	K_i^{CoA}	0.029	mM	
	K_{eq}	0.0281	mM	
Ack	v_{Ack}^{max}	2865.3	mmol/gDCWh	[11]
	K_m^{ACP}	0.16	mM	
	K_m^{ADP}	0.5	mM	
	K_m^{ACE}	7	mM	
	K_m^{ATP}	0.07	mM	
	K_{eq}	174.217	mM	
Acs	v_{Acs}^{max}	17.9068	mmol/gDCWh	[20]
	K_{eq}	0.089971	mmol/gDCWh	
	K_m	0.07	mM	
CS	v_{CS}^{max}	17.36	mmol/gDCWh	
	K_{d1}^{H}	0.00001	mM	
	K_{d2}^{H}	0.0002	mM	
	K_m^{AcCoA}	0.18	mM	
	K_m^{OAA}	0.04	mM	
	K_d^{AcCoA}	0.1	mM	
	K_i^{ATP}	0.58	mM	
	K_{i1}^{2KG}	0.015	mM	
	K_{i2}^{2KG}	0.256	mM	
	K_{i1}^{NADH}	0.00033	mM	
	K_{i2}^{NADH}	0.0084	mM	
	K_{cat0}	1		[12]
ICDH	v_{ICDH}^{max}	24.42	mmol/gDCWh	[15]
	K_{eq}	1000	mM	

Enzyme	Parameter	Value	Unit	References
	k_f	4830	min^{-1}	
	K_m^{2KG}	0.038	mM	
	K_m^{iCiT}	0.0059	mM	
	K_d^{NADP}	0.0013	mM	
	K_m^{NADP}	0.0227	mM	
	K_d^{NADPH}	0.12	mM	
	K_d^{iCiT}	0.03	mM	
	K_{einh}^{NADPH}	0.007	mM	
	K_{eknh}^{2KG}	5.5	mM	
	$K_d^{CO_2}$	1.6	mM	
	$K_{eke}^{CO_2}$	1.6	mM	
	K_{ekn}^{NADP}	0.00016	mM	
	K_m^{NADPH}	0.0036	mM	
	K_{enhe}^{NADPH}	0.028	mM	
αKGDH	v_{2KGDH}^{max}	137.435	mmol/gDCWh	
	K_m^{2KG}	1.0	mM	
	K_I^{2KG}	0.75	mM	
	K_m^{CoA}	0.002	mM	
	K_m^{NAD}	0.07	mM	
	K_m^{SUC}	1.0	mM	
	K_m^{NADH}	0.018	mM	
	K_z	1.5	mM	[16]
SCS	k_f	2.78E+03	min^{-1}	
	k_r	3.99E+03	min^{-1}	
	K_m^{SUCCoA}	0.0077	mM	
	K_m^{ADP}	0.012	mM	
	K_m^{Pi}	2.6	mM	

Enzyme	Parameter	Value	Unit	References
	K_m^{SUC}	0.1	mM	
	K_m^{ATP}	0.02	mM	
	K_m^{CoA}	0.0015	mM	
SDH	K_m^{SUC}	0.1		
	v_{SDH1}	1.11334	mmol/gDCWh	
	v_{SDH2}	1.11334	mmol/gDCWh	
	K_{eq}	10		[16]
Frd	k_f	8.40E+02	min^{-1}	
	k_r	1.06E+04	min^{-1}	
	K_m^{SUC}	0.0015	mM	
	K_m^{FUM}	0.0054	mM	
Fum	v_{FUM1}	1.11334		
	V_{FUM2}	1.11334		
	K_{eq}	10		
	K_m^{FUM}	0.1		[16]
MDH	v_{MDH1}	25.874	mmol/gDCWh	
	v_{MDH2}	25.874	mmol/gDCWh	
	K_{eq}	1.0		
	K_I^{NAD}	0.31	mM	
	K_I^{NADH}	0.04	mM	
	K_I^{MAL}	3.30	mM	
	K_I^{OAA}	0.27	mM	
	K_m^{NAD}	0.1	mM	
	K_m^{NADH}	0.04	mM	
	K_m^{MAL}	1.33	mM	
	K_m^{OAA}	0.27	mM	
	K_{II}^{NAD}	0.31	mM	[16]

Enzyme	Parameter	Value	Unit	References
	K_{II}^{OAA}	0.17	mM	
Icl	$V_{Icl\_f}^{max}$	3.8315	mmol/gDCWh	[25]
	$v_{Icl\_r}^{max}$	2.585/100	mmol/gDCWh	
	K_m^{iCiT}	0.604	mM	
	K_m^{SUC}	0.59	mM	
	K_m^{GOX}	0.13	mM	
	K_I^{ICL}	0.003	mM	
MS	$K_{MS\_f}^{max}$	3.6968	mmol/gDCWh	[25]
	$K_{MS\_r}^{max}$	13.742	mmol/gDCWh	
	K_m^{GOX}	2	mM	
	K_m^{AcCoA}	0.01	mM	
	K_m^{MAL}	1	mM	
	K_m^{CoA}	0.1	mM	
G6PDH	v_{G6PDH}^{max}	0.97922	mmol/gDCWh	[3]
	K_{G6P}	14.4	mM	
	K_{NADP}	0.015	mM	
	K_{NADPH}^{NADP}	0.01	mM	
	K_{NADPH}^{G6P}	0.18	mM	
6PGDH	v_{6GPDH}^{max}	1.81	mmol/gDCWh	[3]
	K_{G6P}	0.1	mM	
	K_{NADP}	0.028	mM	
	$K_{NADPinh}$	0.01	mM	
	K_{ATPinh}	3.0	mM	
Rpe	v_{Rpe}^{max}	18.485	mmol/gDCWh	[3]
	K_{eq}^{Rpe}	1.4	mM	
Rpi	v_{Rpi}^{max}	13.318	mmol/gDCWh	[3]
	K_{eq}^{Rpi}	4.0	mM	

Enzyme	Parameter	Value	Unit	References
TktA	v_{TktA}^{max}	29.348	mmol/gDCWh	[3]
	K_{eq}^{TktA}	1.2	mM	
TktB	v_{TktB}^{max}	316.22	mmol/gDCWh	[3]
	K_{eq}^{TktB}	10.0	mM	
Tal	v_{Tal}^{max}	24.499	mmol/gDCWh	[3]
	K_{eq}^{Tal}	1.05	mM	
Cya	k_f	8.10E+02	min^{-1}	
	K_m^{ATP}	1	mM	
	K_i^{ATP}	1.4	mM	

APPENDIX D: MODEL PARAMETERS USED FOR THE SIMULATIONS OF NITROGEN REGULATION

Enzyme	Kinetic parameter values	Original source
GDH	$GDHcon = 8000*10^{-6}mM$, $K\_GDHcat = 5.3*10^{4}*60$ mmol/gDCW.h; $K\_GDHnh4 = 1.1mM$, $K\_GDH2kg = 0.32mM$, $K\_GDHnadph = 0.04mM$, $K\_GDHglu = 10.0mM$, $K\_GDHnadp = 0.042$, $K\_GDHeq = 1290$, $V\_GDHmax = GDHcon*K\_GDHcat$;	[39]
GS	$GScon = 240*10^{-6}$ mM; $K\_GScat = 1.6*10^{6}$ mmol/gDCW.h: $K\_GSglu = 4.1mM$; $K\_GSnh4 = 0.1mM$; $K\_GSatp = 0.35mM$; $K\_GSgln = 5.56mM$; $K\_GSadp = 0.0585mM$; $K\_GSp = 3.7mM$; $K\_GSeq = 460mM$; $V\_GSmax = GScon*K\_GScat$; h	[39]
GOGAT	$GOGATcon = 130*10^{-6}$ mM; $K\_GOGATcat = 2.1*10^{4}*60/564$ mmol/gDCW.h $K\_GOGATgln = 0.175mM$; $K\_GOGAT2kg = 0.007mM$; $K\_GOGATnadph = 0.0015mM$ $K\_GOGATglu = 11.0mM$; $K\_GOGATnadp = 0.0037mM$; $V\_GOGATmax = GOGATcon*K\_GOGATcat$;	[39]
AST	$V_{ASTmax} = 3793.15$; $K_{ASTglu} = 0.9$; $K_{ASToaa} = 0.58$; $K_{ASTeq} = 6.7$; $K_{AST2kg} = 0.59$; $K_{ASTasp} = 0.45$	[57]
AT	$V_{ATmax} = 0.887$; $K_{ATgs} = 0.0000119$; $KA_{ATpii} = 0.00365$; $K_{AT2kg} = 0.0098$; $K_{ATgln} = 12.2$	[57]

AR	V_{ARmax} =8.09; $K_{ARgsamp}$ =6.97; $K_{ARpiiump}$ =0.00195; K_{ARgln} =12.2	[57]
UT	V_{UTmax} =0.251; K_{UTpii}=0.0191; K_{UTgln} =0.016	[57]
UR	V_{URmax} =0.000109; $K_{URpiiump}$ =0.00755; K_{URgln} =0.016	[57]

APPENDIX E STANDARD COFACTOR CONCENTRATIONS

Cofactor	Value		Unit	Reference
ATP	4.27	mM		[8]
ADP	0.595	mM		[8]
AMP	0.955	mM		[8]
CoA	0.001	mM		[8]
NAD	1.47	mM		[8]
NADH	0.1	mM		[8]
NADP	0.195	mM		[8]
NADPH	0.062	mM		[8]
P_i	10	mM		[43]

DISCLOSURE

Part of this chapter has been previously published in Bacterial Cellular Metabolic Systems Metabolic Regulation of a Cell System with ^{13}C-Metabolic Flux Analysis 2013, Pages 1–54.

REFERENCES

[1]   Nielsen J. It is all about metabolic fluxes. J Bacteriol 2003; 185: 7031-7035.
[2]   Varma A, Palsson DO. Metabolic capabilities of *Escherichia coli*: I. synthesis of biosynthetic precursors and cofactors. J Theor Biol 1993; 165: 477-502.
[3]   Chassagnole C, Noisommit-Rizzi N, Schmid JW, Mauch K, Reuss M. Dynamic modeling of the central carbon metabolism of *Escherichia coli*. Biotechnol Bioeng 2002; 79: 53-73.
[4]   Liao JC, Hou SH, Chao YP. Pathway analysis, engineering and physiological considerations for redirecting central metabolism. Biotechnol Bioeng 1996; 52: 129-140.
[5]   Altintas MM, Eddy CK, Zhang M, McMillan JD, Kompala DS. Kinetic modeling to optimize pentose fermentation in *Zymomonas mobilis*. Biotechnol Bioeng 2006; 94: 273-295.
[6]   Ishii N, Suga Y, Hagiya A, Watanabe H, Mori H, Yoshino M, Tomita M. Dynamic simulation of an *in vitro* multi-enzyme system. FEBS Lett 2006; 581: 413-420.
[7]   Schreyer R, Böck A. Phosphoglucose isomerase from *Escherichia coli* K 10: purification, properties and formation under aerobic and anaerobic. Arch Microbiol 1980; 127: 289-298.
[8]   Karp P, Riley M, Paley S, Pellegrini-Toole A, Krummenacker M. EcoCyc: Encyclopedia of *Escherichia coli* genes and metabolism. Nucleic Acids Res 1997; 25: 43-50.
[9]   Kotlarz D, Garreau H, Buc H. Regulation of the amount and of the activity of phosphofructokinase and pyruvate kinase in *Escherichia coli*. Biochem Biophys Acta 1975; 381: 257-268.

[10]   Kadir TA, Mannan AA, Kierzek AM, McFadden J, Shimizu K. Modeling and simulation of the main metabolism in *Escherichia coli* and its several single-gene knockout mutants with experimental verification. *Microb Cell Fact* 2010; 9: 88.

[11]   Hoefnagel MHN, Starrenburg MJC, Martens DE, Hugenholtz J, Kleerebezem M, Van Swam II, Bongers R, Westerhoff HV, Soep JL. Metabolic engineering of lactic acid bacteria, the combined approach: kinetic modeling, metabolic control and experimental analysis. Microbiology 2002; 148: 1003-1013.

[12]   Mogilevskaya EA, Lebedeva GV, Demin OV. Kinetic model of *E. coli* citrate syntase functioning. *In Proceeding of the 12th Int. Conf. Mathematics Computer Education.* 2006; V3: 934-937.

[13]   Tsuchiya D, Shimizu N, Tomita M (2009) Cooperativity of two active sites in bacterial homodimeric aconitases. Biochem Biophys Res Commun 379: 485-488.

[14]   Doelle HW, Ed. Bacterial metabolism, 2nd edition. New York: Academic Press 1975.

[15]   Mogilevskaya EA, Lebedeva GV, Goryanin II, Demin OV. Kinetic model of functioning and regulation of *Escherichia coli* isocitrate dehydrogenase. Biophysics 2007; 52: 30-39.

[16]   Wright BE, Butler MH, Albe KR. Systems analysis of the tricarboxylic acid cycle in the *Dictyostelium discoideum* I. The basis for model construction. J Biol Chem 1992; 267: 3101-3105.

[17]   Moffet FJ, Bridger WA, The kinetics of Succinyl Coenzyme A Synthetase from *Escherichia coli*. J Biol Chem 1970; 245 (10): 2758-2762.

[18]   Kotte O, Zaugg JB, Heinemann M. Bacterial adaptation through distributed sensing of metabolic fluxes. Mol Sys Biol 2010; 6: 355.

[19]   Henkin J, Abales RH. Evidence against an acyl-enzyme intermediate in the reaction catalyzed by clostridial phosphotransacetylase. Biochemistry 1976; 15: 3475-3479.

[20]   Fung E, Wong WW, Suen JK, Butler T, Lee S-G, Liao JC. A synthetic gene-metabolism oscillator. Nature 2005; 435: 118-122.

[21]   Izui K, Taguchi M, Morikawa M, Katsuki H. Regulation of *Escherichia coli* phosphoenolpyruvate carboxylase by multiple effectors *in vivo*. II. Kinetic studies with a reaction system containing physiological concentrations of ligands. J Biochem 1981; 90: 1321-1331.

[22]   Lee B, Yen J, Yang L, Liao JC. Incorporating qualitative knowledge in enzyme kinetic models using fuzzy logic. Biotechnol Bioeng 1999; 62: 722-729.

[23]   Krebs A, Bridger WA. The kinetic properties of phosphoenolpyruvate carboxykinase of *Escherichia coli*. J Biochem 1980; 58: 309-318.

[24]   Yang C, Hua Q, Baba T, Mori H, Shimizu K. Analysis of *Escherichia coli* anaplerotic metabolism and its regulation mechanisms from the metabolic responses to altered dilution rates and phosphoenolpyruvate carboxykinase knockout. Biotechnol Bioeng 2003, 84: 129-144

[25]   Singh VK, Ghosh I. Kinetic modeling of tricarboxylic acid cycle and glyoxylate bypass in *Mycobacterium tuberculosis*, and its application to assessment of drug target. Theor Biol Med Model 2006; 3: 27.

[26]   Zhao J, Shimizu K. Metabolic flux analysis of *Escherichia coli* K12 grown on ^{13}C-labeled acetate and glucose using GC-MS and powerful flux calculation method. J Biotechnol 2003; 101: 101-117.

[27]   Voegele RT, Sweet GD, Boos W. Glycerol kinase of *Escherichia coli* is activated by interaction with the glycerol facilitator. J Bacteriol 1993; 175: 1087-1094.

[28]   Kleinhans FW. Membrane permeability modeling: Kedem-Katchalsky *vs* a two-parameter formalism. Cryobiology 1998; 37: 271-289.

[29]   Cintolisi A, Clomburg JM, Rigou V, Zygourakis K, Gonzalez R. Quantitative analysis of the fermentative metabolism of glycerol in *Escherichia coli*. Biotechnol Bioeng 2012; 109: 187-198.

[30]   Gonzalez R, Murarka A, Dharmadi Y, Yazdani SS. A new model for the anaerobic fermentation of glycerol in enteric bacteria: Trunk and auxiliary pathways in *Escherichia coli*. Meatb Eng 2008; 10: 234-245.

[31]   Subedi KP, Kim I, Kim J, Min B, Park C. Role of GldA in dihydroxy-acetone and methylglyoxal metabolism of *Escherichia coli* K12. FEMS Microbiol Lett 2008; 279: 180-187.

[32]   Gutknecht R, Beutler R, Garcia-Alles LF, Baumann U, Erni G. The dihydroxyacetone kinase of *Escherichia coli* utilizes a phosphoprotein instead of ATP as phosphoryl donor. EMBO J 2001; 20: 2480-2486.

[33]   Koebman BJ, Westerhoff HV, Snoep JL, Nilsson O, Jensen PR. The glycolytic flux in *E. coli* is controlled by the demand for ATP. J Bacteriol 2002; 184: 3909-3916.

[34]   Knappe J, Blaschkowski HP, Grobner P, Schmitt T. Pyruvate formate lyase of *Escherichia coli*: The acetyl-enzyme intermediate. Eur J Biochem 1974; 50: 253-263.

[35]   Shonn CC, Fromm HJ. Steady-state and pre-steady-state kinetics of coenzyme A linked aldehyde dehydrogenase from *Escherichia coli*. Biochemistry 1981; 20: 7494-7501.

[36]   Bruggeman FJ, Boogerd FC, Westerhoff HV. The multifarious short-term regulation of ammonium assimilation of *Ecsherichia coli*: dissection using an *in silico* replica. FEBS J 2005; 272: 1965-1985.

[37]   Sakamoto N, Kotre AM, Savageau MA. Glutamate dehydrogenase from *Escherichia coli*: purification and properties. J Bacteriol 1975; 124: 775-783.

[38]   Atkinson MR, Blauwkamp TA, Bondarenko V, Studitsky V, Ninfa AJ. Activation of the *glnA*, *glnK*, and *nac* promoters as *Escherichia coli* undergoes the transition from nitrogen excess growth to nitrogen starvation. J Bacteriol 2002; 184: 5358-5363.

[39]   Reitzer L. Nitrogen assimilation and global regulation in *Escherichia coli*. Annu Rev Microbiol 2003; 57: 155-176.

[40]   Lodeiro A, Melgarejo A. Robustness in *Escherichia coli* glutamate and glutamine synthesis studied by a kinetic mode. J Biol Phys 2008; 34: 91-106.

[41]   Ma H, Boogerd FC, Goryanin I. Modelling nitrogen assimilation of *Escherichia coli* at low ammonium concentration. J Biotechnol 2009; 144: 175-183.

[42]   Kalinowski J, Cremer J, Bachmann B, Eggeling L, Sahm H, Ptihler A. Genetic and biochemical analysis of the aspartokinase from *Corynebacterium glutamicum*. Mol Microbiol 1991; 5: 1197-1204.

[43]   Follettie MT, Peoples O, Agoropoulou C, Sinskey AJ. Gene structure and expression of the *Corynebacterium flavum* N13 ask-asd operon: molecular and evolutionary analysis of aspartokinase. J Bacterial 1993; 175: 4096-4103.

[44]   Mlyajima R, Otsuka S, Shiio I. Regulation of aspartate family amino acid biosynthesis in *Brevibacterium flavum*. I. Inhibition by amino acids of the enzymes in threonine biosynthesis. J Biochem 1968; 63: 139-148.

[45]   Yang C, Hua Q, Shimizu K. Development of a kinetic model for L-lysine biosynthesis in Corynebacterium glutamicum and its application to metabolic control analysis, J Biosci Bioeng 1999; 88 (4): 393-403.

[46]   Shiio I, Miyajima R. Concerted inhibition and its reversal by end products of aspartate kinase in *Brevibacterium flavum*. J Biochem 1969; **65**: 849-859.

[47]   Cleland WW. The kinetics of enzyme-catalyzed reactions with two or more substrates or products I. Nomenclature and rate equations. Biochim Biophys Acta 1963; 67: 104-137.

[48]   Yugari Y, Gilvarg C. The condensation step in diaminopimelate synthesis. J Biol Chem 1965; 240: 4710-4716.

[49]   Blickling S, Knäblein J. Feedback inhibition of dihydrodipicolinate synthase enzymes by L-lysine. Biol Chem 1997; 378: 207-210.

[50]   Shedlarski JG. Pyruvate-aspartic semialdehyde condensing enzyme (*Escherichia coli*). In: Tabor H, Tabor CW, Eds. Methods in enzymology, vol. 17, part B. New York, Academic Press 1971; pp. 129-134.

[51]   Schrumpf B, Schwarzer A, Kalinowski J, Ptihler A,Eggeling L, Sahm H. A functionally split pathway for lysine synthesis in *Corynebacterium glutamicum*. J Bacteriol 1991; 174: 4510 4516.

[52]   Sonntag K, Eggeling L, de Graaf AA, Sahm H. Flux partitioning in the split pathway of lysine synthesis in *Corynebacterium glutamicum*: Quantification by "C- and 'HNMR spectroscopy. Eur J Biochem 1993; 213: 1325-1331.

[53]   Misono H, Soda K. Properties of meso-a,s-diaminopimelate dehydrogenase from *Bacillus sphaericus*. J Biol Chem 1980; 255: 10599-10605.

[54]   Kindler SH, Gllvarg C. N-Succinyl-r.-cu,s-diaminopimelate acid deacylase. J Biol Chem 1960; 235: 3532-3535.

[55]   Broer S, Kramer R. Lysine excretion by *Corynebacterium glutamicum*. II. Energetics and mechanism of the transport system. Eur J Biochem 1991; 202: 137-143.

[56]   Kelle R, Laufer B, Carsten B, Weuster-Botz D, Kramer R, Wandrey C. Reaction engineering analysis L-lysine transport by *Corynebacterium glutamicum*. Biotechnol Bioeng 1996; 51: 40-50.

[57]   Yuan J, Doucette CD, Fowler WU, Feng XJ, Piazza M, Rabitz HA, Wingreen NS, Rabinowitz JD. Metabolomics-driven quantitative analysis of ammonia assimilation in *E. coli*. Mol Syst Biol 2009; 5:302

[58]   Hashimoto T, Enzyme kinetics, Kyoritsu pub. Co., Japan, 1971 (in Japanese).

CHAPTER 5

Model Identification, Sensitivity Analysis, and Optimization

Abstract: Model identification method is briefly explained. Then the sensitivity analysis is explained in relation to model parameter identification. In view of sensitivity analysis, the metabolic control analysis (MCA) is explained, followed by its application to find the limiting pathways for lysine fermentation by *Corynebacterium* sp. In relation to flux balance analysis (FBA) and its extension to genome-scale as mentioned in Chapter 3, the linear programming method is explained for the optimization of a single objective function under the constraint of stoichiometric equations. A basic approach for the vector-valued objective function and non-inferior Pareto optimal set is briefly explained. In relation to model parameter identification, various types of direct and gradient-based optimum seeking methods are explained. A global search method such as genetic algorithm (GA) is also explained. Moreover, the optimal operation or optimal control strategies based on the Maximum principle is explained to find the time optimal trajectories with some application to ethanol fermentation.

Keywords: Model parameter identification, sensitivity analysis, metabolic control analysis, MCA, linear programming, non-linear programming, gradient method, vector valued objective function, Pareto optimal, genetic algorithm, GA, maximum principle.

INTRODUCTION

Model identification is made as shown in Fig. **1**, where the type of model or model structure must be first considered based on the available experimental data and based on the purpose such as control, optimization, dynamics, understanding of the process *etc*. Once the model was established, parameter identification with sensitivity analysis may be made based on the experimental data or reference data in the literature. Then the result of computer simulation using the model must be validated by the additional experimental data. This procedure is repeated until satisfied. Sometimes, model uncertainty or the reliability of the model must be assessed based on the statistical analysis. It should be careful to check if the experimental data contain how much noise, if the model structure is appropriate or not, and how much we know about the process systems characteristics.

Here, we first consider the model identification problem, followed by the sensitivity analysis. In relation to sensitivity analysis, we also consider the metabolic control analysis (MCA). We then consider the linear programming

Kazuyuki Shimizu and Yu Matsuoka

problem in relation to flux balance analysis as mentioned in Chapter 3. A vector-valued objective function and non-inferior Pareto optimal set are also explained in relation to FBA and its extension to genome-scale. Then non-linear optimization methods such as direct search method and gradient methods are briefly explained. Noting that those gradient methods are local search methods, the global search method such as genetic algorithm (GA) is also explained. Finally, the basic idea and the approach of the Maximum principle is explained for seeking the time optimal trajectories based on the model developed, and its application to ethanol fermentation is shown. It should be careful that the result obtained highly depends on the accuracy of the model, and the application must be careful with consideration of model uncertainty and robustness.

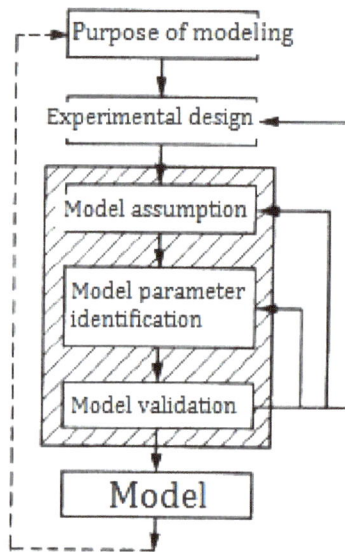

Figure 1: Model identification procedure.

MODEL IDENTIFICATION

Model identification of bio-systems may be made as shown in Fig. **1**, where model structure and model parameters must be appropriately identified. For this, the prior knowledge on the system of concern and the available experimental data must be made clear. The completeness and the accuracy of the model depend on the purpose of using the model. Even a whole cell model does not reflect the real living system *in vivo*, but just the extraction of some feature of the cell system. Thus completeness of the model may not be necessary for it to improve predictions or rationalizations.

The uncertainty of the model comes from the uncertainty of the model structure and model parameters due to ambiguity in the selection of the rate laws underlying the cell system, as well as uncertainty caused by neglecting the metabolic regulation mechanism or improper knowledge on the metabolism *etc.*

Because of the limited experimental data, some model parameters such as kinetic parameters as K_m may be set as those reported in the literature [1]. Care must be taken in such a case, since those may be collected from different sources based on different experimental condition, different physiological state of the cells, different strains, or even different organisms [2]. Some model parameters may be taken from *in vitro* data, but such values may be quite different from those of *in vivo* systems [3]. Some model parameters may be difficult to estimate due to experimental limitation, and may be determined by less accurate ways based on rule of thumb-like considerations, and thus the effect of such model uncertainty on the simulation result must be properly analyzed.

The parameter identification problem may be formulated either as the geometrical problem of minimizing the distance between the model output y(t) and the corresponding experimental data d, or the statistical approach such as the problem of maximizing the likelihood of the observed data given a model that takes the experimental uncertainty into account [4]. In the former case, the parameter identification problem may be expressed as the following minimization problem:

$$J(\theta) = \sum_{i=1}^{N} \frac{[d_i - y(t_i, \theta)]^2}{\sigma_i^2} \tag{1}$$

where d_i (i=1,2,...,N) are the experimental data taken at time points $t_1, t_2, ..., t_N$, and $y(t_i)$ (i=1,2,..., N) are the scalar-valued model outputs. Any distance measure can be considered, where Eq.(1) is a weighted sum of squares, where σ_i^2 is the variance for the i-th measurement taking into account the accuracy of the corresponding data. Here, we consider scalar-valued output y for simplicity. The extension to vector-valued outputs may be considered in the similar way [5]. The identification of parameter vector $\hat{\theta}$ may be made as

$$\hat{\theta} = \arg \min_{\theta} J(\theta) \tag{2}$$

where the optimization methods are explained in the later sections of the present Chapter.

Another approach of parameter estimation may be made based on the statistical consideration regarding the experimental observations as realizations of random variables [6]. Let ε_i be the measurement error for $y(t_i)$, then d_i is expressed as

$$d_i = y(t_i) + \varepsilon_i \tag{3}$$

The model output may be expressed as

$$y(t) = h(x(t), u(t), \theta) + \varepsilon \tag{4}$$

where y, h, x, u, θ, and ε are shown in vectors of appropriate dimension, where x is the state vector, u is the input vector, and θ is the parameter vector. If the measurement errors are assumed to be independent, and normally distributed with zero mean and variance σ_i^2 for the i-th data point, the likelihood of observation by θ can be expressed as

$$\mathcal{L}(\theta) = c \prod_{i=1}^{N} \left[-\frac{\{d_i - y(t_i, \theta)\}^2}{2\sigma_i^2} \right] \tag{5}$$

where c is the constant. The parameter vector θ that minimizes $\mathcal{L}(\theta)$ is called as the maximum likelihood estimate.

In relation to model parameter estimation, the identifiability of the model parameters must be correctly analyzed. The models are often over-parameterized and too complex in their structures in contrast to the available experimental data [7, 8]. In such a case, several parameter sets may equally satisfy the measured experimental data, where the model structure is unidentified [9]. A simple way of evaluating the identifiability of parameters in practice may be to look at the parameter confidential intervals probably based on likelihood-based confidence intervals.

The identifiability analysis may guide to the experimental design so that enough data are obtained for model parameter estimation. If it is difficult to do so in practice, some model simplification may be made, where the model reduction method for the case with different time scales is explained in Chapter 3.

SENSITIVITY ANALYSIS AND METABOLIC CONTROL ANALYSIS

Local Sensitivity Analysis

The local parameter sensitivity analysis is to determine the degree of the state or output variables in response to a change in model parameters such that

$$S_{ij} = \frac{\Delta x_i / x_i}{\Delta p_j / p_j} \tag{6}$$

where x_i is the i-th state variable, p_j is the j-th model parameter. The sensitivity analysis can be used to validate the model and to see which parameters affect the most on the specific metabolite production *etc*. Some practical application may be seen for acetone-butanol-ethanol fermentation by *Clostridium* sp., where different fermentation patterns in response to each model parameter were quantitatively evaluated toward higher butanol production [10]. Another application of sensitivity analysis may be also seen for lactate fermentation by *Lactococcus lactis* by finding the targets of genetic manipulation [11].

Metabolic Control Analysis

It is of practical interest to identify the rate-limiting steps and the influence of participating metabolites on the metabolic regulation for metabolic engineering [12]. A theoretical framework for this purpose has been investigated based on the sensitivity analysis as **metabolic control analysis (MCA)**. The basic idea has been independently proposed by Heinrich and Rapoport [13], and Kacser and Burns [14]. This method has been shown to be useful for the theoretical analysis [15, 16], and the method has been further extended [17].

In MCA, three types of coefficients such as **flux control coefficient (FCC)**, **concentration control coefficient (CCC)**, and **elasticity coefficient (EC)** play important roles for the analysis. FCC is defined as

$$C_{i,k} = \frac{\partial J_k}{\partial e_i} \frac{e_i}{J_k} = \frac{\partial \ln J_k}{\partial \ln e_i} \tag{7}$$

where J_k is the k-th flux of the new (pseudo) steady state after the perturbation of enzyme E_i, and e_i denotes the concentration of E_i. FCC measures how much E_i affects J_k. The following **summation theorem** holds for such normalized sensitivities:

$$\sum_{i=1}^{N} C_{ik} = 1 \tag{8}$$

Likewise CCC is defined as

$$C_{i,k} = \frac{\partial x_k}{\partial e_i}\frac{e_i}{x_k} = \frac{\partial \ln x_k}{\partial \ln e_i} \tag{9}$$

where x_k is the k-th metabolite concentration of the new (pseudo) steady state after perturbation of e_i. Namely CCC denotes how much e_i affects x_k. Moreover, EC is defined as

$$\varepsilon_{ik} = \frac{\partial v_k}{\partial x_i}\frac{x_i}{v_k} = \frac{\partial \ln v_k}{\partial \ln x_i} \tag{10}$$

where v_k is the reaction rate as a function of x_k. Unlike FCC and CCC, EC is the local property without changing other x_j ($j \neq i$). There is a relationship between FCC and EC called **connection theorem** such that

$$\sum_{i=1}^{N} C_{ik}\varepsilon_{ki} = 0 \quad (i=1,2,\ldots,K) \tag{11}$$

In practice, FCC can be computed by matrix method based on the above summation theorem and connection theorem [17].

Identification of Rate-Limiting Steps in Lysine Pathway

Consider next how MCA can be used for identifying the rate-limiting pathway for the specific metabolite production. For this, consider the lysine production pathways as given in Fig. **2.** The mathematical equations for computer simulation for lysine fermentation are given in Chapters 4. Fig. **3** shows the typical lysine fermentation result in the batch culture of *Corynebacterium glutamicum* ATCC 21253 [18], where FCCs with respect to several enzymes of interest are shown in Fig. **4**. As can be seen from Fig. **4**, the lysine flux is controlled mainly by aspartokinase (Ask) and lysine permease (PERM). The FCC for Ask is higher than those for other enzymes in the early period of lysine production. Ask activity is inhibited by threonine, and Ask activity increases after threonine concentration becomes low, where threonine was initially added in the medium, and it is assimilated as cultivation proceeded, and its concentration becomes low less than 1.5 mM after 18 h of cultivation [18]. Although the concentration of intracellular threonine decreased to such low level in the lysine production phase, Fig. **4** indicates that Ask reaction remained the rate-limiting step in the lysine synthesis pathway.

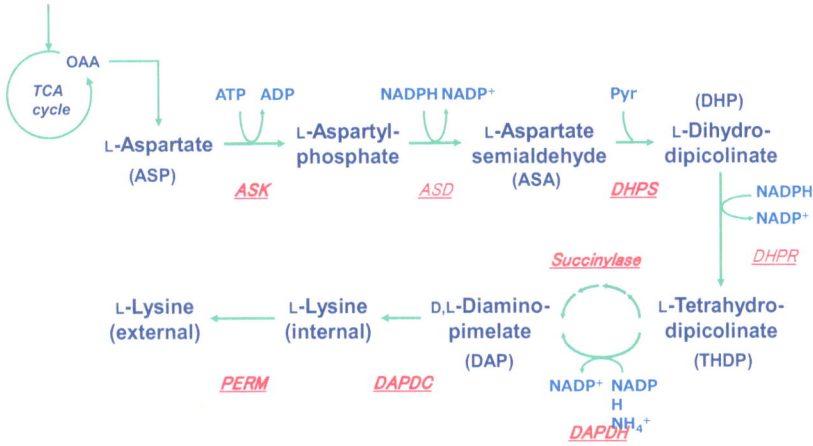

Figure 2: Lysine synthetic pathways.

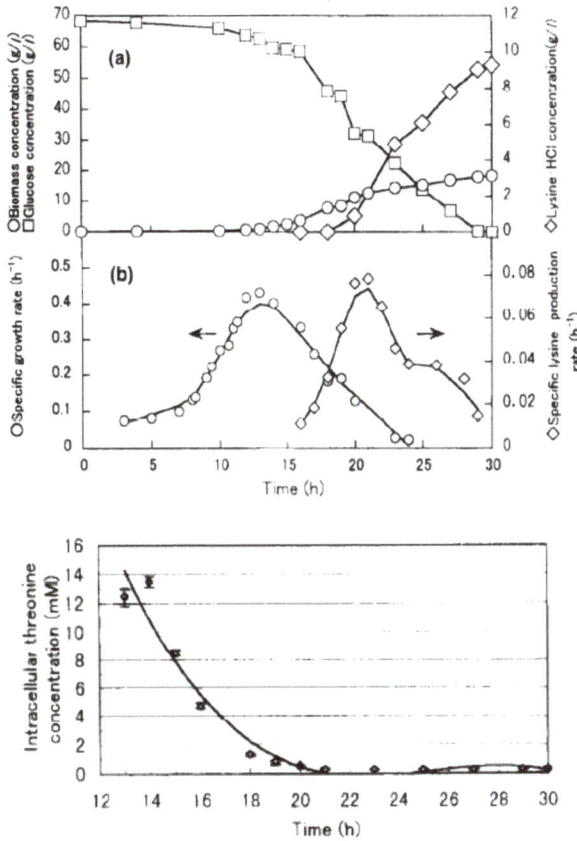

Figure 3: Typical batch lysine fermentation using *C.glutamicum* ATCC 21253 [18].

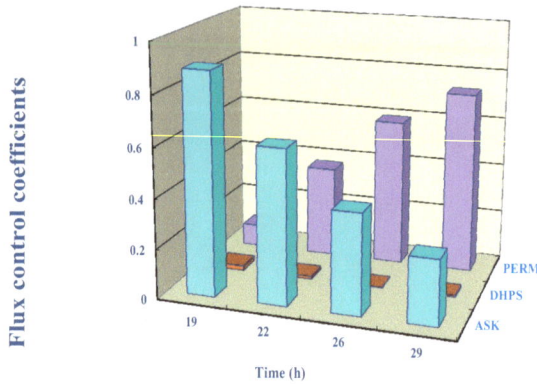

Figure 4: The changes of FCCs with respect to time for three enzymes [19].

With the accumulation of extracellular lysine, however, the FCC for lysine permease increased markedly, and the rate-limiting step shifted to the excretion catalyzed by permease in the late production phase (after approximately 25 h). The FCC for dihydrodipicolinate synthase (DHPS) is very low, which may be due to the nature of the homoserine auxotroph of this strain. For those enzymes associated with diaminopimelate (DAP), such as DAPDH-SUC (diaminopimelate dehydrogenase and enzymes involved in succinylase pathway) and diaminopimelate decarboxylase (DAPDC), the FCCs are close to zero during the entire period of lysine production, suggesting that these enzymatic reactions do not markedly influence the lysine synthesis. For the above MCA analysis, concentrations of cofactors and precursors were assumed to be constant. FCC of Ask was calculated by perturbation of Ask activity, and finding the new steady state.

Based on the above MCA result, the recombinant *C. glutamicum* 21253-33 was constructed by over-expressing *lysC* gene by plasmid pJC33, thus enhancing Ask activity [19]. In addition, another recombinant strain, *C. glutamicum* 21253-23 was also constructed by over-expressing *dapA* gene encoding DHPS [19]. It is difficult for another rate-limiting enzyme PERM to be coordinated by gene manipulation due to a dearth of molecular information regarding this export carrier. In order to verify the over-expression of such genes as *lysC* and *dapA*, the activities of Ask and DHPS were compared to the parent strain. As shown in Table **1**, the specific activities of Ask in 21253-33 and DHPS in 21253-23 were enhanced considerably compared to those in the parent strain ATCC 21253. Furthermore, to study the influence of the increased activities of Ask and DHPS on lysine production, several cultivations were carried out using these strains [19].

As shown in Fig. **5**, the cell growth rate, and the glucose consumption and lysine production rates between the parent strain and the recombinant strain 21253-23 are similar, while for the recombinant strain 21253-33 over-expressing Ask, a significant increase in lysine production can be observed. Approximately 20% increase in lysine synthesis flux could be achieved for strain 21253-33 compared to the parent strain (Fig. **5**) [19]. It also indicates that increasing the activity of an enzyme with low FCC had little effect on improving the flux through the lysine pathway as suggested by MCA.

Table 1: Enzyme activities of the mutants as compared to the wild type strain

Enzyme	Specific activity (U/mg of protein)		
	21253	**21253-33**	**21253-23**
ASK	**0.03±0.01**	**0.35±0.05**	ND
DHPS	**0.07±0.01**	ND	**0.61±0.08**

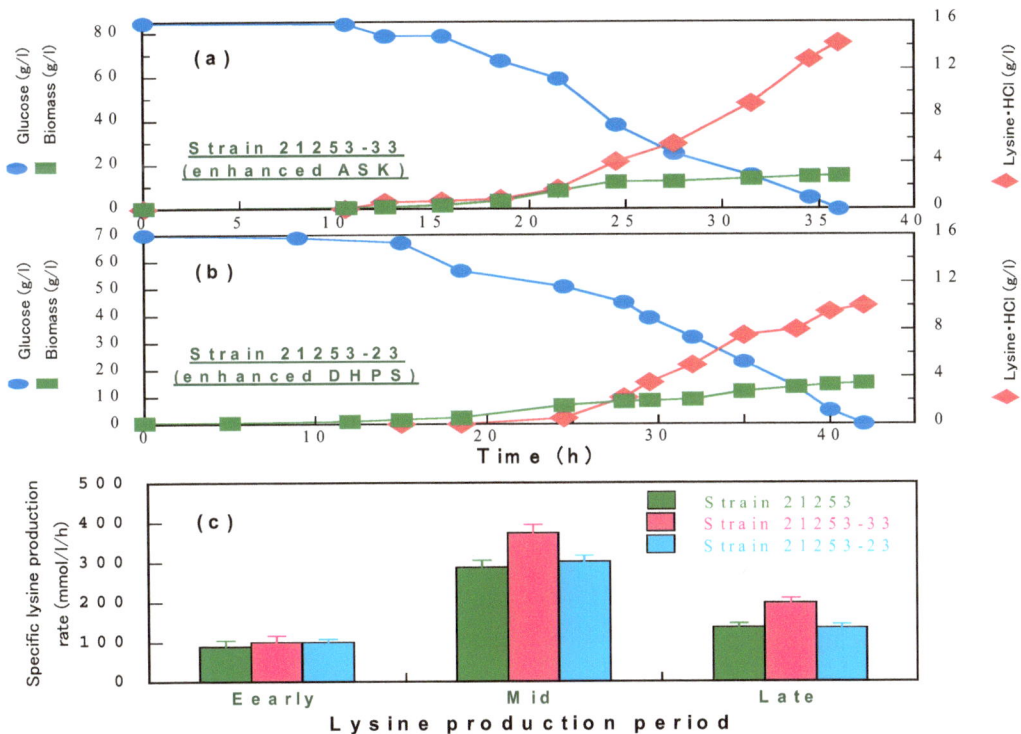

Figure 5: Lysine fermentation using *C.glutamicum* mutant with amplified *lysC* as compared to the parent strain and the mutant with amplified *dapA* [19].

OPTIMIZATION BY LINEAR PROGRAMMING METHOD

The optimization problem often appears in FBA or genome-scale FBA. The **optimization** of the system is to find the condition that determines the maximum or minimum with respect to the performance measure. In general, this measure is called as **objective function** or **performance index**, and the variables to be operated are called **operating variable** or **control variable**.

Consider the **linear programming** problem as formulated as

$$\max : J = c^T x \text{ (objective function)} \tag{12a}$$

$$Ax \le b \text{ (constraints)} \tag{12b}$$

$$x \ge 0 \text{ (non-negative constraints)} \tag{12c}$$

or

$$\min : J = c^T x \text{ (objective function)} \tag{13a}$$

$$Ax \ge b \text{ (constraints)} \tag{13b}$$

$$x \ge 0 \text{ (non-negative constraints)} \tag{13c}$$

where the above formulation is called as canonical form. In the above formulation, c is the cost vector, A is the coefficient matrix, x is the variable vector, and b is the constant vector. The linear programming (LP) problem has been formulated by Danzig in 1947 [20].

Consider the following example to understand the algorithm of LP:

$$\max : J = 4x_1 + 3x_2$$

$$x_1 + 2x_2 \le 2$$

$$6x_1 + 4x_2 \le 7$$

$$x_1 \ge 0, \ x_2 \ge 0$$

where the **feasible region** which satisfies the constraints is shown as the shaded region in Fig. **1**. The objective function can be expressed as $x_2 = -(4/3)x_1 + J/3$.

Thus the problem is to draw the lines of having the slope of -4/3 and choose the one that gives the highest y-axis intersect as shown in Fig. **6**. As seen in Fig. **6**, B extreme point is the optimal point. Note that another extreme point may become the optimum point depending on the employed objective function. In the practical application, the optimal condition may be different depending on the objective function such as the productivity or yield.

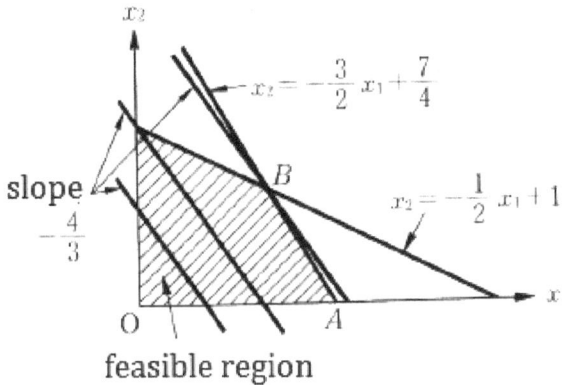

Figure 6: Feasible region and the change in the optimal point depending on the objective function.

In the case where the search space is more than 2-dimensional space, it is not easy to draw such figure as Fig. **6**, but the algorithm for finding the optimum point has been established based on the simplex tableau.

VECTOR-VALUED OBJECTIVE FUNCTION

In the flux balance analysis (FBA), some appropriate objective functions such as the maximization of the cell growth rate, the specific substrate consumption rate, and/or the metabolite production rate must be introduced due to excess degrees of freedom. It was, however, shown that no single objective function can accurately represent the flux data for the different culture condition [21]. Rather, a **vector-valued objective function** or multiple objective functions must be considered, resulting in Pareto optimal set to represent the metabolic fluxes [22], where the influential objective function may be the maximum ATP yield, maximum biomass yield, and minimum sum of absolute fluxes (which corresponds to minimum enzyme investment).

Consider the typical objective function such as the productivity (J_1), product titer (J_2), and the yield (J_3), where the proper performance evaluation must be made.

Although the above objective functions are important in practice, it is not possible to maximize all of them at the same time in general. Instead, we have to consider the **non-inferior set** (or **Pareto optimal set**) to assess the sensitivity of the trade-offs among objective functions. The non-inferior set is the collection of the points where there are no other superior feasible points [23].

Fig. **7** shows the schematic illustration for the non-inferior set (shaded region), where the contour map of J_1 is depicted onto J_2-J_3 plane. The non-inferior set can be obtained by using the constraint method in which one-objective programming problem is solved repeatedly for all the feasible values by fixing the other objective functions to be constant values within the feasible region.

Note that it is not easy to determine the optimal point on the Pareto optimal set in practice depending on the variety of culture conditions. The optimal operation point may be determined somewhere on the non-inferior set if additional data such as the overall cost and profit are available.

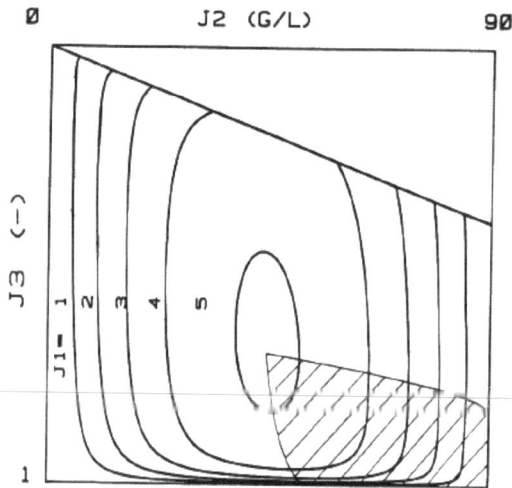

Figure 7: Non-inferior (Pareto optimal) set in J_1-J_2-J_3 space.

MODEL PARAMETER IDENTIFICATION

Once the model was developed, the model parameters must be identified [24]. Some parameters in the model may be directly obtained from the specific experiments such as the identification of kinetic parameters by *in vitro* experiments, or from the literature references. Others must be inferred by comparing the output of the model (or simulation result) with the corresponding experimental data, where the **identifiability** and the **reliability** of the identified model parameters must also be evaluated.

The parameter estimation problem can be formulated as the following minimization problem. Namely, if the data set with N measured points, and the corresponding simulation result as model outputs were given, the following objective function to be minimized can be defined for some distance measure of the vector of residuals.

$$J(\theta) = \sum_{i=1}^{N} \|d_i - y_i(\theta)\| \tag{14}$$

where $\| \bullet \|$ is the norm or distance measure such as ℓ_1, ℓ_2, or ℓ_∞, where ℓ_1 is the absolute value, ℓ_2 is the least square, and ℓ_∞ is the maximum distance. Any distance may be used as far as the following condition (axiom) is satisfied:

$$(i) \quad d_{ii} = 0$$
$$(ii) \quad d_{ik} > 0$$
$$(iii) \quad d_{ik} = d_{ki}$$
$$(iv) \quad d_{is} + d_{sk} \geq d_{ik}$$

where d_{ij} is the distance between i and j.

The typical parameter estimation is made based on a weighted sum of squares as a measure of distance such that

$$J(\theta) = \sum_{i=1}^{N} \frac{(d_i - y_i(\theta))^2}{\sigma_i^2} \tag{15}$$

where σ_i^2 is the weight for the i-th data point by considering the variance of the data point. Namely, if the i-th data vary large, the contribution of the corresponding term will be small. Then the parameter values can be obtained by solving the minimization problem such that

$$\theta = \arg\min_{\theta} J(\theta) \tag{16}$$

The experimental data reflect the true system, but the experimentally observed data may be fluctuated and deviated by a measurement error such that

$$d_i = y_i + \varepsilon_i \ (i=1,2,\ldots,N) \tag{17a}$$

or

$$y = h(x(t), u(t), \theta) + \varepsilon \tag{17b}$$

If the measurement errors are assumed to be independent and normally distributed with zero mean with variance of σ_i^2, the likelihood for parameter estimation may be expressed as

$$L(\theta) = k \prod_{i=1}^{N} \exp\left(-\frac{d_i - y_i(\theta)}{2\sigma_i^2}\right) \tag{18}$$

where k is some constant, and this is called the maximum likelihood estimation.

NON-LINEAR OPTIMIZATION

To find the optimal solution to Eq.(16) with (15) in the parameter space, the non-linear optimization method must be considered, where both local and global methods may be considered, since the objective function is in general multimodal, and the local minimum may not be the global minimum.

Local Optimization Methods

If the gradient of the objective function can be computed with respect to θ, the optimal solution is searched until the gradient vanishes using gradient-based methods, while if it is not available, or difficult to obtain, the direct-search methods can be used to find the optimal solution without explicit derivatives, where the gradient may be indirectly computed numerically.

Direct Search Methods

The **direct-search method** has been introduced by Hooke and Jeeves (1961) for the pattern search purpose [25]. Direct search methods select a finite number of points in the search space, and approach to the optimal point without using explicitly derived derivatives [26, 27].

Among the direct search methods, the **simplex method** developed by Nelder and Mead [28] has been shown to be powerful and often used in practice [24, 29]. This method is to find the optimal point based on an adaptive simplex, where it is the simplest polytope with n+1 vertices in n-dimensional solution space. For example, it is a triangle in the 2-dimensional space, and tetrahedron in the 3-dimensional space and so on. As shown in Fig. **8**, the objective function is evaluated at all vertices, and the vertices are ordered according to the value. Let f_1, f_2, and f_3 be the values of the objective function at the vertices x_1, x_2, and x_3, respectively, where x_1 is assumed to be the worst vertex. Then the better point may exist at the opposite side of x_1 over the line drawn between x_2 and x_3. Let the middle point between x_2 and x_3 be x_{23}. Then foind the new vertex x_4 on the vector directing from x_1 to x_{23} such that

$$x_4 = (1+\alpha)x_{23} - \alpha x_1 \tag{19}$$

where α is the parameter. The same procedure is repeated next for x_2, x_3, and x_4 until the optimal point is searched. If the convergence is slow, α can be increased, while if the optimum point is in the interior of the simplex, α can be reduced, and the search procedure is repeated until the optimum point is reached within the predetermined tolerance.

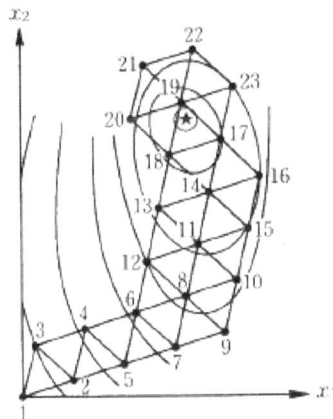

Figure 8: Optimum searching by simplex method.

This method can be also applied if the optimum point is on the boundary of the feasible region, where this method is called as **complex method** developed by Box [30].

Gradient-Based Method

Let $f=f(\mathbf{x})$ be the objective function for n-dimensional search space such as $\mathbf{x}=[x_1,x_2,\ldots,x_n]^T$, where the function is assumed to be explicitly defined. If there is no constraint on **x**, the following necessary condition for the optimality must be satisfied at the stationary point:

$$\frac{\partial f}{\partial x} = \nabla f(x) = [\frac{\partial f}{\partial x_1}, \frac{\partial f}{\partial x_2}, \bullet\bullet\bullet, \frac{\partial f}{\partial x_n}] = 0 \tag{20}$$

Consider how to search the optimal point from the arbitrarily chosen initial point x_0 by hill climbing or descending sequentially. Let x_i be the point obtained by the i-th searching step, and derive the expression for determining the point of the next searching point. For this, consider the Taylor series expansion of f around x_i and truncate higher terms than the first order term such that

$$df = df(x) - df(x_i) = \nabla f(x_i)dx = \sum_{k=1}^{n} \frac{\partial f(x_i)}{\partial x_k}dx_k \tag{21}$$

where $dx=x-x_i$. The problem is to find the direction of steepest descent or ascent among the equal distance direction. This problem can be solved by using Lagrange multiplier, resulting in

$$dx = -\eta_i \nabla^T f(x_i) \tag{22}$$

Therefore, the algorithm for sequentially searching the optimum point is expressed as

$$x_{i+1} = x_i + dx = x_i - \eta_i \nabla^T f(x_i) \tag{23}$$

where η_i is the positive constant, and is the step size. This method is called as the **gradient method or steepest descent (or ascent) method**.

Now the problem is how to choose the value of η_i. The direction is determined, and x_i is already determined. Therefore, x_{i+1} or dx is a function of η_i, and $f(x_{i+1})$ becomes a function of only η_i. Namely,

$$f(x_{i+1}) = f[x_i - \eta_i \nabla^T f(x_i)] = f(\eta_i)$$

from which the optimal value for η_i is determined by solving the following necessary condition for the stationary point:

$$\frac{df}{d\eta_i} = \frac{\partial f(x_{i+1})}{\partial x_{i+1}} \frac{\partial x_{i+1}}{\partial \eta_i} = -\nabla f(x_{i+1})\nabla^T f(x_i) = 0 \tag{24}$$

This method is sometimes called as **optimal gradient method**, and its converging nature is shown in Fig. **9**.

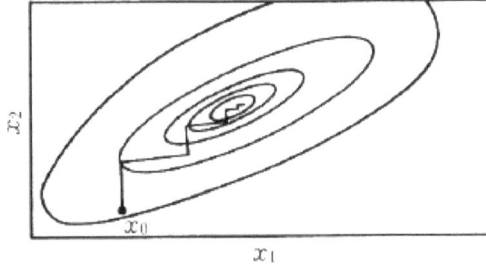

Figure 9: Optimum gradient method.

As expected, the convergence is high based on the gradient-based method. However, as approaching to the optimal point, its converging rate becomes significantly low as can be seen in Fig. **9**, where the reason is that the moving step becomes small, where the second order term in the Taylor series expansion of f around $\mathbf{x}_i$ becomes dominant around the optimal point. Consider then the Taylor series expansion of f with respect to $\mathbf{x}_i$ up to the second order terms.

$$df = f(x_i + dx) - f(x_i) = \nabla f(x_i)dx + \frac{1}{2}dx^T \left[\frac{\partial^2 f(x_i)}{\partial x^2}\right]dx \tag{25}$$

where $[\partial^2 f(x_i)/\partial x^2]$ is the Hessian matrix evaluated at $\mathbf{x}_i$. Let us take the derivative of the above equation with respect to $d\mathbf{x}$, and set it to be zero. Then the following equation is obtained.

$$dx = -\eta_i \left[\frac{\partial^2 f(x_i)}{\partial x^2}\right]^{-1} \nabla f(x_i) \tag{26}$$

Thus the algorithm for searching the optimal point is expressed as

$$x_{i+1} = x_i - \eta_i \left[\frac{\partial^2 f(x_i)}{\partial x^2}\right]^{-1} \nabla f(x_i) \tag{27}$$

where η_i is the positive constant, and this algorithm is called as the **second order gradient method**.

There are several other similar methods such as conjugate gradient method, Fletcher-Reeve's method, Davidon-Fletcher-Powell's method, Powell's method *etc*. In the case where some constraints are imposed on the search space, the optimal condition must satisfy Kurn-Tucker theorem for non-linear programming problem.

Global Search Method

It is in general not easy to find the global optimum as the search space becomes large without information about the shape of the objective function. In certain cases, the random search method such as Monte-Carlo method may be considered. It has been recognized that the **evolutionary method** such as **genetic algorithm** is powerful for the global search [31].

Consider the simple **genetic algorithm (GA)**, where the GA algorithm is based on the evolution process of the population. Each organism (or living system) is characterized by **genotype**, from which **phenotype** is defined. Among the variety of population, each individual is evaluated by the fitness value, and some are selected to survive based on the fitness values, where the higher the fitness value is, the higher the probability of survive. The survived individual is made crossover with appropriate opponent, and sometimes change by mutation with low probability. These operations are repeated over generation, and find the strains having the highest fitness value.

Consider GA algorithm by referring Fig. **10**. The GA algorithm starts with generating a set of individuals constituted of the strings having only 1 or 0 sequences randomly, where the size of the population must be predetermined. Once genome or the character sequence is composed of plural genes, the location of gene is called **locus**. In the present case, either 1 or 0 is the gene, where its candidate for the locus is called **allele**.

Next, choose two strains in the population, and apply **crossover** operation by the predetermined crossover probability. Moreover, some gene is replaced by its allele (namely from 1 to 0 or from 0 to 1) based on the predetermined mutation probability. Then consider mapping from each genotype to phenotype, where this operation is called as **decoding**, while the inverse mapping is called as **coding**.

After above operation, compute fitness value for each individual, and select those which have higher fitness values. This is one **generation**, and the above procedure is repeated until some criterion is satisfied, or until the pre-specified generation is established.

Coding

⇩

Generation of genes

⇩

Adjustment of the number of individuals   Crossover

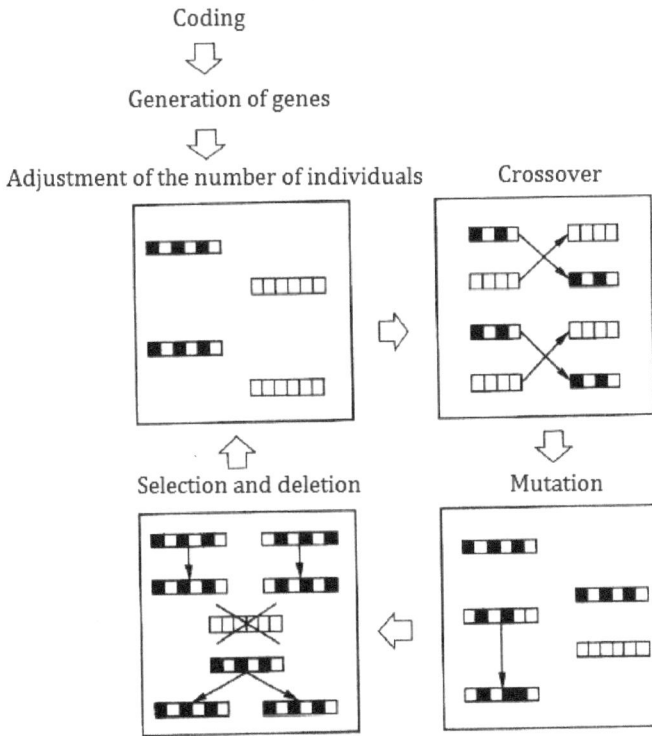

Selection and deletion   Mutation

Figure 10: Simple GA.

Consider the simple example to understand the idea of GA as stated above. For this, consider the example of minimizing $f(x) = x^2 - 1$ on $x \in [0,1]$. Let the value of x be expressed by 4 digits such that 0 is expressed as [0 0 0 0], and 1 is expressed as [1 1 1 1]. Any value between 0 and 1 can be expressed as

$$x = a + \frac{n(b-a)}{2^4 - 1}$$

where a=0, and b=1, and n takes the values from 0 to 15 (=2^4-1). For example, s_1=[1 0 1 0] corresponds to

$$x_1 = 0 + \frac{10(1-0)}{15} = \frac{2}{3}$$

This operation is decoding, where the phenotype is the value of the function such that

$$f(x_1) = f(2/3) = (2/3)^2 - 1 = -5/9$$

which is the fitness value.

For the crossover operation, the following three parameters must be predetermined: Namely, probability of crossover P_c, crossover points CP, and generation gap C, where generation gap means the ratio of how much individuals are replaced by the newly generated strains by crossover operation.

For the crossover operation, choose two strains from N individuals randomly, and generate the number on [0,1] randomly, and if its value is more than P_c, crossover operation is conducted between the two. Then choose two strains from the rest of N-2 individuals and make the similar operation. In this way, NC strains are replaced among N individuals. There are several crossover operations such that

a) One-point crossover or simple crossover

Interchange all the genes at randomly chosen crossover point CP as illustrated below.

*Parent*1	110|0100	$\rightarrow$	*Daughter*1	110|1001
*Parent*2	000|1001		*Daughter*2	000|0100

b) Multipoint crossover

This is the extension of one-point crossover, where the exchange is made at multiple points such that

*Parent*1	11|0010|0	$\rightarrow$	*Daughter*1	11|0100|1
*Parent*2	00|0100|1		*Daughter*2	00|0010|0

c) Uniform crossover

First, develop mask pattern by the random generation of {1,0}, and then place the gene of parent 1 if the mask pattern is 1, while the gene of parent 2 is retained if the mask patter is 0 for one daughter, while the other daughter is the reversed pattern such that

Parent 1 1100100

Parent 2 0001001

Mask 1011011

Daughter 1 1000000

Daughter 2 0101101

Consider next the mutation. With only crossover operation, the variety of new population is limited, which results in the local optimum. In order to generate the variety of population so that the global optimum is obtained, the mutation is considered. The mutation is made as follows: Namely, for each locus, generate the random number on [0,1], and if the number is larger than the predetermined mutation probability P_m, then its value is replaced by the allele such that

$$11 \mid 0 \mid 0100 \rightarrow 11 \mid 1 \mid 0100$$

where P_m must be smaller than Pc, since mutation may lose or destroy the promising genes obtained by crossover operation. Note that if P_m becomes larger, the optimization becomes close to random search.

The idea of natural selection in nature is realized as **reproduction** or **selection** in GA, where the most representative one is the **roulette selection**. Let s_i be the i-th individual, and $f(s_i)$ be the fitness value of s_i. Then the selection probability is determined by

$$Ps_i = \frac{f(s_i)}{\sum_{i=k}^{N} f(s_i)}$$

In this way, the probability of selection becomes higher if the fitness value is high, while the individual with low fitness value can be selected even if the selection probability is low. There are other several selection strategies such as expected-value selection, ranking selection, tournament selection *etc.*

As implied before, some strain with high fitness value may be lost or destroyed by crossover or mutation, and therefore, it may be considered to retain some strains with high fitness values without any operation. This is called as elite preserving selection.

OPTIMAL CONTROL FOR BIOPROCESSES

In the typical batch and fed-batch cultures, the process state such as cell concentration changes with respect to time. In such a situation, it is often required to find the optimal operation or optimal culture condition with respect to time that maximizes some objective function. In the present section, we consider how to find such optimal operation theoretically with some application examples. Although the optimal solution may be obtained as explained in the followings, the treatment of the result must be careful due to model uncertainty inherent in bioprocesses.

Consider first how to find the optimal operation that maximizes the objective function by the **variational method** [32, 33]. Let J be the **functional** objective function such that

$$J = \int_0^\tau \varphi(\dot{x}(t), x(t), t)dt \tag{28}$$

where the functional means that J is determined for some function x(t). Consider the problem of finding the optimal trajectory x*(t) that maximizes J under the constraints that $x(0)=x_0$ and $x(\tau)=x_\tau$. As shown in Fig. **11**, consider the perturbation from the optimal trajectory x*(t) such that

$$x(t) = x^*(t) + \varepsilon\eta(t) \tag{29}$$

where ε is the small constant parameter, and $\eta(t)$ is an arbitrary function which satisfies

$$\eta(0) = \eta(\tau) = 0 \tag{30}$$

Substituting Eq.(29) into Eq.(28), we have

$$J = \int_0^\tau \varphi(\dot{x}^*(t) + \varepsilon\dot{\eta}^*(t), x^*(t) + \varepsilon\eta(t), t)dt \tag{31}$$

By taking the derivative with respect to ε, we have

$$\frac{dJ(\varepsilon)}{d\varepsilon} = \int_0^\tau \left(\frac{\partial\varphi}{\partial\dot{x}}\frac{\partial\dot{x}}{\partial\varepsilon} + \frac{\partial\varphi}{\partial x}\frac{\partial x}{\partial t} \right)dt = \int_0^\tau \left(\frac{\partial\varphi}{\partial\dot{x}}\dot{\eta} + \frac{\partial\varphi}{\partial x}\eta \right)dt$$

$$= \left[\frac{\partial \varphi}{\partial \dot{x}} \eta \right]_0^\tau - \int_0^\tau \frac{d}{dt} \left(\frac{\partial \varphi}{\partial \dot{x}} \right) \eta \, dt + \int_0^\tau \left(\frac{\partial \varphi}{\partial x} \eta \right) dt$$

$$= -\frac{\partial \varphi}{\partial \dot{x}} \bigg|_{t=0} \eta(0) + \frac{\partial \varphi}{\partial \dot{x}} \bigg|_{t=\tau} \eta(\tau) + \int_0^\tau \left[\frac{\partial \varphi}{\partial x} - \frac{d}{dt} \left(\frac{\partial \varphi}{\partial \dot{x}} \right) \right] \eta \, dt \qquad (32)$$

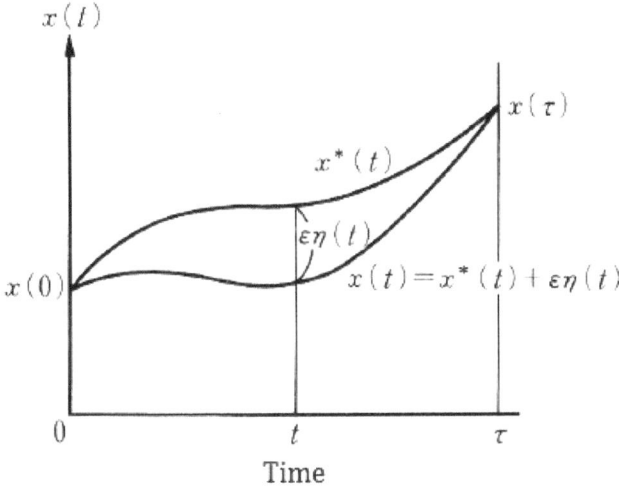

Figure 11: Optimal trajectory and its variation.

By using the condition of Eq.(30), the 1st and the 2nd terms of the last equation become 0 due to B.C.s of Eq.(30). Moreover, the following necessary condition must be satisfied for J to be maximized with respect to ε :

$$\frac{dJ(\varepsilon)}{d\varepsilon} \bigg|_{\varepsilon=0} = 0 \qquad (33)$$

Therefore, the following equation is obtained:

$$\int_0^\tau \left[\frac{\partial \varphi}{\partial x} - \frac{d}{dt} \left(\frac{\partial \varphi}{\partial \dot{x}} \right) \right] \eta \, dt = 0 \qquad (34)$$

For any function of $\eta(t)$, Eq.(34) must be satisfied, and thus the following equation must be satisfied:

$$\frac{\partial \varphi}{\partial x} - \frac{d}{dt}\left(\frac{\partial \varphi}{\partial \dot{x}}\right) = 0 \tag{35}$$

This is called as **Euler-Lagrange equation**. If the boundary condition is not given, the following equation must be satisfied from Eq.(32):

$$\left.\frac{\partial \varphi}{\partial \dot{x}}\right|_{t=0} = 0 \text{ (if initial condition is not specified)} \tag{36a}$$

$$\left.\frac{\partial \varphi}{\partial \dot{x}}\right|_{t=\tau} = 0 \text{ (if final condition is not specified)} \tag{36b}$$

Consider the simple example to understand the above derivation, and find the optimal x(t) that maximize the following objective function:

$$J = \int_0^1 \sqrt{1+\left(\dot{x}\right)^2}\, dt$$

with boundary conditions such that x(0)=0 and x(1)=1. The Euler-Lagrange equation becomes as

$$\frac{d}{dt}\left[\frac{\dot{x}}{\sqrt{1+\left(\dot{x}\right)^2}}\right] = 0 \quad \Rightarrow \quad \frac{\dot{x}}{\sqrt{1+\left(\dot{x}\right)^2}} = C_1 \text{ (constant)}$$

Then $\dot{x}$ is obtained as

$$\dot{x} = \pm\frac{C_1}{\sqrt{1+C_1^2}} = C_2 \text{ (new constant)}$$

Then by integration, we have

$$x(t) = C_2 t + C_3 \text{ (}C_3\text{: integral constant)}$$

By considering the B.C.s, we have

$$x(0) = C_3 = 0 \text{ and } x(1) = C_2 + C_3 = 1$$

Therefore, the solution becomes as

$$x*(t) = t$$

The variational method as mentioned above has some limitation in practice, where the more general approach known as the **Maximum principle** is considered next [32,33]. Consider the general state equation as expresses as

$$\frac{dx}{dt} = f(x, u, t) \tag{37}$$

where x=$[x_1, x_2, \ldots, x_n]^T$ is the n-dimensional state variable vector, u=$[u_1, u_2, \ldots, u_m]^T$ the m-dimensional operation variable vector. Consider the optimal control u(t) that maximizes the following objective function:

$$J = \varphi[x(\tau), \tau] + \int_0^\tau g(x, u, t) dt \tag{38}$$

The problem is to find the optimal solution to Eq.(38) under the constraint of Eq.(37). This may be equivalent to maximize the following objective function by introducing Lagrange multiplier (function) λ (t):

$$J = \varphi[x(\tau), \tau] + \int_0^\tau [g(x, u, t) + \lambda^T \{f(x, u, t) - \dot{x}\}] dt \tag{39}$$

Introduce a new function called **Hamilton function** as

$$H(x, u, \lambda, t) = g(x, u, t) + \lambda^T f(x, u, t) \tag{40}$$

By substituting Eq.(40) into Eq.(39), we have

$$J* = \varphi[x(\tau), \tau] + \int_0^\tau \{H(x, u, \lambda, t) - \lambda^T \dot{x}\} dt = \varphi[x(\tau), \tau] - \lambda^T(\tau)x(\tau) + \lambda^T(0)x(0) + \int_0^\tau \{H(x, u, \lambda, t) - \dot{\lambda}^T x\} dt \tag{41}$$

Suppose that the initial time 0 and the final time τ are specified. Then, consider the perturbation of J with respect to the change in u(t) by δ u(t), where x(t) also changes to x(t)+ δ x(t) such that

$$\delta J* = \varphi[x(\tau) + \delta x(\tau), \tau] - \varphi[x(\tau), \tau] - \lambda^T(\tau)[x(\tau) + \delta x(\tau)] + \lambda^T(\tau)x(\tau) + \lambda^T(0)[x(0) + \delta x(0)] - \lambda^T(0)x(0)$$

$$+ \int_0^\tau [H(x + \delta x, u + \delta u, \lambda, t) - \lambda^T(x + \delta x)]dt - \int_0^\tau [H(x, u, \lambda, t) - \dot{\lambda}^T x]dt$$

$$= \left[\left(\frac{\partial \varphi}{\partial x} - \lambda^T\right)\delta x\right]_{t=\tau} + [\lambda^T \delta x]_{t=0} + \int_0^\tau \left[\left(\frac{\partial H}{\partial x} + \dot{\lambda}^T\right)\delta x + \frac{\partial H}{\partial u}\delta u\right]dt \qquad (42)$$

Consider choosing λ such that the coefficient associated with δ x to be 0 such that

$$\dot{\lambda}^T = -\frac{\partial H}{\partial x} \qquad (43a)$$

$$\dot{\lambda}^T(\tau) = \frac{\partial \varphi}{\partial x}\bigg|_{t=\tau} \qquad (43b)$$

where λ is called as **adjoint variable**, and the above equations are called as **adjoint equations**. If Eq.(43) is satisfied, Eq.(42) becomes as

$$\delta J* = \lambda^T(0)\delta x(0) + \int_0^\tau \left(\frac{\partial H}{\partial u}\delta u\right)dt \qquad (44)$$

If x(0) is specified, δ x(0)=0, and for J* to be maximized (or minimized), δ J must be 0. Thus the following equation must hold.

$$\frac{\partial H}{\partial u} = 0 \qquad (0 \le t \le \tau) \qquad (45)$$

Eqs.(43) and (45) are the similar equations as those derived by the method of variation. In summary, the **canonical form** is as follows:

$$\dot{x} = f(x, u, t) = \left(\frac{\partial H}{\partial \lambda}\right)^T \qquad (46a)$$

$$\dot{\lambda} = -\left(\frac{\partial H}{\partial x}\right)^T \qquad\qquad (46b)$$

with

$$x(0)=x_0 \text{ (given)} \qquad\qquad (46c)$$

and

$$\lambda(\tau) = \left(\frac{\partial \varphi}{\partial x}\right)^T\Bigg|_{t=\tau} \qquad\qquad (46d)$$

The optimal u is obtained from Eq.(45). Note that the initial condition is given for x, while the terminal condition is specified for λ, and thus the **two-point boundary value problem** must be solved in practice.

Consider taking the derivative of H with respect to t such that

$$\frac{dH}{dt} = \frac{\partial H}{\partial t} + \frac{\partial H}{\partial x}\dot{x} + \frac{\partial H}{\partial u}\dot{u} + \frac{\partial H}{\partial \lambda}\dot{\lambda}$$

By substituting Eqs.(46a) and (46b) for $\dot{x}$ and $\dot{\lambda}$, and using Eq.(45) for $\partial H/\partial u$, we have

$$\frac{dH}{dt} = \frac{\partial H}{\partial t} + \frac{\partial H}{\partial x}\left(\frac{\partial H}{\partial \lambda}\right)^T + \frac{\partial H}{\partial u}\dot{u} - \frac{\partial H}{\partial \lambda}\left(\frac{\partial H}{\partial x}\right)^T = \frac{\partial H}{\partial t} \qquad (47)$$

which means that if t is not explicitly shown in H, or f and g, then Eq.(47) becomes as dH/dt=0, and thus H is constant over the optimal trajectory.

Consider reformulating the Maximum principle from the application point of view, and consider the following state equations:

$$\frac{dx_i}{dt} = f_i(x_1, x_2, ..., x_n, u) \qquad (i = 1, 2, ..., n) \qquad\qquad (48a)$$

$$x_i(0) = x_{i0} \qquad (i = 1, 2, ..., n) \qquad\qquad (48b)$$

where u is the **control input**. Consider then the following objective function:

$$J = C_1 x_1(\tau) + C_2 x_2(\tau) + ... + C_m x_m(\tau) \tag{49}$$

where J is the function of the set of state variables at t= τ . In the case where the objective function contains the followings such as utility cost *etc.* in practice:

$$J = \int_0^\tau f_0(x,u)dt \tag{50}$$

This can be also treated in the similar manner by introducing a new variable x_0 such that

$$\frac{dx_0}{dt} = f_0(x,u) \tag{51}$$

and by modifying J of Eq.(49) as

$$J = x_0(\tau) + \sum_{i=1}^{m} C_i x_i(\tau) \tag{52}$$

Consider the above problem by the Maximum principle. For this, define Hamilton function as

$$H = \lambda_1 f_1 + \lambda_2 f_2 + \bullet \bullet \bullet + \lambda_n f_n \tag{53}$$

where λ_i (i=1,2,...,n) are the adjoint variables which satisfy

$$\frac{d\lambda_i}{dt} = -\frac{\partial H}{\partial x_i} \qquad (i = 1,2,...,n) \tag{54}$$

The terminal conditions for λ_i (i=1,2,...,n) are given as

$$\lambda_i(\tau) = \frac{\partial J}{\partial x_i}\bigg|_{t=\tau} = C_i \qquad (i = 1,2,...,n) \tag{55}$$

If $x_i(\tau)$ is specified, $\lambda_i(\tau)$ cannot be determined from Eq.(55). Moreover, we set $\lambda_0=1$ for J to be maximized.

Suppose that the control input is constrained as

$$u_{min} \leq u \leq u_{max} \tag{56}$$

Then the optimal u*(t) is obtained from the Maximum principle as

$$u*(t) = \begin{cases} u_{min} & if \ \dfrac{\partial H}{\partial u} < 0 & (57a) \\[2mm] u^0 & if \ \dfrac{\partial H}{\partial u} = 0 & (57b) \\[2mm] u_{max} & if \ \dfrac{\partial H}{\partial u} > 0 & (57c) \end{cases}$$

In the case where H is a linear function with respect to u, the optimal solution cannot be obtained from the above equation, but the following equation can be used:

$$\frac{d^i}{dt^i}\left(\frac{\partial H}{\partial x_i}\right) \neq 0 \tag{58}$$

Namely, the differentiation is continued until nonzero term appears for the LHS of Eq.(58), where the optimal solution obtained by Eq.(58) is called as **singular control**.

To understand the Maximum principle, consider the simple example as

$$\frac{dx}{dt} = f(x,u) = x + u$$

$$x(0) = 1$$

$$J(u) = \int_0^\tau f_0(x,u)dt = \int_0^1 -(x^2 + u^2)dt$$

where $\tau = 1$ is specified, and x(0) is specified. The Hamilton function is expressed as

$$H(x,\lambda,u) = f_0 + \lambda f = -(x^2 + u^2) + \lambda(x + u)$$

$$\frac{d\lambda}{dt} = -\frac{\partial H}{\partial x} = 2x - \lambda$$

$$\lambda(1) = 0$$

The optimal u which maximizes J is obtained by

$$\frac{\partial H}{\partial u} = -2u + \lambda = 0$$

from which we have u*= λ /2. Substituting this into the state equation, we have

$$\frac{dx}{dt} = x + \frac{\lambda}{2}$$

By taking the derivative with respect to time, we have

$$\frac{d^2x}{dt^2} = \frac{dx}{dt} + \frac{1}{2}\frac{d\lambda}{dt} = x + \frac{\lambda}{2} + \frac{1}{2}(2x - \lambda) = 2x$$

Therefore, the solution to this equation becomes as

$$x*(t) = C_1 e^{-\sqrt{2}t} + C_2 e^{\sqrt{2}t}$$

From the initial condition

$$x*(0) = C_1 + C_2 = 0$$

The equation for λ becomes as

$$\lambda*(t) = 2[-(1+\sqrt{2})C_1 e^{-\sqrt{2}t} + (1+\sqrt{2})C_2 e^{\sqrt{2}t}]$$

Using the terminal condition for λ (λ (1)=0), we have

$$\lambda*(1) = 0 = 2[-(1+\sqrt{2})C_1 e^{-\sqrt{2}} + (1+\sqrt{2})C_2 e^{\sqrt{2}}]$$

where C_1 and C_2 can be determined from the above two equations, and then the optimal solution u*(t) is obtained.

Consider next the practical application example. Fig. 12 shows the batch culture of *Saccharomyces cerevisiae* cultivated at different temperatures, where it indicates that the higher the culture temperature, the initial cell growth and the ethanol production rate is higher, whereas it becomes declined soon. On the other hand, at lower temperature, although the initial cell growth rate is lower, the final values of cell and ethanol concentrations are higher. Let us consider the optimal temperature profile which gives the highest ethanol concentration.

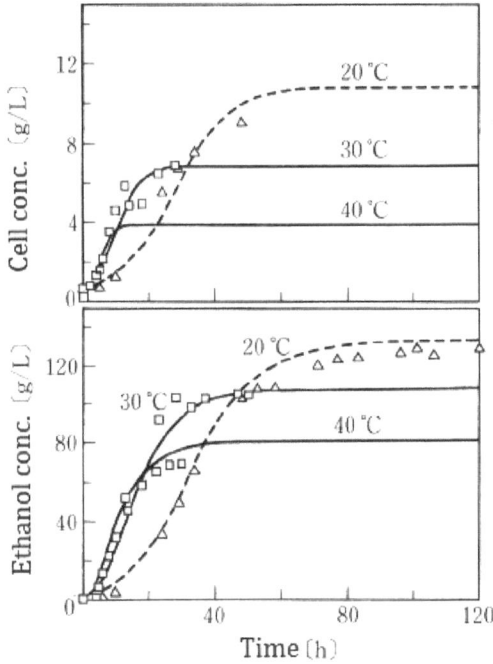

Figure 12: Effect of temperature on the cell growth and ethanol formation.

Consider the following equation for the state equations:

$$\frac{dX}{dt} = \mu(T,P)X \tag{59a}$$

$$\frac{dP}{dt} = \rho(T,P)X \tag{59b}$$

Where X and P are the cell and ethanol concentrations, respectively. T is the culture temperature, and μ and ρ are the specific cell growth rate, and the ethanol production rate, respectively. In the above equation, the substrate concentration is not considered, where the substrate is assume to be enough supplied. Let ρ be expressed as the Ludeking-Piret model as

$$\rho(T,P) = \alpha\mu(T,P) + \beta(T,P) \tag{60}$$

Consider then the following equations for μ and β :

$$\mu(T,P) = \mu_0(T,P)e^{-k_1(T)\int_0^t P^n dt'} \tag{61a}$$

$$\beta(T,P) = \beta_0(T,P)e^{-k_2(T)\int_0^t P^n dt'} \tag{61b}$$

where μ_0, β_0, k_1, and k_2 are assumed to be expressed by the **Arrhenius relationship** such as for example,

$$\mu_0(T) = A_1 e^{-\frac{E1}{R(273+T)}} \tag{62}$$

where A_1 and E_1 are the constant, and the similar equations can be also considered for the others.

Now the objective is to maximize the final ethanol concentration at the specified time τ , and thus the objective function is as follows:

$$J = P(\tau) \tag{63}$$

There is an integral term for μ and β as shown by Eq.(61). This can be treated by introducing a new variable ς such as

$$\frac{d\varsigma}{dt} = P^n \tag{64}$$

In order to find the optimal time profile for T, the Maximum principle can be applied. Although the solution cannot be obtained analytically in this case and in many practical applications, this may be obtained by numerically finding optimal T(t) that maximize the Hamilton function such that

$$T_i^{(n)} = T_i^{(n-1)} + \delta \left(\frac{\partial H}{\partial T_i} \right)^{(n-1)} \tag{65}$$

where T_i is the temperature of the i-th segment among the n-divided sub-sections, and the superfix "(n)" means the n-th iterative computation. Moreover, δ is the step-size for the hill climbing algorithm. Fig. **13** shows the resulting optimal temperature, where the temperature was constrained between the two boundary values set from the practical application point of view. The performance could be improved by about 6-7 % as compared to the case of constant temperature.

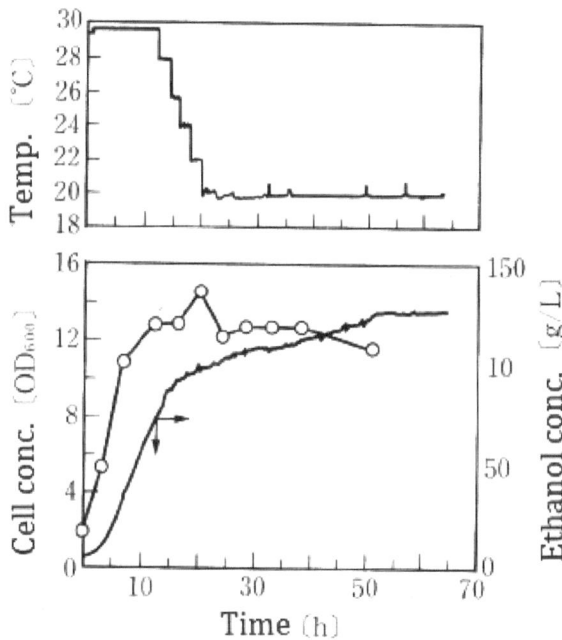

Figure 13: Optimal temperature pattern and the improvement of ethanol fermentation.

Consider yet another example such as fed-batch cultivation. In this case, the substrate F is often the control variable to maximize the final product concentration. In this case, the Hamilton function is expressed as a liner function with respect to F such that

$$H = A + BF \tag{66}$$

where the optimal F that maximizes H becomes as

$$F = F_{\text{max}} \qquad if \qquad B > 0 \tag{67a}$$

$$F = F_{\text{min}} \qquad if \qquad B < 0 \tag{67b}$$

Where this operation is called **Bang-Bang control**. In the case where B=0, the optimal solution becomes **singular control** as mentioned before.

OPTIMAL OPERATION BY GREEN'S THEOREM

As mentioned in Chapter 3, the state equations for the cell growth and substrate consumption rate may be expressed as

$$\frac{dXV}{dt} = \mu XV \tag{68a}$$

$$\frac{dSV}{dt} = FS_F - vXV \tag{68b}$$

Eq.(68a) may be re-expressed as

$$\frac{dx}{\mu x} = dt \tag{68a'}$$

where x≡XV. The integration of both sides of this equation gives the following equation:

$$\tau = \int_0^\tau dt = \int_{x(0)}^{x(\tau)} \frac{dx}{\mu x} \tag{69}$$

In general, the problem of maximizing the productivity is equivalent to minimize the time τ to attain the specified cell concentration.

Consider the general case with two variables such as x_1 and x_2, where the initial state I and final state F are specified (Fig. **14**). Consider then the two routes Γ_1 and Γ_2, where both start from I and reach to F as shown in Fig. **14**. Let Γ be the closed loop moving counter-clockwise, and let Σ be the area surrounded by Γ. Then the following equation holds for any function of U and V, where those are the functions of x_1 and x_2 from the Green's theorem [34,35]:

$$\oint_\Gamma \{U(x_1,x_2)dx_1 + V(x_1,x_2)dx_2\} = \iint_\Sigma \left(\frac{\partial V}{\partial x_1} - \frac{\partial U}{\partial x_2}\right)dx_1 dx_2 \tag{70}$$

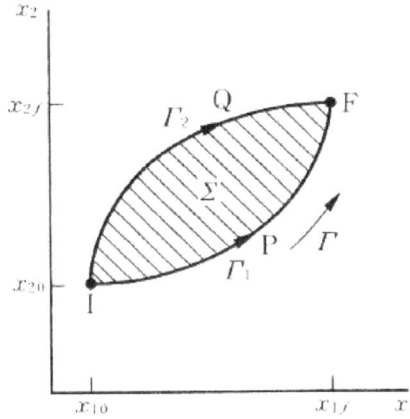

Figure 14: The two routes in x_1-x_2 space for the optimal operation by Green's theorem.

Now consider the original problem again as mentioned before, and let $U \equiv 1/\mu x$ and $V=0$, then the above equation becomes as

$$\int_{\Gamma_1} \frac{1}{\mu x}dx - \int_{\Gamma_2} \frac{1}{\mu x}dx = \iint_\Sigma -\frac{\partial}{\partial S}\left(\frac{1}{\mu x}\right)dxdS \tag{71}$$

If τ_1 and τ_2 are the time required to move from I to F *via* Γ_1 (IPF) and Γ_2 (IQF), respectively, then the following equation holds from Eq. (69):

$$\tau_1 - \tau_2 = \iint_\Sigma \frac{1}{x\mu^2}\left(\frac{\partial \mu}{\partial S}\right)dxdS \tag{72}$$

where μ is assumed to be a function of S. In the above equation, $\tau_1 > \tau_2$ holds if $\partial\mu/\partial S > 0$, while $\tau_1 < \tau_2$ if $\partial\mu/\partial S < 0$, since the term $(1/x\mu^2)$ is always positive. In the case where $\mu(S)$ is expressed as Monod type model such as $\mu(S) = \mu_m/(K_s + S)$, $\tau_2 < \tau_1$ since $\partial\mu/\partial S = \mu_m/(K_s + S)^2 > 0$. Namely, the minimum time is attained by taking the upper route between the two routes, and thus the optimal operation is attained by IAF as shown in Fig. **15**, where IA implies the supply of the substrate at time 0, followed by the batch cultivation from A to F.

In the case where μ is expressed as the substrate inhibition model such that $\mu(S) = \mu_m S/(K_s + S + S^2/K_i)$, find the value of S that maximize μ, where it is shown as S_m. The optimal operation becomes as IMNF as shown in Fig. **15**. This route indicates such operation that the initial supply of substrate to S_m at time 0, followed by the fed-batch culture from M to N, and then batch culture from N to F.

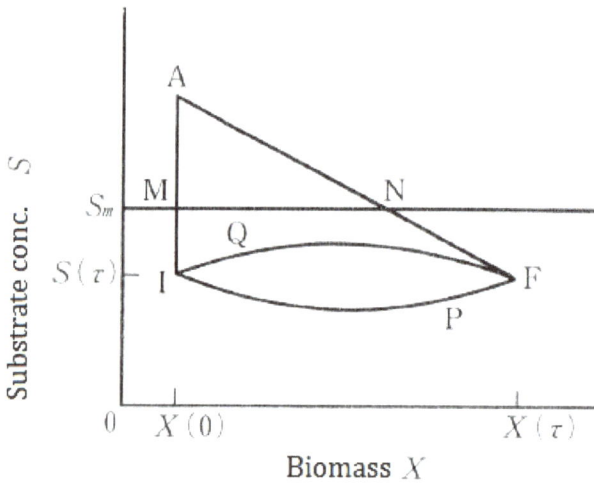

Figure 15: Optimal operation by Green's theorem.

The above analysis is based on the unstructure model. Since the model contains uncertainty, the optimal operation obtained above may be backed up by on-line optimization control *etc*. It may be, however, useful to get insight into the optimal operation.

CONCLUDING REMARKS

In the present Chapter, the model identification, sensitivity analysis, and some typical optimization procedures are explained. In relation to sensitivity analysis, metabolic control analysis (MCA) is explained with some application to lysine fermentation. Although it might be useful in detecting the limiting pathway for the specific metabolite production, the result highly relies on the accuracy of the model.

Liner programming problem may be considered for the FBA and its genome-scale model, where a vector valued objective function must be incorporated in general, and may be difficult to find the optimal point on the Pareto optimal non-inferior set. In the case of finding the optimal operation, the application may be limited in practice to unstructure model, but it may be useful to get insight into the optimal time trajectories.

DISCLOSURE

Part of this chapter has been published in [36].

REFERENCE

[1] Alvarez-Vasquez F, Sims KJ, Hannun YA, Voit EO. Integration of kinetic information on yeast sphingolipid metabolism in dynamical pathway models. J. Theor Biol 2004; 226: 265-291.

[2] Costa RS, Machado D, Rocha I, Ferreira EC. Critical perspective on the consequences of the limited availability of kinetic data in metabolic dynamic modeling. IET Syst Biol 2011; 5: 157-163.

[3] Minton AP. How can biochemical reactions within cells differ from those in test tubes ? J Cell Sci 2006; 119: 2863-2869.

[4] Almquist J, Cvijovic M, Hatzimanikatis V, Nielsen J, Jirstrand M. Kinetic models in industrial biotechnology-Improving cell factory performance. Metab Eng 2014; 24:38-60

[5] Raue A, Kreutz C, Maiwald T, Bachmann J, Schilling M, Klingmuller U, Timmer J. Structural and practical identifiability analysis of partially observed dynamical models by exploiting the profile likelihood. Bioinformatics 2009; 25: 1923-1929.

[6] Ljumg L. System identification: Theory for the user, Prentice-Hall, Englewood Cliffs, NJ, USA 1987.

[7] Schmidt H, Madsen MF, Dano S, Cedersund G. Complexity reduction of biochemical rate expressions. Bioinformatics 2008; 24: 848-854.

[8] Sunnåker M, Schmidt H, Jirstrand M, Cedersund G. Zooming of states and parameters using a lumping approach including back-translation. BMC Syst Biol 2010; 4: 28.

[9] Bellman R. On structural identifiability. Math Biosci 1970; 7: 329-339.

[10] Shinto H, Tashiro Y, Yamashita M, Kobayashi G, Sekiguchi T, Hanai T *et al.* Kinetic modeling and sensitivity analysis of acetone-butanol-ethanol production. J Biotechnol 2007; 131: 45-56.

[11] Oshiro M, Shinto H, Tashiro Y, Miwa N, Sekiguchi T, Okamoto M *et al.* Kinetic modeling and sensitivity analysis of xylose metabolism in Lactococcus lactis IO-1. J Biosci Bioeng 2009; 108 (5): 376-384.

[12] Stephanopoulos G, Aristidou AA, Nielsen J. Metabolic engineering-Principles and methodorogies-(Chapter 11), Academic press, USA 1998.

[13] Heinrich R, Rapoport TA. A linear steady state treatment of enzymatic chains. General properties, control and effecter strength. Eur J Biochem 1974; 42: 89-95.

[14] Kacser H, Burns J. The control of flux. In: Davies DD, Ed. Rate control of biological processes. Cambridge, Cambridge Univ Press 1973; pp. 65-104.

[15] Fell D, Understanding the control of metabolism, Portland press, London 1997.

[16] Heinrich R, Schuster S, Eds. The regulation of cellular systems. New York: Chapman and Hall 1996.

[17] Liao JC, Delgado J. Dynamic metabolic control theory. A methodology for investigating metabolic regulation using transient metabolic data. Ann N Y Acad Sci 1992; 665: 27-38

[18] Yang C, Hua Q, Shimizu K. Development of a kinetic model for L-lysine biosynthesis in *Corynebacterium glutamicum* and its application to metabolic control analysis. J Biosci Bioeng 1999; 88: 393-403.

[19] Hua Q, Yang C, Shimizu K. Metabolic control analysis for lysine synthesis using *Corynebacterium glutamicum* and experimental verification. J Biosci Bioeng 2000; 90: 184-192.

[20] Hadley G, Linear programming, Addison-Wesley Publ. Co., London 1962

[21]   Schuetz R, Kuepfer, Sauer U. Systematic evaluation of objective functions for predicting intracellular fluxes in *Escherichia coli*. Mol Syst Biol 2007; 3:119.

[22]   Schuetz R, Zamboni N, Zampieri M, Heinemann M, Sauer U. Multidimensional optimality of microbial metabolism. Science 2012; 336:601-604.

[23]   Cohon JL. Multiobjective programming and planning, Acad Press, New York 1978.

[24]   Ashyraliyev M, Fomekong-Nanfack Y, Kaandorp JA, Blom JG. Systems biology: parameter estimation for biochemical models,FEBS J 2009; 276: 886–902

[25]   Hooke R, Jeeves TA. Direct search solution of numerical and statistical problems. J Assoc Comput March 1961; 8, 212-229.

[26]   Powell MJD. Direct search algorithm for optimization calculations. Acta Numerica 1998; **7**, 287-336

[27]   Kolda TG, Lewis RM, Torczon V. Optimization by direct search: new perspectives on some classical and modern methods. SIAM Rev 2003; **45**, 385-482.

[28]   Nelder JA, Mead R. A simplex method for function minimization. Comput J 1965; 7, 308-313.

[29]   Lagarias JC, Reeds JA, Wright MH, Wright PE. Convergence properties of the Nelder-Mead simplex method in low dimensions. SIAM J Optim 1998; 9, 112-147.

[30]   Box, MJ. A New Method of Constrained Optimization and a Comparison With Other Methods. The Computer J 1965; 8 (1): 42-52

[31]   Goldberg DE. Genetic algorithm in search, optimization, and machine learning, Addison-Wesley 1989.

[32]   Bolchanski BG. The mathematical theory of optimal processes, Interscience Publ. 1962

[33]   Fan LT. The maximum principle and its application, 1969

[34]   Hildebrand FB. Advanced calculus for applications, 2nd edition, Prentice-Hall, New Jersey 1976

[35]   Wylie CR Advanced engineering mathematics-4th edition, McGraw-Hill, Tokyo 1975

[36]   Shimizu K, Bioprocess systems analysis method, Corona pub. Co., Japan, 1997 (in Japanese).

CHAPTER 6

Steady-State and Dynamic Characteristics

Abstract: Once the appropriate model was developed, the steady-state and dynamic characteristics such as multiple steady states and stability *etc.* can be analyzed. Multiple steady states are analyzed from the point of view of Catastrophy. The dynamic characteristics around the steady state are analyzed based on the perturbation of the state and the stability analysis with phase plane analysis. The bifurcation analysis is briefly explained for the limit cycle bifurcating from the steady state as Hopf-bifurcation. The dynamics of continuous stirred tank fermenter (CSTF) for the different expression of the cell growth rate is explained. The dynamics of the Lotka-Volterra population model is explained. The chaotic behavior is also explained for the simple logistic model.

Keywords: Dynamics, multiple steady states, Catastrrophy theory, phase plane, limit cycle, stability, Hopf-bifurcation, chaos.

INTRODUCTION

Once the appropriate model was established, one can analyse the steady state and dynamic characteristics based on the model developed. It is important to understand such characteristics for analysing the bioprocess systems from the points of view of operation, optimization, and control. It is useful for analyzing the phenomena. The structure of the steady states and stability of the states are closely related, and the stability and the oscillating and/ or chaotic behaviours are also closely related.

Here, we consider the catastrophy theory and its application to understand the structure of the steady states. Then the local dynamic characteristics such as stability and oscillating behaviour are explained by perturbing the nonlinear system. The application may be considered for the continuous stirred tank fermentor and the Lotka-Volterra population dynamic model.

MULTIPLE STEADY STATES AND CATASTROPHY

Basis for Catastrophy theory

The **Catastrophy theory** has been proposed by René Tom in 1972 in his publication entitled "Stabilité Structurelle et Morphogenèse" paying particular attention to the abrupt or catastrophic change of the state by differential geometry of modern mathematics [1-3]. This has been recognized to be useful for the modelling and the analysis of morphology, evolutional behaviour, animal

Kazuyuki Shimizu and Yu Matsuoka

behaviour as well as psychology and social science, where those modelling has been believed to be difficult.

Let F be the potential energy or potential, and is expressed as

$$F = \frac{x^4}{4} + u\frac{x^2}{2} + vx \tag{1}$$

where x is the state variable, and u and v are the parameters. Let u=-3 for convenience, and consider how F changes with respect to v as shown in Fig. **1**, where the local minimum represents the feasible state in practice. As shown in Fig. **1**, the catastrophic change occurs at v=-2, where one of the local minimum disappears, and the state abruptly changes.

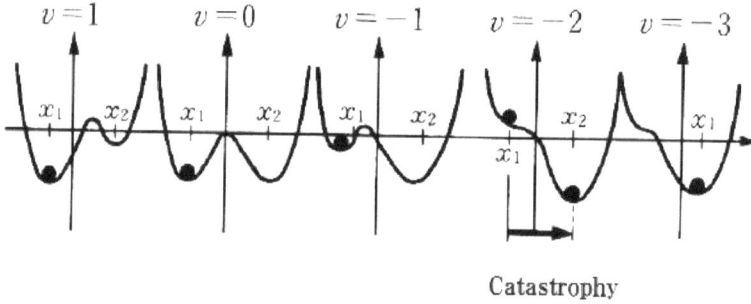

Figure 1: Change of potential.

Let the set {u,v,x} which minimizes F such that $\partial F / \partial x = x^3 + ux + v = 0$ be called as equilibrium space. Namely,

$$M_F = \{(u,v,x)|x^3 + ux + v = 0\} \tag{2}$$

where M is called as manifold.

If we plot the point (u,v,x) of M_F onto the 3-dimensional space R^3, we may obtain the plane as shown in Fig. **2**. This plane satisfies $v = -x^3 - ux$, and thus this may be expressed as v(x,u). Consider then mapping u-x plane onto M_F as shown in Fig. **3** such that

$$\phi : R^2 \to M_F \qquad (u,x) \to (u, -x^3 - ux, x)$$

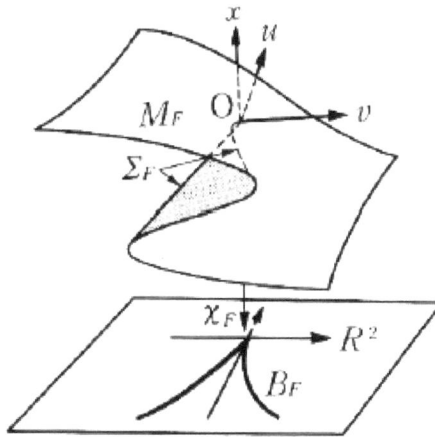

Figure 2: Equilibrium space.

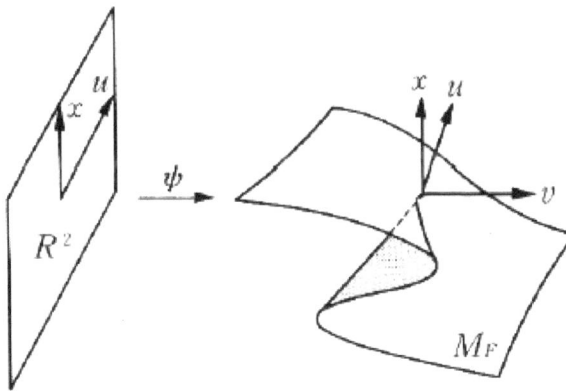

Figure 3: Mapping of u-x plane into M_F.

As can be seen from the figure, R^2 and M_F show one-to-one mapping. This is called as differmorphic in differential geometry, and ϕ is called as the diffeomorphism from R^2 to M_F. Namely, R^2 and M_F are the same in the sense of differential topology point of view.

Let G_F be the set of stable equilibrium points such that

$$G_F = \{(u,v,x) \mid \frac{\partial F}{\partial x} = x^3 + ux + v = 0, \qquad \frac{\partial^2 F}{\partial x^2} = 3x^2 + u > 0\} \tag{3}$$

Namely, the state is stable when the 2nd derivative of F with respect to x is positive. Let Σ_F be the boundary between the points which satisfy $\partial^2 F / \partial x^2 > 0$, and those that do not satisfy, then

$$\Sigma_F = \{(u,v,x) \mid x^3 + ux + v = 0, \qquad 3x^2 + u = 0\} \qquad (4)$$

As seen from Fig. **2**, this set is the line of folding M_F.

Next, consider the point which satisfies $\partial^3 F / \partial x^3 = 0$. Namely,

$$\frac{\partial F}{\partial x} = \frac{\partial^2 F}{\partial x^2} = \frac{\partial^3 F}{\partial x^3} = 0 \qquad (5)$$

The point which satisfies these equations corresponds to O in Fig. **2**, where u=0, v=0, x=0, and the point is called as **cusp point**.

Let χ be the projection of R^3 onto u-v plane such that

$$\chi : R^3 \rightarrow R^2, \qquad (u,v,x) \rightarrow (u,v)$$

Moreover, let χ_F be the projection of M_F onto u-v plane such that $\chi_F : M_F \rightarrow R^2$, and this is called as **cusp catastrophe mapping**.

The point on M_F where Jacobian is 0 is called as **singular point** for $\chi_F : M_F \rightarrow R^2$, while the other point is called as **non-singular point**.

Consider the above example again to understand the above nature. First, consider the projection of any point (u,v,x) on M_F onto (u,v) in R^2. As mentioned before, the original coordinate on M_F is expressed as (u, x^3 ux, x). Since $\chi_F(u, -x^3 - ux, x) = (u, -x^3 - ux)$, the necessary and sufficient condition for (u,x) to be the singular point of χ_F is that the Jacobian of this mapping becomes 0 such that

$$\begin{vmatrix} \dfrac{\partial u}{\partial u} & \dfrac{\partial u}{\partial x} \\ \dfrac{\partial(-x^3 - ux)}{\partial u} & \dfrac{\partial(-x^3 - ux)}{\partial x} \end{vmatrix} = \begin{vmatrix} 1 & 0 \\ -x & -3x^2 - u \end{vmatrix} = -3x^2 - u = 0 \qquad (6)$$

where the set of singular points of χ_F is the folded line Σ_F.

Let the set of the singular points of χ_F be $B_F \subset R^2$, where it is called as **bifurcation set**. In the present example, this is expressed as

$$B_F = \{(u,v) \mid \frac{\partial F}{\partial x} = x^3 + ux + v = 0, \qquad \frac{\partial^2 F}{\partial x^2} = 3x^2 + u = 0\} \tag{7}$$

From $\partial F / \partial x = 0$ and $\partial^2 F / \partial x^2 = 0$, $v = -x(u + x^2)$ and $u = -3x^2$ are obtained. If we eliminate x from these equations, we have $4u^3 + 27v^2 = 0$. Namely,

$$B_F = \{(u,v) \mid 4u^3 + 27v^2 = 0\} \qquad \subset R^2 \tag{8}$$

In the similar way, the 2nd order bifurcation set $B_F{}^2$ is obtained by imposing another constraint such as $\partial^3 F / \partial x^3 = 0$ as

$$B_F{}^2 = \{(0,0)\} \qquad \subset B_F \subset R^2 \tag{9}$$

Consider the shape of F, where the inside of B_F or $4u^3 + 27v^2 < 0$ is satisfied. F has two local minima, and one local maximum which is called **unstable equilibrium point**, while outside of B_F is stable equilibrium space. The above example is for the cusp Catastrophy. There are other types of Catastrophy, and those are summarized in Table **1**, where the corresponding figure is also given in Fig. **4**.

Table 1: Elementary catastrophy

1.	$\dfrac{x^3}{3} + ux$	fold
2.	$\pm\dfrac{x^4}{4} + u\dfrac{x^2}{2} + vx$	cusp
3.	$\dfrac{x^5}{5} + u\dfrac{x^3}{3} + v\dfrac{x^2}{2} + wx$	swallow tail
4.	$\pm\dfrac{x^6}{5} + u\dfrac{x^4}{4} + v\dfrac{x^3}{3} + w\dfrac{x^2}{2} + tx$	butterfly
5.	$x^3 + y^3 + wxy - ux - vy$	hyperbolic umbilic
6.	$\dfrac{1}{3}x^3 - xy^2 + w(x^2 + y^2) - ux - vy$	elliptic umbilic
7.	$x^2 y \pm y^4 + ux^2 + vy^2 + wx + ty$	parabolic umbilic

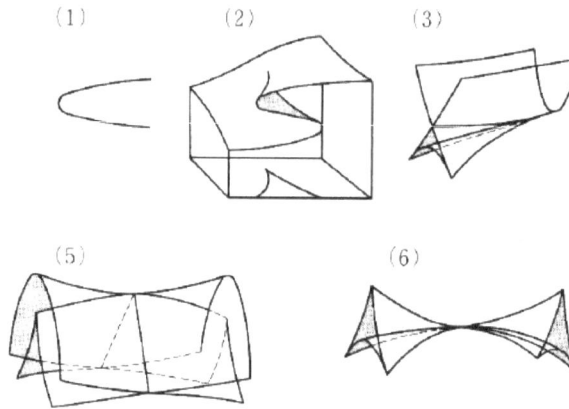

Figure 4: Elementary Chatastrophy.

Application of Catastrophy Theory to the Phase Transition in Thermodynamics

Consider how the Catastrophy theory can be applied, and how the Catastrophy is observed in the natural phenomena. For this, consider the phase transition by the phase plane of volume-pressure curves at constant temperature as shown in Fig. **5** [4]. Consider increasing the pressure by keeping the temperature constant at T_3, where the state at A abruptly changes from gas phase to liquid phase by reducing the volume to the state B. This is the Catastrophy in phase transition. Fig. **5** also illustrates how the P-V curve changes when the temperature was increased, and the points A and B connect each other at temperature T_c, where it becomes an inflexion point in the P-V curve. This is the critical point (T_c, P_c, V_c), and no more drastic phase change occurs at higher temperature than T_c.

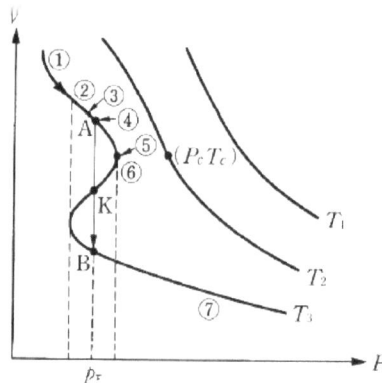

Figure 5: Volume-pressure curve for the fluid.

Consider analyzing this phenomenon by Catastrophy theory as mentioned in the previous section. Let U be the **internal energy**, and let S be the **entropy**, then the **free energy** can be expressed from thermodynamics as

$$E = U - TS \tag{10}$$

Then the **Gibb's function** can be defined as

$$G = E + PV \tag{11}$$

From the thermodynamics, it has been known that the **Gibb's free energy minimum principle** holds, where given P and T, the state of the system (P,T,V) is determined, and thus V is determined so that G(P,T,V) is minimized.

Consider the following **van der Waals equation**:

$$(P + \frac{a}{V^2})(V - b) = RT \tag{12}$$

where a and b are the constants depending on the type of gas, and R is the gas constant.

Let M_G be the stable equilibrium state (P,V,T) of the gas of concern, then M_G is expressed as

$$M_G = \{(P,V,T) \in R^3 \Big| \frac{\partial G(P,V,T)}{\partial V} = 0\} \tag{13}$$

Namely, Eq.(12) is determined by $\partial G / \partial V = 0$. Then by taking the derivative of Eq.(12) with respect to V, we have

$$\frac{\partial^2 G}{\partial V^2} = 0 \quad \Rightarrow \quad \frac{V^3 P - aV + 2ab}{V^3} = 0 \tag{14}$$

Further derivative with respect to V gives

$$\frac{\partial^3 G}{\partial V^3} = 0 \quad \Rightarrow \quad 3V^2 P - a = 0 \tag{15}$$

from which we have $P = a / 3V^2$. By substituting this into Eq.(14), we have $V = 3b$, which in turn gives that $P = a / 27b^2$. By substituting this into Eq.(12), we have

$$T = \frac{8a}{27bR} \tag{16}$$

from which the cusp point or the critical point is obtained as $(P_c, V_c, T_c) = (a/27b^2, 8a/27bR, 3b)$. Therefore, the Gibb's function is the potential with cusp point at (P_c, V_c, T_c).

The appropriate coordinate with (P_c, V_c, T_c) to be the origin makes the state equation as expressed by Eq.(12) become as

$$x^3 + ux^2 + vx = 0$$

Let us change the coordinate. For this, consider the following change of variables:

$$\overline{P} = \frac{27b^2 P}{a}, \qquad \overline{T} = \frac{27bRT}{8a}, \qquad \overline{V} = \frac{V}{3b} \tag{17}$$

Then Eq.(12) becomes as

$$(P + \frac{a}{V^2})(V - b) = \left(\frac{a\overline{P}}{27b^2} + \frac{a}{9b^2 V^2} \right)(3b\overline{V} - b) = RT$$

$$= \frac{a}{9b}\left(\overline{P} + \frac{3}{\overline{V}^2} \right)\left(\overline{V} - \frac{1}{3} \right) = \frac{8aR\overline{T}}{27bR}$$

$$\therefore \quad \left(\overline{P} + \frac{3}{\overline{V}^2} \right)\left(\overline{V} - \frac{1}{3} \right) = \frac{8\overline{T}}{3} \tag{18}$$

If the density as expressed by $\overline{\rho}(= 1/\overline{V})$ is used instead of $\overline{V}$, Eq.(18) becomes as

$$\left(\overline{P} + 3\overline{\rho}^2 \right)\left(\frac{1}{\overline{\rho}} - \frac{1}{3} \right) = \frac{8\overline{T}}{3} \tag{19}$$

where the change of variables such that $p = \overline{P} - 1$, $x = \overline{\rho} - 1$, $t = \overline{T} - 1$ gives the origin of $(\overline{P}, \overline{\rho}, \overline{T})$ to $(0,0,0)$, and the above equation can be expressed as

$$x^3 + \frac{(8t + p)x}{3} + \frac{(8t - 2p)}{3} = 0 \tag{20}$$

Consider further the change of variables such that

$$u = \frac{8t + p}{8}, \qquad v = \frac{8t - 2p}{3} \tag{21}$$

Then Eq.(20) can be expressed as the standard form as given in Table 1 as

$$x^2 + ux + v = 0$$

Namely, the Gibb's function is expressed as

$$G(u, v, x) = \frac{x^4}{4} + u\frac{x^2}{2} + vx + C \ \ (\text{constant}) \tag{22}$$

which corresponds to the cusp catastrophy as given in Table **1**. M_G in this case can be expressed as

$$M_G = \{v, v, x) \,|\, x^3 + ux + v = 0\}$$

where

$$u = \frac{1}{3}\left(\frac{27bRT}{a} + \frac{27b^2 p}{a} - 9\right)\left(= \frac{8t - p}{3}\right)$$

$$v = \frac{1}{3}\left(\frac{27bRT}{a} + \frac{54b^2 p}{a} - 10\right)\left(= \frac{8t - 2p}{3}\right)$$

$$x = \frac{3b}{V} - 1$$

Consider again the case of increasing the pressure P in the P-V plane while keeping the temperature at T_3. Fig. **6** shows how the potential G changes when the state changes from ① to ⑦ along the line corresponding to the temperature at T_3. At point ③, new local minimum appears, but it is not the global minimum, and thus the state does not change, while at point ④, the local minimal point becomes minimum, and thus the state abruptly changes from ④ to ⑤ based on the Gibb's free energy minimum principle [4]. This rule is called as **Maxwell convention**.

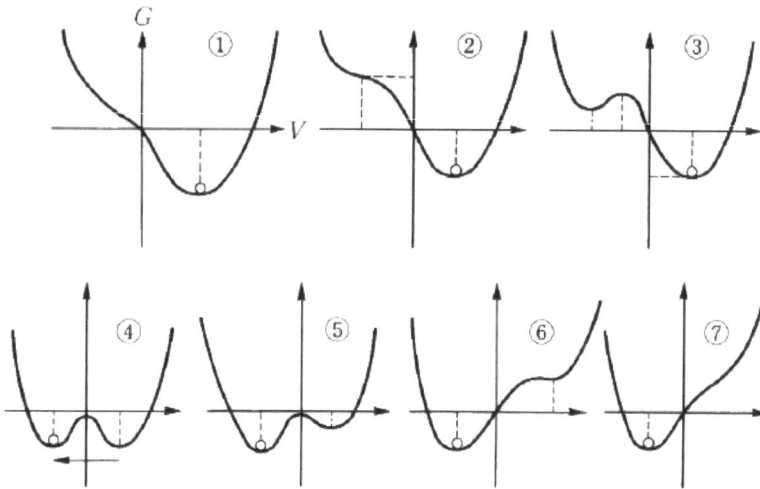

Figure 6: Change of the potential.

DYNAMIC ANALYSIS AND PHASE PLANE ANALYSIS FOR THE PERTURBED LINEAR SYSTEM

Basic Equations for the Perturbed Local Analysis

Consider the n-th order non-linear system such as

$$\frac{dx}{dt} = f(x) \qquad x \in R^n, \qquad f \in R^n \tag{23}$$

where x is the n-dimensional state vector. Consider the steady-state that satisfies

$$f(x) = 0 \tag{24}$$

where $\bar{x}$ is also called as **singular point**, or **equilibrium point**. Consider then the perturbation of **f** around the steady state, where this can be made by Taylor series expansion of **f** around $\bar{x}$ such that

$$f(x) = f(\bar{x} + \Delta x) = f(\bar{x}) + \frac{\partial f}{\partial x}\bigg|_{\bar{x}} \Delta x + \bullet\bullet\bullet \tag{25}$$

Substituting $\mathbf{x} = \bar{x} + \Delta \mathbf{x}$ into Eq.(23), and approximating the RHS of Eq.(25) by the liner term for $\Delta \mathbf{x}$, the following equation is obtained:

$$\frac{d(\Delta x)}{dt} = A\Delta x \tag{26}$$

where A is defined as

$$A \equiv \left.\frac{\partial f}{\partial x}\right|_{\bar{x}} \tag{27}$$

and is called as **Jacobi matrix** or **Jacobian**, and the (i,j)th component of A is given as

$$a_{ij} \equiv \left.\frac{\partial f_i}{\partial x_j}\right|_{\bar{x}} \tag{28}$$

Consider then the following **characteristic equation**:

$$\left| A - \lambda I \right| = 0 \tag{29}$$

where λ is called **eigen value**, and I is the unit matrix.

Simple Example

Consider the following simple example to understand the dynamic characteristics of the system:

$$\frac{dx_1}{dt} = f_1 \equiv 2x_1^2 + 2x_2$$

$$\frac{dx_2}{dt} = f_2 \equiv 5x_1x_2 + 5$$

At steady state

$$2(\bar{x}_1^2 + \bar{x}_2) = 0$$

$$5(\bar{x}_1\bar{x}_2 + 1) = 0$$

Solving these equations, we have

$$\bar{x} = (\bar{x}_1, \bar{x}_2) = (1, -1)$$

Moreover, a_{ij} can be obtained as

$$a_{11} \equiv \left.\frac{\partial f_1}{\partial x_1}\right|_{\bar{x}} = 4\bar{x}_1 = 4 \qquad\qquad a_{12} \equiv \left.\frac{\partial f_1}{\partial x_2}\right|_{\bar{x}} = 2$$

$$a_{21} \equiv \left.\frac{\partial f_2}{\partial x_1}\right|_{\bar{x}} = 5\bar{x}_2 = -5 \qquad\qquad a_{22} \equiv \left.\frac{\partial f_2}{\partial x_2}\right|_{\bar{x}} = 5\bar{x}_1 = 5$$

Therefore, A-matrix is obtained as

$$A = \begin{bmatrix} 4 & 2 \\ -5 & 5 \end{bmatrix}$$

The characteristic equation becomes as

$$|A - \lambda I| = \left\| \begin{bmatrix} 4 & 2 \\ -5 & 5 \end{bmatrix} - \lambda \begin{bmatrix} 1 & 0 \\ 0 & 1 \end{bmatrix} \right\| = \begin{vmatrix} 4-\lambda & 2 \\ -5 & 5-\lambda \end{vmatrix}$$

$$= (4-\lambda)(5-\lambda) - (-5)2 = \lambda^2 - 9\lambda + 30 = 0$$

The eigen values are obtained by solving this equation as

$$\lambda = \frac{9 \pm \sqrt{81-120}}{2} = \frac{9 \pm \sqrt{39}i}{2}$$

Now the dynamic patterns may be classified based on the eigen values as shown in Table **2**, where the dynamic behaviour around the steady state is illustrated as Fig. **7**.

Hopf Bifurcation

Consider the non-linear equation as expressed as Eq.(23). Paying particular attention to the model parameter, and consider how the dynamics change when parameter changes. Let us rewrite Eq.(23) by explicitly showing parameter p such that

$$\frac{dx}{dt} = f(x, p) \tag{30}$$

where **f** is assumed to be smooth enough, and thus continuously differentiable. Let $\bar{x}$ be the steady state solution to Eq.(30) such that

$$0 = f(\bar{x}, p) \tag{31}$$

Table 2: Classification of the dynamics

	Eigen value	Stability	Type of singular point
(1)	$\lambda_1 \neq \lambda_2$ で $\lambda_1 < 0,\ \lambda_2 < 0$	A. S.	S. N.
(2)	$\lambda_1 = \lambda_2 < 0$	A. S.	S. N.
(3)	$\lambda_1 \neq \lambda_2$ で $\lambda_1 > 0,\ \lambda_2 > 0$	U.	U. N.
(4)	$\lambda_1 = \lambda_2 > 0$	U.	U. N.
(5)	$\lambda_1 < 0 < \lambda_2$	U.	S. P.
(6)	$\lambda_1 = a + bi,\ \lambda_2 = a - bi,\ a < 0$	S.	S. F.
(7)	$\lambda_1 = a + bi,\ \lambda_2 = a - bi,\ a > 0$	U.	U. F.
(8)	$\lambda_1 = bi,\ \lambda_2 = -bi$	S.	C.
(9)	$\lambda_1 = 0,\ \lambda_2 < 0$	S.	—
(10)	$\lambda_1 = 0,\ \lambda_2 > 0$	U.	—

A. S. : asymptotically stable   , U. : unstable

S. : stable   S. N. : stable node

U. N. : unstable node   S. P. : saddle point

S. F. : stable focus   U. F. : unstable focus

     C. : center

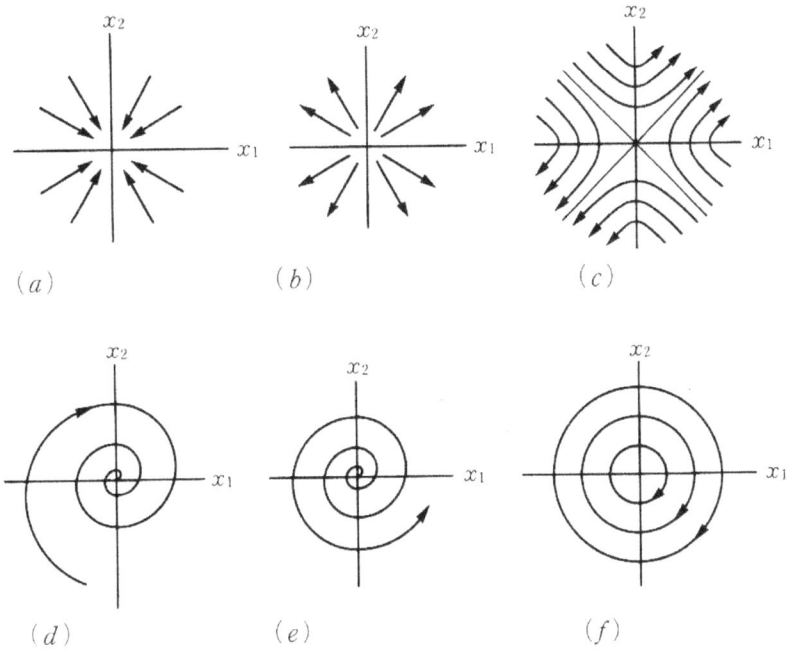

(a)   (b)   (c)

(d)   (e)   (f)

Figure 7: Dynamics around the equilibrium point

The perturbed linear equation at the equilibrium point is expressed as

$$\frac{d(\Delta x)}{dt} = \frac{\partial f}{\partial x}\Big|_{\bar{x}} \Delta x \tag{32}$$

Suppose that $A(\mathbf{p}_0)$ has the eigenvalues of pure imaginary pair $\pm i\omega_0$ at $\mathbf{p} = \mathbf{p}_0$, and let $\mathbf{v}_1$ and $\mathbf{v}_1^*$ (conjugate of $\mathbf{v}_1$) be the corresponding eigenvectors. Then Eq.(32) has the solution such that $\Delta x = v_1 e^{i\omega_0 t}$ and $\Delta x^* = v_1^* e^{-i\omega_0 t}$. Assume that other eigenvalues do not have pure imaginary pair. Then the periodic solution bifurcates from $x = x_0$ around $(\mathbf{x}, \mathbf{p}) = (0, \mathbf{p}_0)$. Namely, by considering the small positive parameter ε, we have

$$\eta(t) = \eta(t; \varepsilon) = \bar{x} + \varepsilon(v e^{i\omega_0 t} + v^* e^{-i\omega_0 t}) + o(\varepsilon) \tag{33}$$

where $o(\varepsilon)$ denotes the higher order terms than ε, and $o(\varepsilon) \to 0$ as $\varepsilon \to 0$. The period of $\eta(t; \varepsilon)$ is $2\pi\omega(\varepsilon)$, and among the eigenvalues of $A(\mathbf{p}_0)$, if all the real part of eigenvalues other that $\pm i\omega_0$ is negative, and also $\text{Re}[d\lambda(p_0)/dp] \bullet p_2 > 0$ is satisfied, then the bifurcated periodic solution is asymptotically stable. On the other hand, if $A(\mathbf{p}_0)$ has some positive real part of the eigenvalue, or $\text{Re}[d\lambda(p_0)/dp] \bullet p_2 < 0$ is satisfied, then the bifurcated periodic solution becomes unstable. The bifurcation of such periodic solution is called as **Hopf bifurcation**.

Dynamics of the Continuous Culture System

Consider the dynamics of the continuous stirred tank fermentor (CSTF) as mentioned in Chapter 3 for the modelling. The unstructure model may be expressed

$$\frac{dX}{dt} = f_1 \equiv (\mu(S) - D)X \tag{34a}$$

$$\frac{dS}{dt} = f_2 \equiv D(S_F - S) - \nu(S)X \tag{34b}$$

where X is the cell concentration, S and S_F are the substrate concentrations of the fermentor and the feed, respectively, D is the dilution rate, μ is the specific growth rate, and ν is the specific substrate consumption rate.

The steady state can be obtained by

$$0 = (\mu(\overline{S}) - D)\overline{X} \tag{35a}$$

$$0 = D(S_F - \overline{S}) - v(\overline{S})\overline{X} \tag{35b}$$

There are two solutions which satisfy these equations such that

a) Trivial solution

$$\overline{X} = 0 \text{ and } \overline{S} = S_F \tag{36}$$

where this corresponds to the washout.

b) Non-trivial solution

$$\mu(\overline{S}) = D \tag{37a}$$

$$D(S_F - \overline{S}) = v(\overline{S})\overline{X} \tag{37b}$$

If v is expressed as μ (S)/Y, Eq.(37b) can be expressed by using the relationship of Eq.(37a) as

$$Y(S_F - \overline{S}) = \overline{X} \tag{37b'}$$

Consider first the simple case when μ (S) is expressed as μ (S)=kS to understand the dynamics of CSTF. Then from Eq.(37a) and (37b)', we have

$$\overline{S} = D/k \text{ and } \overline{X} = Y(S_F - D/k)$$

where X>0 and thus $D<kS_F$ must be satisfied.

Let A be the Jacobian matrix. Then

$$a_{11} \equiv \left.\frac{\partial f_1}{\partial X}\right| = \mu(\overline{S}) - D = k\overline{S} - D$$

$$a_{12} \equiv \left.\frac{\partial f_1}{\partial S}\right| = \frac{\partial \mu(\overline{S})}{\partial S}\overline{X} - k\overline{X}$$

$$a_{21} \equiv \frac{\partial f_2}{\partial X}\bigg| = -\frac{\mu(\overline{S})}{Y} = -\frac{k\overline{S}}{Y}$$

$$a_{22} \equiv \frac{\partial f_2}{\partial S}\bigg| = -D - \frac{\overline{X}}{Y}\frac{\partial \mu(\overline{S})}{\partial S} = -D - \frac{k\overline{X}}{Y}$$

At the trivial solution, A-matrix becomes as

$$A = \begin{bmatrix} kS_F - D & 0 \\ -\dfrac{kS_F}{Y} & -D \end{bmatrix}$$

and the characteristic equation becomes as

$$|A - \lambda I| = \begin{vmatrix} kS_F - D - \lambda & 0 \\ -\dfrac{kS_F}{Y} & -D - \lambda \end{vmatrix} = (kS_F - D - \lambda)(-D - \lambda) = 0$$

from which we have

$$\lambda_1 = -D < 0, \qquad \lambda_2 = kS_F - D$$

Thus

$$\lambda_2 > 0 \text{ when } D < kS_F$$

$$\lambda_2 \leq 0 \text{ when } D \geq kS_F$$

In summary, at the trivial steady state (washout),

unstable (saddle point) if $0<D<kS_F$ since $\lambda_1< 0$ and $\lambda_2> 0$

stable (node)   if $D>kS_F$ since $\lambda_1< 0$ and $\lambda_2< 0$

At the non-trivial steady state, since A-matrix becomes as

$$A = \begin{bmatrix} 0 & k\overline{X} \\ -\dfrac{kS_F}{Y} & -D - \dfrac{k\overline{X}}{Y} \end{bmatrix} = \begin{bmatrix} 0 & kY(S_F - D/k) \\ -D/Y & -kS_F \end{bmatrix}$$

the characteristic equation becomes as

$$|A - \lambda I| = \begin{vmatrix} -\lambda & kY(S_F - D/k) \\ -\dfrac{D}{Y} & -kS_F - \lambda \end{vmatrix} = \lambda^2 + kS_F\lambda + kD(S_F - D/k) = 0$$

This can be solved as

$$\lambda = \frac{-kS_F \pm (2D - kS_F)}{2} \quad \Rightarrow \quad \lambda_1 = D - kS_F, \qquad \lambda_2 = -D < 0$$

As seen before, the non-trivial steady state is meaningful when $D < kS_F$ holds, and both eigenvalues become negative, which indicates that the dynamics around the non-trivial steady state is always asymptotically stable (node).

Fig. **8a** shows how $\overline{X}$ changes with respect to D for the case of Monod model, where the solid line indicates stable state, while broken line indicates the unstable state. Fig. **8b** shows the phase plane diagram at the dilution rate D_1 as given in Fig. **8a**.

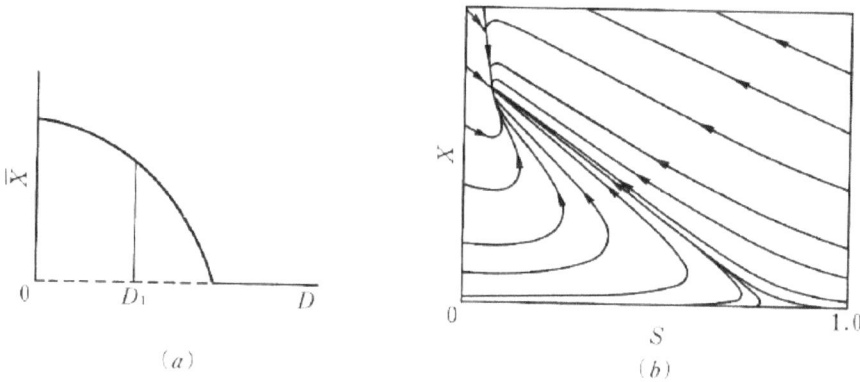

Figure 8: Dynamics of CSTF.

Dynamic of CSTF in Detail

Basic Formulation

Consider the dynamic analysis for **continuous stirred tank fermentor (CSTF)** in more detail, or in more general way. Consider again Eq.(34), and re-express with dimensionless variables such that

$$\frac{dx_1}{d\tau} = (DaM(x_2) - 1)x_1 \tag{38a}$$

$$\frac{dx_2}{d\tau} = -x_2 + \frac{DaM(x_2)x_1}{y(x_2)} \tag{38b}$$

where $x_1 \equiv X / Y(S_F)S_F$, $x_2 \equiv (S_F - S) / S_F$, $\tau \equiv Dt$, $Da \equiv \mu(S_F) / D$,
$M(S) \equiv \mu(S) / \mu(S_F)$, $y(S) \equiv Y(S) / Y(S_F)$. Here, Da is called as **Damköhler number**.

The steady-state solution to Eq.(38) becomes as

a) Trivial solution (washout): $\bar{x}_1 = \bar{x}_2 = 0$ $\tag{39}$

b) Non-trivial solution: $\bar{x}_1 = \dfrac{y(\bar{x}_2)\bar{x}_2}{DaM(\bar{x}_2)}$ $\tag{40a}$

$$M(\bar{x}_2) = \frac{1}{Da} \tag{40b}$$

where " $\bar{}$ " means the steady state.

Stability of the Steady State

Let det A and tr A be the determinant and the trace of A, respectively, where tr A is the summation of the diagonal elements such that tr A = a_{11} + a_{22}. At trivial steady state, these are expressed as

$$A = \begin{bmatrix} -1 + Da & 0 \\ Da & -1 \end{bmatrix}, \ \det A = 1 - Da, \ trA = -2 + Da$$

The condition for the local stability is expressed as follows:

$$\det A > 0 \ \text{and} \ trA < 0$$

where the proof may be found elsewhere. By applying this condition to the trivial steady state, we have

$$Da < 1 \tag{41}$$

At non-trivial steady state, we have

$$A = \begin{bmatrix} 0 & DaM'(\overline{x_2})\overline{x_1} \\ \dfrac{\overline{x_2}}{\overline{x_1}} & -1 + Da\overline{x_1}\left\{\dfrac{M(\overline{x_2})}{y(\overline{x_2})}\right\}' \end{bmatrix}$$

where " ` " means the derivative. By applying the local stability condition as mentioned above, we have

$$M'(\overline{x_2}) < 0 \tag{42a}$$

$$-1 + \left\{\dfrac{M(\overline{x_2})}{y(\overline{x_2})}\right\}' \dfrac{y(\overline{x_2})\overline{x_2}}{M(\overline{x_2})} < 0 \tag{42b}$$

Bifurcation of Limit Cycle

The necessary condition for bifurcation of the limit cycle from the steady state (Hopf bifurcation) is that the eigenvalues have pure imaginary parts as mentioned before (proof may be found elsewhere). Namely,

$$\det A > 0 \tag{43a}$$

and

$$trA = 0 \tag{43b}$$

If we apply this to the trivial solution, we have

$$\det A = 1 - Da = -1 < 0$$

and thus bifurcation will not occur at the trivial steady state.

At the non-trivial steady state, we have the following condition.

$$M'(\overline{x_2}) < 0 \tag{44a}$$

$$\left\{\dfrac{M(\overline{x_2})}{y(\overline{x_2})}\right\}' \dfrac{y(\overline{x_2})\overline{x_2}}{M(\overline{x_2})} = 1 \tag{44b}$$

If the yield coefficient is constant (where this is usually satisfied in practice), $y'(x_2) = 0$, we have $M'(x_2) = M(x_2)/x_2 > 0$ from Eq.(44b). However, this contradicts to (44a), and thus Hopf bifurcation will not occur in such a case.

The sufficient condition for the occurrence of Hopf bifurcation is given as

$$trB^* \neq 0 \tag{45a}$$

where

$$B^* \equiv \left.\frac{\partial A^\varepsilon}{\partial x}\right|_{\substack{x=x_0 \\ \varepsilon=0}} \left.\frac{dx}{dDa}\right|_{\substack{Da=Da^0 \\ \varepsilon=0}} + \left.\frac{\partial A^\varepsilon}{\partial Da}\right|_{\substack{Da=Da^0 \\ \varepsilon=0}} \tag{45b}$$

where Da^0 is that satisfies the necessary condition (43), x^ε is the non-trivial solution when Da was perturbed by ε, and A^ε is the corresponding Jacobi matrix. If the above condition is applied to CSTF, we have

$$\left\{\frac{M(\bar{x}_2)}{y(\bar{x}_2)}\right\}'' \neq 0 \tag{46}$$

Moreover, the bifurcated limit cycle becomes asymptotically stable based on the Friedlich's theory as

$$3\left\{\frac{M(\bar{x}_2^0)}{y(\bar{x}_2^0)}\right\}'' \bar{x}_2^0 < \left\{\frac{M(\bar{x}_2^0)}{y(\bar{x}_2^0)}\right\}'' \left\{1 + \frac{4M''(\bar{x}_2^0)\bar{x}_2^0}{3M'(\bar{x}_2^0)}\right\} \tag{47}$$

The Dynamic Analysis for the Substrate Inhibition Model

Consider the dynamic analysis of CSTF where the cell growth rate is expressed as the substrate inhibition model such as

$$\mu(S) = \frac{\mu_m}{\dfrac{K_s}{S} + 1 + \dfrac{S}{K_i}} \tag{48}$$

where μ_m, K_s, and K_i are constant model parameters, and this equation may be reduced to Monod model when $K_i \to \infty$. Moreover, let us assume the yield coefficient to be a linear function of S such that

$$Y(S) = a + bS \tag{49}$$

Where a and b are the constant model parameters. Above two equations can be expressed as the non-dimensional variables as mentioned before as

$$M(\overline{x_2}) = \frac{(1+\gamma+\alpha\gamma)(1-\overline{x_2})}{(1-\overline{x_2})^2 + \gamma(1-\overline{x_2}) + \alpha\gamma} \tag{50a}$$

$$y(\overline{x_2}) = \frac{1+\beta}{1+\beta-\overline{x_2}} \tag{50b}$$

$$\text{where } \alpha \equiv K_s / S_F, \quad \beta \equiv a/bS_F, \quad \gamma \equiv K_i / S_F.$$

The condition for the non-trivial solution to have multiple steady states is

$$\alpha\gamma < 1 \tag{51}$$

The stability condition is expressed as

$$\overline{x_2} > 1 - \sqrt{\alpha\gamma} \tag{52}$$

Another stability condition is tr A <0 as mentioned before. Fig. **9** shows how tr A changes when parameters α, β, and γ changed with respect to $\overline{x_2}$. Consider the curve of tr A going through point P in Fig. **9**, where the following equation is satisfied at P:

$$trA = 0, \quad (trA)' = 0 \tag{53}$$

The set of parameters which satisfies this equation is the boundary, from which the unstable trivial solution appears. Let the numerator of tr A be $g(x_2)$, then

$$g(\overline{x_2}) = -3(1-\overline{x_2})^4 + 2\varsigma(1-\overline{x_2})^3 + \eta(1-\overline{x_2})^2 - \alpha\beta\gamma \tag{54}$$

where $\varsigma \equiv 1-\beta-\gamma$, $\eta \equiv \beta+\gamma-\alpha\gamma-\beta\gamma$. Since the denominator of tr A is positive at $0 < x_2 < 1$, the condition of (50) is equivalent to

$$g(\overline{x_2}) = 0, \quad g'(\overline{x_2}) = 0 \tag{55}$$

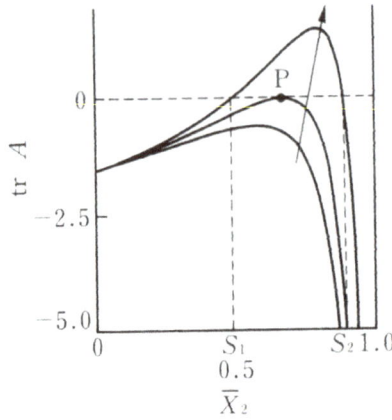

Figure 9: Change the root for tr A.

After some manipulations, the parameter set which satisfies the condition becomes as

$$F \equiv \frac{\varsigma^4}{32} + \frac{\varsigma\eta}{8} + \frac{\eta^2}{12} - \alpha\beta\gamma + \varsigma(\frac{\varsigma^2}{96} + \frac{\eta}{36})\sqrt{9\varsigma^2 + 24\eta} = 0 \qquad (56)$$

The curve which goes through point P does not satisfy the sufficient condition of (45), and thus the necessary and sufficient condition for the limit cycle to be bifurcated from the steady state is as follows:

$$F < 0 \qquad (57)$$

The condition for the limit cycle to be stable can be expressed from (47) as

$$\{1 - S_i + \theta_i(v_i\theta_i + \xi_i)\}\left[\frac{9S_i}{\theta_i} - \{(1-S_i)^2 + \gamma(1-S_i) + \alpha\gamma\} - \frac{8v_iS_i}{3\xi_i}\right] + 9\lambda_iS_i\theta_i^2 < 0 \qquad (i = 1, 2)$$

$$(58)$$

where S_i is the value of $\overline{x}_2$ which gives $trA = 0$ in Fig. **9**, and other parameters are defined as

$$\theta_i \equiv \frac{1 - S_i + \beta}{(1 - S_i)^2 + \gamma(1 - S_i) + \alpha\gamma}$$

$$v_i \equiv (1-Si)^3 - 3\alpha\gamma(1-S_i) - \alpha\gamma^2$$

$$\lambda_i \equiv (1-Si)^4 - 6\alpha\gamma(1-S_i)^2 + 4\alpha\gamma^2(1-S_i) + (\alpha\gamma)^2 - \alpha\gamma^3$$

$$\xi_i \equiv (1-Si)^2 - \alpha\gamma$$

Fig. **10** shows the steady states with respect to Da. The dynamics change depending on the values of m and n in Fig. **10**, and S_1 and S_2 in Fig. **9**, and then consider the boundary of the change in the dynamics. The boundary for $m<S_1<S_2$, $S_1<m<S_2$, and $S_1<S_2<m$ becomes as

$$g(m) = g(1-\sqrt{\alpha\gamma}) = 0$$

and the following equation is obtained from Eq.(54) as

$$2\varsigma\sqrt{\alpha\gamma} + \gamma - 4\alpha\gamma - \beta\gamma = 0 \qquad (59)$$

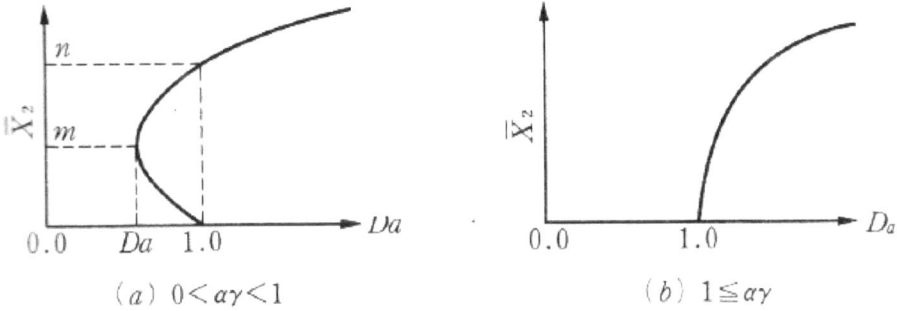

Figure 10: Multiplicity of the non-trivial steady states.

In the similar way, the following equation defines the relative magnitude of m, S_1, and S_2:

$$-3(\alpha\gamma)^3 + 2\varsigma(\alpha\gamma)^2 + \eta(\alpha\gamma) - \beta = 0 \qquad (60)$$

Fig. **11** shows the curves which satisfy Eqs.(55), (58), (59), and (60), where each region surrounded by these curves shows different dynamics. Fig. **12** shows how the dynamic pattern changes, while Fig. **13** shows the phase plane diagram for the region of A~H in Fig. **12**.

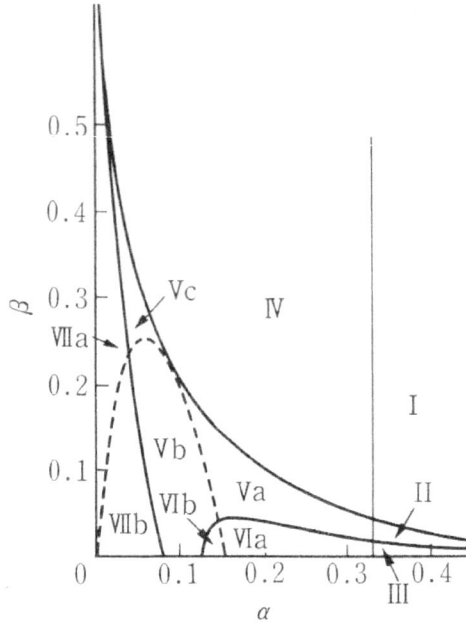

Figure 11: Classification of the dynamics in the parameter space ($\gamma = 3$).

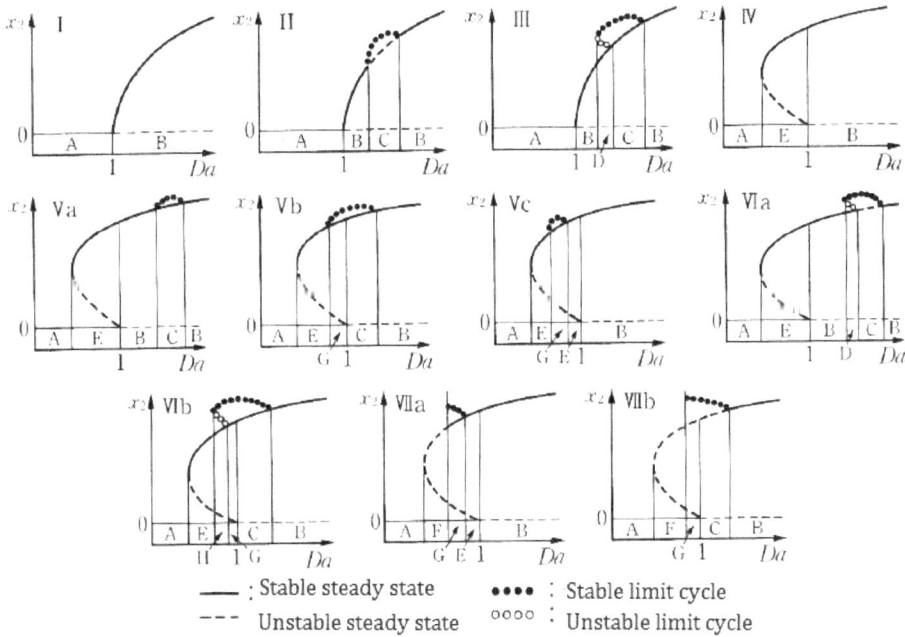

Figure 12: Classification of dynamics on Da-x_2 plane.

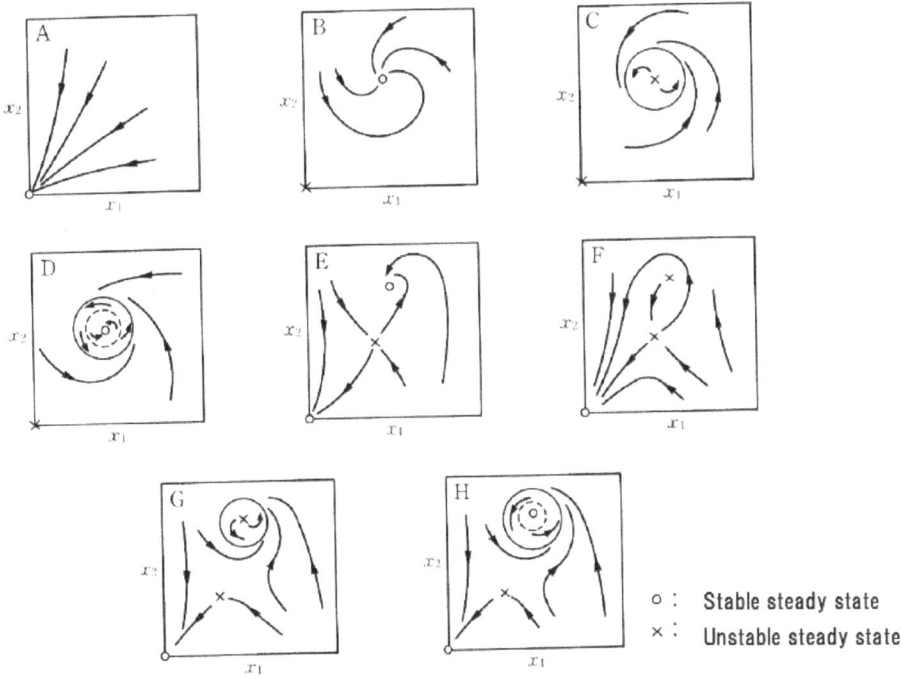

Figure 13: Classification of the dynamics in the phase plane

Effect of the Feedback Control on the Dynamics

The dynamics as mentioned above can be changed by introducing the feedback control, where the dilution rate is corrected by measuring the cell concentration as

$$D = D\{1 + \frac{K_c(X - X_s)}{X_s}\}$$ (61)

where D is the dilution rate at the steady state, X_s is the set point for the cell concentration, K_c is the proportional gain of the feedback controller. Substituting this into Eq.(34), and using non-dimensional variables, we have the basic equation for the closed loop system as

$$\frac{dx_1}{d\tau} = -\{1 + k_c(x_1 - x_{1s})\}x_1 + DaM(x_2)x_1$$ (62a)

$$\frac{dx_2}{d\tau} = -\{1 + k_c(x_1 - x_{1s})\}x_2 + \frac{DaM(x_2)x_1}{y(x_2)}$$ (62b)

where

$$k_c \equiv K_c Y(S_F) S_F / X_s .$$

The stability condition for the trivial solution is

$$k_c x_{1s} < 1 - Da_1 \tag{63}$$

which indicates that the trivial solution can be un-stabilized by appropriately choosing k_c even when Da is less than 1. Moreover, the necessary condition for the limit cycle to be bifurcated is not satisfied at the trivial solution even if the proportional feedback control is applied.

The stability condition for the non-trivial solution is as follows:

$$-k_c \overline{x}_1 \left[-1 + Da \overline{x}_1 \left\{ \frac{M(\overline{x}_2)}{y(\overline{x}_2)} \right\}' \right] - Da M'(\overline{x}_2) \overline{x}_2 (1 - k_c \overline{x}_1) > 0 \tag{64a}$$

and

$$-1 - k_c \overline{x}_1 + Da \overline{x}_1 \left\{ \frac{M(\overline{x}_2)}{y(\overline{x}_2)} \right\}' < 0 \tag{64b}$$

The necessary condition for Hopf bifurcation can be expressed as

$$-(k_c \overline{x}_1)2 + Da M'(\overline{x}_2) \overline{x}_2 (k_c \overline{x}_1) - Da M(\overline{x}_2) \overline{x}_2 > 0 \tag{65}$$

This is satisfied if

$$M'(\overline{x}_2) < 0 \tag{66a}$$

or

$$M'(\overline{x}_2) > \frac{4}{Da \overline{x}_2} \tag{66b}$$

Moreover, the followings must be satisfied:

$$\frac{2 - DaM'(\overline{x}_2)\overline{x}_2 - \sqrt{\varphi}}{2} < \frac{-\overline{x}_2 y'(\overline{x}_2)}{y(\overline{x}_2)} < \frac{2 - DaM'(\overline{x}_2)\overline{x}_2 + \sqrt{\varphi}}{2} \qquad (67)$$

where

$$\varphi \equiv \{DaM'(\overline{x}_2)\overline{x}_2\}^2 - 4DaM'(\overline{x}_2)\overline{x}_2$$

Fig. **14** shows the effect of feedback control on the dynamics.

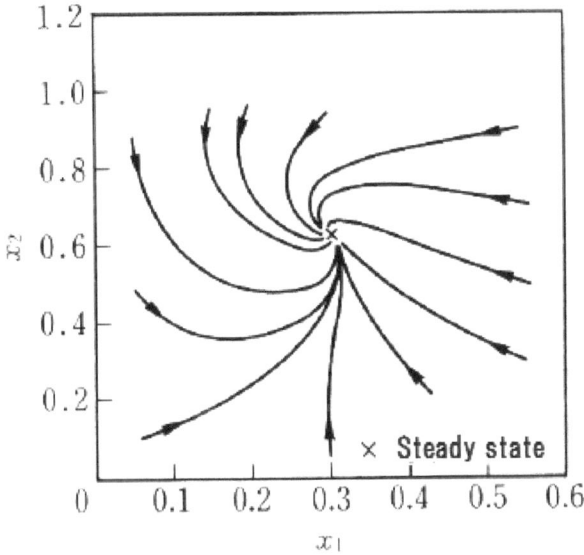

Figure 14: Phase plane of the closed loop system.

Population Dynamics of Lotka-Volterra Model

Dynamics of Lotka-Volterra Two Competing Model

The Lotka-Volterra two competing model as explained in Chapter 3 may be expressed using the non-dimensional variables as

$$\frac{dx_1}{d\tau} = (1 - x_1 - x_2)x_1 \qquad (68a)$$

$$\frac{dx_2}{d\tau} = (a - bx_1 - cx_2)x_2 \qquad (68b)$$

where

$$x_1 \equiv n_1 / K_1, \quad x_2 \equiv s_2 n_2 / K_1, \quad \tau \equiv a_1 t, \quad b \equiv a_2 s_1 K_1 / a_1 K_2, \quad c \equiv a_2 K_1 / a_1 s_2 K_2.$$

There exist four steady states for this system, and these are

$$\bar{x}_{00} = [0,0]^T, \quad \bar{x}_{10} = [1,0]^T, \quad \bar{x}_{01} = [0, a/c]^T, \quad \bar{x}_{11} = [(c-a)/(c-b), (a-b)/(c-b)]^T$$

The stability can be checked by the method as explained in the previous sections for the following cases: (a) a<b,c, (b) b<a<c, (c) c<a<b.

Dynamics of Lotka-Volterra Prey Model

Lotka-Volterra prey model can be also expressed using the dimensionless variables as

$$\frac{dx_1}{d\tau} = (1 - x_1 - kx_2)x_1 \tag{69a}$$

$$\frac{dx_2}{d\tau} = (-\gamma + akx_1)x_2 \tag{69b}$$

where $x_1 \equiv n_1 / K$, $x_2 \equiv n_2$, $\tau \equiv at$, $\alpha \equiv Kd/ab$, $k \equiv b$, $\gamma \equiv c/a$. This system

has three steady states such as $\bar{x}_{00} = [0,0]^T$, $\bar{x}_{10} = [1,0]^T$, $\bar{x}_{11} = [\dfrac{\gamma}{\alpha k}, \dfrac{1}{k}\left(1 - \dfrac{\gamma}{\alpha k}\right)]^T$

a)   In the case where $\gamma \geq \alpha k$

There exist two steady states $\bar{x}_{00}$ and $\bar{x}_{10}$, where $\bar{x}_{00}$ is unstable, and $\bar{x}_{10}$ is globally asymptotically stable. In this case, the death rate c of the prey becomes large, and thus the predator cannot get enough prey to survive.

b)   In the case where $\gamma < \alpha k$

There exist two steady states $\bar{x}_{00}$ and $\bar{x}_{11}$, where $\bar{x}_{00}$ is unstable, and the coexisting steady state is globally asymptotically stable

Dynamics of Three Competing Prey-Predator Lotka-Volterra Model

Consider next the three prey-predator Lotka-Volterra model, and this can be expressed by the non-dimensional variables as

$$\frac{dx_1}{d\tau} = (1 - x_1 - x_2 - kx_3)x_1 \qquad\qquad\qquad (70a)$$

$$\frac{dx_2}{d\tau} = (a - bx_1 - cx_2 - x_3)x_2 \qquad\qquad\qquad (70b)$$

$$\frac{dx_3}{d\tau} = (-\gamma + \alpha kx_1 + \beta x_2)x_3 \qquad\qquad\qquad (70c)$$

The steady state for the three organisms to coexist is expressed as

$$\overline{x}_{+++} = \left[\frac{\gamma c^0 - \beta a^0}{|B|}, \frac{-(\gamma b^0 - \alpha k a^0)}{|B|}, \frac{\beta(a-b) + \gamma(b-c) + \alpha k(c-a)}{|B|}\right]^T$$

where

$$a^0 \equiv ak - 1, \quad b^0 \equiv bk - 1, \quad c^0 \equiv ck - 1$$

$$B \equiv \begin{bmatrix} 1 & 1 & k \\ b & c & 1 \\ -\alpha k & -\beta & 0 \end{bmatrix}, \; |B| \equiv \alpha k c^0 - \beta b^0 = \alpha k(ck - 1) - \beta(bk - 1)$$

The stability condition becomes as

a)   In the case where $|B| > 0$

For γ which satisfies $\beta(a^0 / c^0) < \gamma < \alpha k(a^0 / b^0)$, there exists $\overline{x}_{+++}$, and asymptotically stable.

b)   In the case where $|B| < 0$

For γ which satisfies $\alpha k(a^0 / b^0) < \gamma < \beta(a^0 / c^0)$, there exists $\overline{x}_{+++}$, but unstable. In this case, if the following condition is satisfied, the chaotic behaviour emerges:

$$\frac{1}{k} < a < c < b, \qquad |B| > 0$$

CHAOS

Chaos Phenomenon

In 1961, American meteorologist Edward Norton Lorenz (1917–2008) noticed that the subtle change in the initial condition gives disastrous change in the weather in his computer simulation. This kind of phenomenon is called as **butterfly effect**, which implies that only a slight swing of the butterfly causes the big storm in another city far away from there. This problem is called as **initial value sensitivity** or **initial value dependency**.

The above phenomenon indicates that the detailed deterministic model using super computer does give undetermined or unpredicted chaotic behaviour in the simulation. This is called as **chaos**. This phenomenon is originated in the non-lineality of the system equations, and is considered to be the acronym to **cosmos** in the universe. Moreover, this phenomenon is different from the random behaviour, and this may be called rather as **deterministic chaos**.

Tent Mapping

Consider the triangular function (tent mapping) as shown in Fig. **15**, and consider the following operation:

$$x_{n+1} = f(x_n) \tag{71}$$

where starting with x(0) somewhere on the x-axis, draw straight line upward until the function f, where this value is set as x(1), and repeat this operation as indicated in Fig. **15**. Then it turns out that the following statement holds: Namely, the point started form [0,1] on the x-axis does not go outside of the range. Moreover, the point started with the slightly different point x_0' ($x_0 \neq x_0'$) will be magnified to 2^n times larger by the n-times mappings in [0,0.5] or [0.5,1]. This indicates that the slight deviation of the initial condition gives exponentially different values, and this phenomenon is called as **chaos**. In the tent mapping, the interval [0,0.5] is magnified to [0,1] by one mapping, and thus the original domain is expanded. Moreover, since f(0.5)=1 and f(1)=0, the domain [0.5,1] is overturned, or refolded. This **stretch** and **folding** may be the essential feature of chaos.

In order to trace the point in the phase plane, one way is to observe continuously, while another method is to observe discretely or in the digital fashion, where the latter is called as **Poincare mapping**.

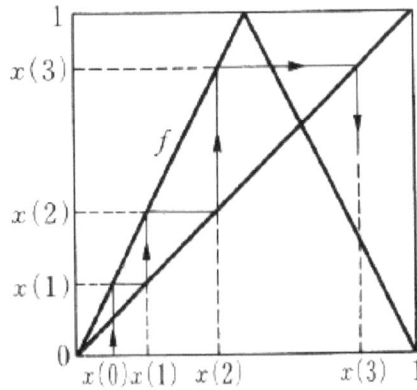

Figure 15: Tent mapping.

Chaos Emerged from Logistic Equation

In 1976, the mathematical bio-scientist Robert Mey analyzed how the behaviour of insect population changes. Let X_{n+1} be the number of population of the insect of concern at the (n+1)th year, and it is a function of the number of the population at n-th year such that

$$X_{n+1} = f(X_n) \tag{72}$$

This function may change depending on the type of insect, but Robert Mey considered the following logistic equation:

$$X_{n+1} = aX_n(1-X_n) \tag{73}$$

This indicates that the population will keep increasing by the 1st term in the RHS of the above equation, while the population will converge or reach to some value by incorporating the 2nd term. Consider how $X_1, X_2, ..., X_n$ changes starting with X_0. For this, consider X-Y plane, where X_n is on the X-axis, and X_{n+1} is taken in the Y-axis direction. First, prepare the line Y=X such that $X_n = X_{n+1}$. Starting with X_0 on the X-axis, X_1 is given by the above function. Next draw the horizontal line from there and find the point crossing with the diagonal line $X_{n+1} = X_n$, then this value of f at this point becomes X_2.

In the case of a=0.9 for example, the population decreases, and eventually all die as seen in Fig. **16**a. If the value of a was increased to 1.5, the population keeps increasing and reach to X=1-1/a=0.333.... If a becomes 2.5, the dynamics show overshoot and converge to X=1-1/a=0.6. If a=3.2, the population becomes oscillatory as shown in the figure. If the value of a is larger than 3.449, the population is oscillatory, but returned to the original population every 4 years, while a is more than 3.56, it returns to the original population in every 8 years. If the value of a is 3.5699, the original state will not be returned. This is the **chaos**.

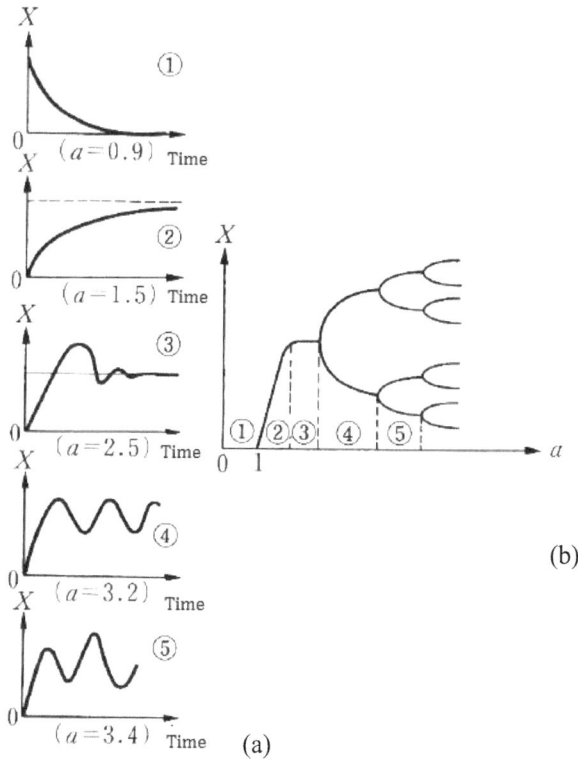

Figure 16: Change of the dynamics with respect to a in the logistic equation (a), and the bifurcation diagram (b).

Fig. **16b** shows the schematic illustration of **attracter**, and called as **bifurcation diagram**, where x-axis shows the value of a, and y-axis shows the value of X. In the figure, ①-⑤ correspond to the transient behaviour as shown in Fig. **16a**. Namely, the parameter value in ④ gives oscillation with one period, while the value at ⑤ gives **period doubling**, and so on. Fig. **17** shows the simulation result for Fig. **16b**, where the dense portion is called as **strange attracter**, and this is chaos. Namely, chaos indicates that the state at some point will be never returned to the same position, but keep changing the state in the attracter.

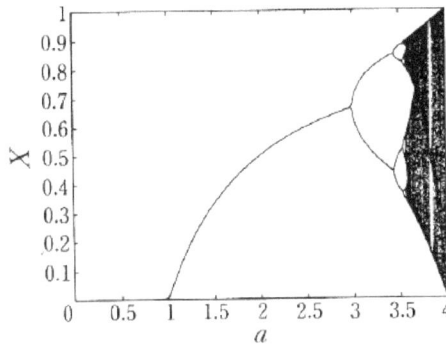

Figure 17: Bifurcation diagram.

There are several condition to check if the state is chaos or not such as Lee and York's theorem, and Sarkovskii's theorem. Moreover, Liapounov number can be also used to check the chaos easily, where details may be found elsewhere.

It is important to notice that the simple equation such as the logistic equation can give complicated behaviour, which indicates that the complex phenomena observed in nature may be governed by the simple rules.

CONCLUDING REMARKS

The steady state and dynamic characteristics depend on the model developed. In general, the simple model that can express the essential feature of the system is preferred. However, the simplified lumped model may miss the steady state and dynamic characteristics inherent in the original system. The experimental data that are obtained by the observation of the phenomena at certain condition are limited, while the global essential nature may be analyzed based on the model developed as explained in the present chapter.

DISCLOSURE

Part of this chapter has been previously published in [5].

REFERENCES

[1]   Gilmore R. Catastroph theory for scientists and engineers, Wiley Inter-Science 1981.
[2]   Poston T, Stewart I. Catastroph theory and its applications, Pitman, London 1978.
[3]   Lu Y-C. Singularity theory and introduction to catastrophe theory, Springer-Verlag, Berlin, Germany 1976.
[4]   Noguchi H. The Catastrophy (in Japanese) Science Co. 1980.
[5]   Shimizu K, Mathematical approaches to the analysis of biosystems, Corona Publ. Co., Japan, 1999 (in Japanese)

CHAPTER 7

Metabolic Regulation of the Main Metabolism

Abstract: It is quite important to understand the overall regulation mechanism of a cell system for the proper modeling. Namely, it is desirable to incorporate how the cell detect environmental signals, integrate such information, and how the cell system is regulated. In particular, carbon catabolite regulation (CCR) is of primal importance for the analysis of overflow metabolism and for the selective assimilation of multiple carbon sources. Metabolic regulation is made by enzyme level regulation and transcriptional regulation *via* the transcription factors. The effects of feed-forward and feed-back regulations on the metabolism is also explained based on the simple linear system. For CCR, proper understanding on the phosphotransferase system (PTS) and the transcriptional regulation by cAMP-Crp and Cra is important. The coordinated regulation between catabolic and anabolic (nitrogen source-assimilation) metabolism may be made by the keto acid such as αKG. The effect of oxygen level on the metabolism is explained in terms of global regulators such as ArcA/B and Fnr. Moreover, the nitrogen regulation in response to nitrogen limitation is explained. The modeling may be made by taking into account the metabolic regulation mechanisms of the central metabolism.

Keywords: Metabolic regulation, catabolite regulation, nitrogen regulation, transcription factors, Cra, cAMP-Crp, acetate overflow, catabolite repression, multiple carbon sources, ArcA, Fnr, feed-forward regulation, feed-back regulation, nitrogen regulation.

INTRODUCTION

The living organisms must survive in response to the variety of environmental perturbations. For this, living organisms sense environmental changes by detecting extracellular signals such as the concentrations of nutrients, and the growth condition such as oxygen availability *etc*. These signals eventually feed into the transcriptional regulatory systems, which affect the physiological and morphological changes that enable organisms to adapt effectively for survival [1].

In order to understand the cell system in response to culture environment, the coupling between the recognition or sensing of the environmental condition and adjustment of the metabolic system must be properly understood. Moreover, although local regulation mechanisms are known to exist, it is not clear how those local regulation systems are coordinated on the systems level, where this may be made by "distributed sensing of the intracellular metabolic fluxes" [2].

In bacterial adaptation to the culture environment, the global regulators detect the change in culture environment, and control the metabolic pathway genes [3-5].

Kazuyuki Shimizu and Yu Matsuoka

Here, the metabolic regulation is considered focusing on catabolite regulation, nitrogen regulation, and the effect of oxygen level on the metabolism of *E. coli* (but not limited to *E. coli*).

PTS AND CATABOLITE REGULATION

Glucose Uptake by PTS

The first step in the metabolism of carbohydrates is the transport of these molecules from periplasm into cytosol. In bacteria, various carbohydrates are taken up by several mechanisms [6]. The most common mechanism is the transport by phosphotransferarse system (PTS) [7], where PTS is widespread in bacteria and absent in archaea and eukaryotic organisms [8]. PTS is composed of soluble and nonsugar-specific components, Enzyme I (EI) encoded by *ptsI* and phosphohistidine carrier protein (HPr) encoded by *ptsH*, where they transfer phosphoryl group from phosphoenol pyruvate (PEP) produced in the main metabolism to the sugar-specific enzymes IIA and IIB. Another component of PTS is enzyme IIC (in some cases also IID) which is an integral membrane protein permease that transports sugar molecules, where it is phosphorylated by EIIB (Fig. **1**). So far, 21 different EII complexes have been identified in *E. coli*, and those are involved in the transport of about 20 different carbohydrates such as glucose, fructose, mannose, mannitol, galactitol, sorbitol *etc.* [7, 9]. EII^{Glc} is composed of soluble $EIIA^{Glc}$ encoded by *crr* and of integral membrane permease $EIICB^{Glc}$ encoded by *ptsG*.

In *ptsG* mutant, glucose can still be transported by EII^{Man} complex and other transporters such as GalP and MglCAB, where the cell can grow with less growth rate than the wild-type strain [10]. In such a case, *galP* is induced, where it codes for low-affinity galactose: H^+ symporter GalP. The genes in the *mglBCA* operon encode an ATP-binding protein, a galactose/glucose periplasmic binding protein, and an integral membrane transporter protein, respectively, forming Mgl system for galactose/glucose import [11]. The glucose molecule transported either by GalP or Mgl system must be phosphorylated by glucokinase (Glk) encoded by *glk* from ATP to become glucose 6-phosphate (G6P) [12] (Fig. **1**).

Uptake of Various Carbon Sources in *E. coli*

In the case of using fructose, it is transported by fructose-PTS, which has its own HPr like protein domain called FPr. Namely, the phosphate of PEP is first transferred to EI (as EI-P), but then this phosphate is transferred to FPr instead of HPr, and in turn the phosphate is transferred *via* fructose specific $EIIA^{Frc}$,

E II BC^{Frc}, and phosphorylate fructose, where phosphorylated fructose becomes fructose 1-phosphate (F1P) [13] (Fig. **2**). The *fruBKA* operon is under control of cAMP-Crp, and thus glucose is preferentially consumed by glucose PTS when glucose co-exists, while this operon is repressed by Cra [14, 15]. Because of this, *cra* gene knockout enables co-consumption of glucose and fructose with fructose to be consumed faster as compared to glucose [16], where activated FruB in *cra* mutant competes with HPr (for glucose phosphorylation) for the phosphate of EI-P. Since phosphorylation of E II A^{Glc} *via* HPr becomes lower [17], the glucose uptake rate decreases as compared to the wild-type strain [16].

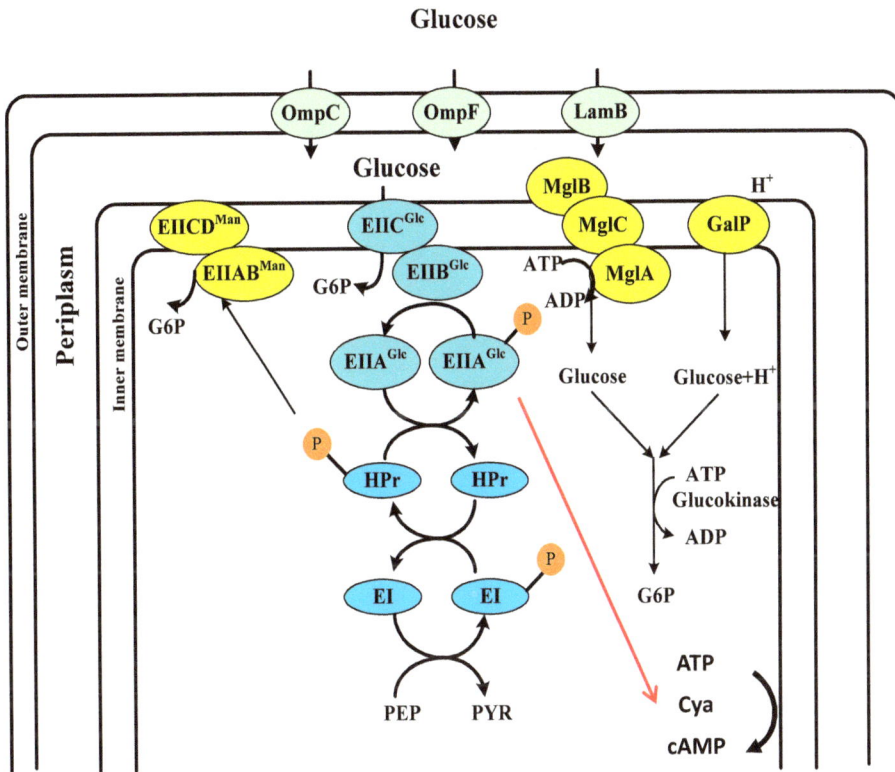

Figure 1: Glucose transport by PTS and non-PTS

In the case of using xylose as a carbon source, it is transported either by an ATP dependent high affinity ABC transporter encoded by *xylFGH*, or ATP independent low affinity proton symporter encoded by *xylE* [18, 19]. In the case of xylose utilization, the transcription factor XylR regulates *xylAB/xylFGH* [20], where *xylR* is under control of cAMP-Crp (Fig. **2**), and then catabolite repression occurs when glucose co-exists, where glucose is preferentially consumed first.

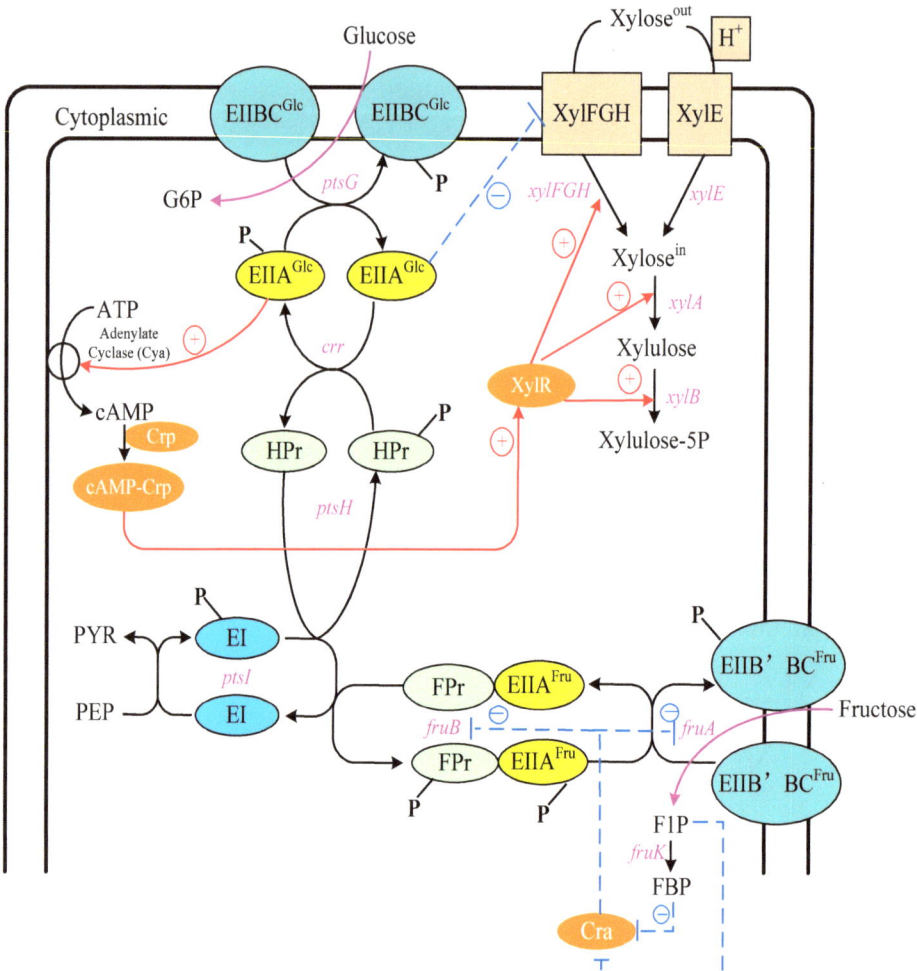

Figure 2: Glucose PTS and fructose PTS, and xylose assimilating pathways.

Glycerol has been paid recent attention for the production of biofuels and biochemicals, since it is a by-product of the biodiesel production [21-25]. In *E. coli*, glycerol is transported, and phosphorylated to produce dihydroxy acetone phosphate (DHAP) of the central metabolism *via* the pathway encoded by *glpF*, *glpK*, and *glpD*, where ATP (or in certain cases PEP) is used for the phospholylation at GlpK reaction, while NADH is produced at GlpD reaction under aerobic condition (Fig. **3**). These genes are under catabolic regulation by cAMP-Crp, so that glycerol is assimilated after glucose was depleted if glucose co-exists. NADH production at GlpD becomes important for the biofuels production under anaerobic condition affecting NADH/NAD$^+$ balance for dehydrogenase reactions.

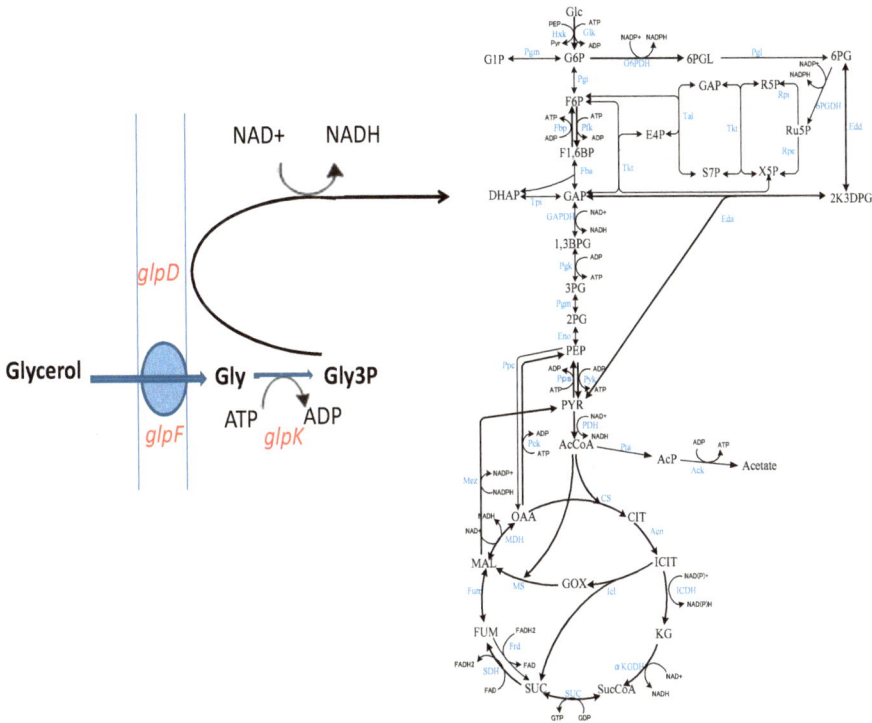

Figure 3: Glycerol assimilation pathways.

In the case of using glycerol as a single carbon source, cAMP-Crp increases due to the increase in the phosphorylated E II A^{Glc}, where cAMP-Crp induces *glpFKD* genes. Since FBP concentration decreases in the case of using glycerol as a carbon source, Cra is activated, and this together with up-regulation of cAMP-Crp causes *pckA* gene as well as TCA cycle genes to be up-regulated [26, 27].

The main drawback of using glycerol as a carbon source is the slow glycerol uptake rate, resulting in the lower cell growth rate. The slow up-take rate of glycerol is due to allosteric inhibition of GlpK by FBP [28]. The glycerol uptake rate can be increased by modulating GlpK by evolutional mutation with relaxing of feedback inhibition of GlpK by FBP [28]. However, as the glycerol uptake rate is increased, and the cell growth rate is increased, the phosphate of PEP or EI-P may be used for the phosphorylation at GlpK reaction competing with HPr of the glucose-PTS, and thus the phosphorylation level of E II A^{Glc} decreases, and in turn cAMP level decreases, and cause acetate overflow metabolism [29].

Catabolite Regulation in *E. coli*

Among the culture environment, carbon sources are by far important for the cell from the point of view of energy generation and biosynthesis. Living organisms can use various compounds as carbon sources, where these can be either cometabolized or selectively used with preference for the specific carbon sources among available carbon sources. One typical example of selective carbon-source usage is the diauxie phenomenon observed in many bacteria such as *E. coli* when a mixture of glucose and other carbon sources is used as a carbon source, where this phenomenon was first observed by Monod [30]. Subsequent investigation on this phenomenon has revealed that selective carbon source utilization is common, and that glucose is the preferred carbon source in many organisms, where the presence of glucose often prevents the use of other carbon sources. This preference of glucose over other carbon sources has been named as **glucose repression**, or more generally **carbon catabolite repression (CCR)** [31].

CCR is one of the most important regulatory phenomena in many bacteria [32-34]. CCR is important for the cells to compete with other organisms in nature, where it is crucial to select a preferred carbon source in order to improve the cell growth rate, which then results in survival among other competing organisms. The ability to select the appropriate carbon source that allows fastest growth may be the driving force for the evolution of CCR [35].

The central players in CCR in *E. coli* are the transcriptional activator Crp, the signal metabolite cyclic AMP (cAMP), adenylate cyclase (Cya), and PTSs. The cAMP-Crp complex and the repressor Mlc are involved in the regulation of *ptsG* gene and *pts* operon expression. In contrast to Mlc, where it represses the expression of *ptsG*, *ptsHI,* and *crr* [36], cAMP-Crp complex activates *ptsG* gene expression [37] (Fig. 1) (Appendix A). These two antagonistic regulatory mechanisms guarantee a precise adjustment of *ptsG* expression levels under various conditions [38]. Unphosphorylated EIIA^{Glc} inhibits the uptake of other non-PTS carbohydrates by the so-called **inducer exclusion** [39], while phosphorylated EIIA^{Glc} (EIIA^{Glc}-P) activates Cya, which generates cAMP from ATP and leads to an increase in the intracellular cAMP level [40]. In the absence of glucose, Mlc binds to the upstream of *ptsG* gene and prevents its transcription. If glucose is present in the medium, the amount of unphosphorylated EIICB^{Glc} increases due to the phosphate transfer to glucose. In this situation, Mlc binds to EIICB^{Glc}, and thus it does not bind to the operator region of *pts* genes [38, 41, 42]. If the concentration ratio between PEP and PYR (PEP/PYR) is high, EIIA^{Glc} is predominantly phosphorylated, whereas if this ratio is low, then EIIA^{Glc} is predominantly dephosphorylated [43]. EIIA^{Glc} is preferentially dephosphorylated when *E. coli* cells grow rapidly with glucose as a carbon source [39, 43].

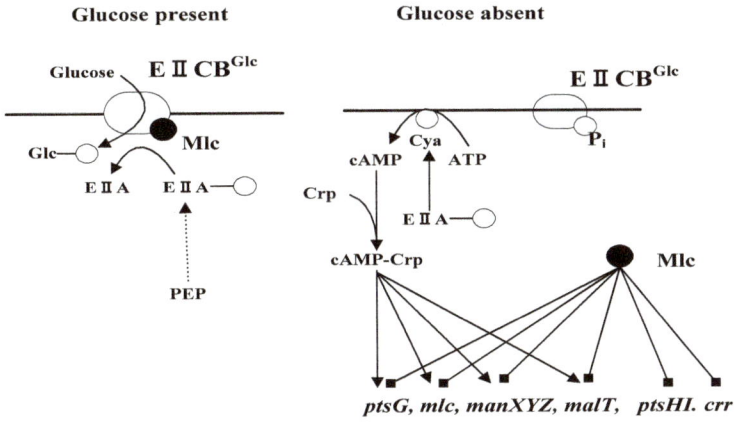

Figure 4: Regulation of PTS genes by cAMP-Crp and Mlc.

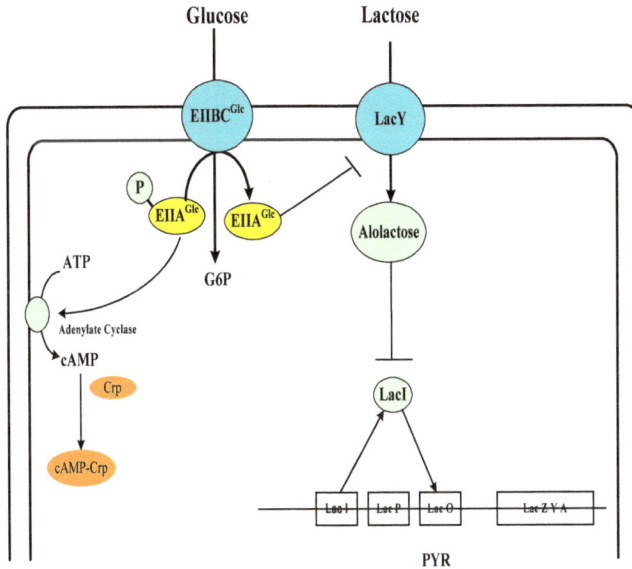

Figure 5: Inducer exclusion and the activation of Cya for cAMP formation in the glucose-lactose system.

In *E. coli,* the *lac* operon is only expressed if allolactose (a lactose isomer formed by β-galactosidase) binds and inactivates the *lac* repressor. Lactose cannot be transported into the cell in the presence of glucose, because the lactose permease, LacY is inactive in the presence of glucose [44]. As shown in Fig. **5,** phosphorylated EIIA^{Glc} is dominant when glucose is absent, and does not interact with LacY, whereas unphosphorylated EIIA^{Glc} can bind and inactivates LacY when glucose is present [45, 46]. This occurs if lactose co-exists [47]. The similar mechanism may be seen for the transport of other secondary carbon sources such as maltose, melibiose, raffinose, and galactose [48, 49].

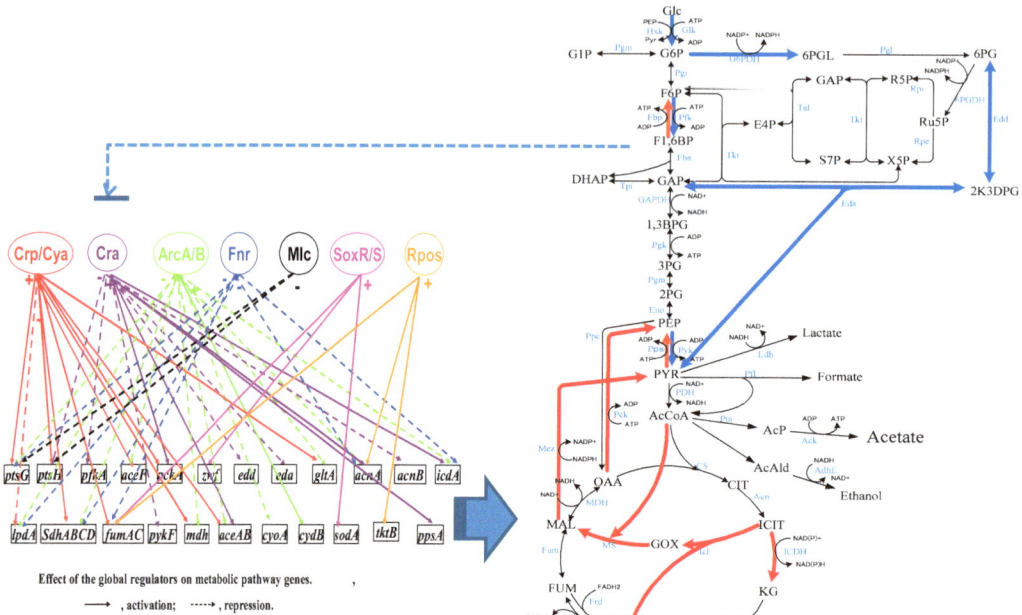

Figure 6: Control of metabolic pathway genes by Cra.

Carbon Flow Control

In addition to cAMP-Crp, which acts depending on the level of glucose concentration, the **catabolite repressor/activator protein (Cra)** plays an important role in the control of carbon flow in *E. coli* [13, 50-52]. The carbon uptake and glycolysis genes such as *ptsHI*, *pfkA*, *pykF*, *zwf*, and *edd-eda* are repressed, while gluconeogenic genes such as *ppsA*, *fbp*, *pckA*, *icdA*, *aceA*, and *aceB* are activated by Cra in *E. coli* (Fig. 6) (Appendix A). The mutant defective in *cra* gene is unable to grow on gluconeogenic substrates such as pyruvate, acetate, and lactate. This appears to be due to deficiency in the gluconeogenic enzymes such as Pps, Pck, some TCA cycle enzymes, two glyoxylate-shunt enzymes, and certain electron transport carriers.

The gluconeogenic pathway is deactivated by the knockout of *cra* gene, and the carbon flow toward glycolysis, and thus the glucose consumption rate is expected to increase, since glycolysis pathway genes such as *ptsHI*, *pfkA*, and *pykF* are activated by *cra* gene knockout, where *cra* gene knockout indeed enabled the increase of the glucose consumption rate, and thus improved the metabolite production rate [53]. However, the regulation mechanism is complex, and it must be careful, since *icdA, aceA, B,* and *cydB* genes are repressed, while *zwf* and *edd* gene expression are activated, and thus the Entner-Doudoroff (ED) pathway is activated by *cra* gene knockout [54].

Enzyme Level and Transcriptional Regulations in the Central Metabolism

The cell system achieves the coupling between recognition and adjustment through transcription factors (TFs), whose activities respond to the culture environment, and regulate the expression of the associated genes. This combined recognition and adjustment forms the reaction networks that overarch the metabolic and genetic layers [2]. In general, fast action is made by the enzyme level regulation such as the feed-forward activation of pyruvate kinase (Pyk) by fructose 1,6-bisphosphate (FBP), and the feedback inhibition of phosphofruct kinase (Pfk) by phosphoenol pyruvate (PEP), a motif that enables a high level of the upstream metabolite to lower the level of the downstream metabolite [55] (Fig. **7a**). The slow action is made through the transcriptional regulation, where lower cAMP-Crp level represses the expression of TCA cycle genes, while (FBP-inhibited) Cra activates the expression of gluconeogenic pathway genes as well as some of the TCA cycle genes and the glyoxylate pathway genes, while represses the expression of the glycolysis genes in the case of *E. coli* (Appendix A) (Fig. **7b**).

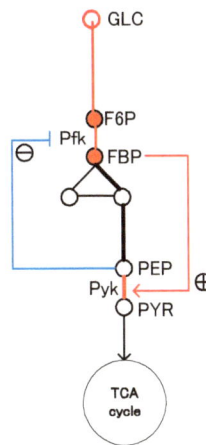

(a)

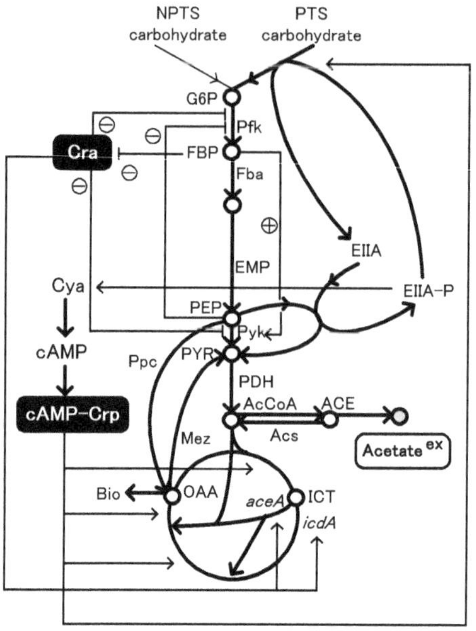

(b)

Figure 7: Enzyme level regulation (a) and transcriptional regulation (b) of the main metabolism.

The levels of the flux-signaling metabolites become coupled, enabling a robust, coherent response of the TFs. The coherent behavior of the overall system is, therefore, not established by a common transcriptional master regulator, but arises from the molecular interactions within the system itself [2]. It may be considered that the system of reactions of the lower glycolysis and the feed-forward activation of Pyk by FBP translate flux information into the concentration of FBP, and that this feed-forward activation affects the linearization of the glycolytic kinetics [56] as will be explained later. In fact, feed-forward regulation has been known to ensure the structural robustness against perturbations [55]. This mechanism may be conserved in many organisms. For example, in *Saccharomyces cerevisiae*, sugar uptake rate is well correlated with the respiratory and fermentative pathways, or the specific ethanol production rate, and the similar relationship may be seen between the glycolysis flux and FBP [57, 58], where Pyk is also feed-forward activated by FBP in *S. cerevisiae* [59].

Moreover, the increased pyruvate (PYR) goes down through pyruvate dehydrogenase (PDH) to acetyl CoA (AcCoA), where AcCoA is homeostatic. Namely, the increase in AcCoA activates Ppc activity [60, 61], thus reducing the upcoming Pyk-PDH fluxes, and increases oxaloacetate (OAA) and activates citrate synthase (CS) reaction, thus activating the outgoing flux from AcCoA. In this way, AcCoA concentration decreases, forming the feed-back regulation against the initial increase in AcCoA (Fig. **8**). In another view, this phenomenon may be considered as the feed-forward regulation in the sense that the repression of TCA cycle activity is detected by the increase in AcCoA, which causes the activation of the anaplerotic Ppc pathway, and backs up the precursor metabolite such as OAA for biosynthesis, where it is expected to be decreased due to deactivated TCA cycle.

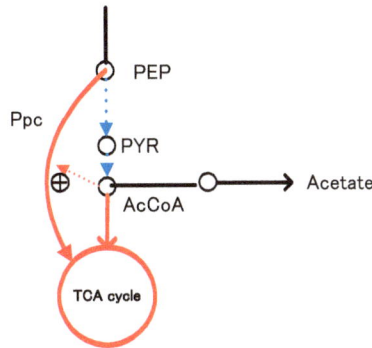

Figure 8: Homeostasis of AcCoA in relation to Ppc.

As the glucose uptake rate increases, the TCA cycle flux tends to increase by the increased OAA and AcCoA, and then NADH is overproduced. The accumulated NADH inhibits citrate synthase (CS) and isocitrate dehydrogenase (ICDH) allosterically [62], forming feedback regulation, and thus results in AcCoA accumulation, which in turn causes acetate overflow metabolism. This enzyme level regulation by NADH in the TCA cycle can be verified by incorporating NADH oxidase (NOX) [63], or by adding nicotinic acid [64], whereby activating TCA cycle. This effect is more enhanced under oxygen limitation [65]. In the long run, the expression of TCA cycle genes is eventually repressed by the transcriptional regulation by cAMP-Crp toward steady state as will be explained later.

The typical growth condition changes from glucose-rich to acetate-rich condition in the batch culture. This requires a significant reorganization of the central metabolism from glycolysis to gluconeogenesis. Although the molecular mechanism underlying the metabolic transition from glucose to acetate has been extensively investigated in *E. coli* [66], its dynamics have been poorly understood. Since it is critical for the cell to efficiently and quickly reprogram the metabolism under the changing environmental condition, the cell must have the elaborate managing system.

The anaplerotic enzyme Ppc is allosterically activated by both FBP and AcCoA as mentioned above, which then enables the ultrasensitive regulation of anapleurosis

[67]. Namely, after glucose depletion, FBP concentration decreases accordingly, where Ppc and Pyk activities decrease in turn by the allosteric regulation, and PEP consumption is almost completely turned off. These make PEP concentration to be increased, and this buildup of PEP is kept nearly constant during certain period, and this may serve to quickly uptake the glucose by PTS if it becomes available again [67]. This mechanism is important for the fed-batch culture compensated by DO-stat or pH-stat, where carbon limitation often occurs periodically, and the uptake of carbon source can be made quickly and efficiently without delay by the above mechanism.

Moreover, after glucose depletion, FBP level drops, and thus Ppc activity decreases, while PEP carboxykinase (Pck) activity is activated by the activated Cra caused by the decreased FBP. This reveals the mechanism of avoiding the futile cycling caused by Ppc and Pck during gluconeogenic phase [67], where ATP generation becomes important. During the active glycolysis with enough sugars available, this futile cycling occurs, and loses ATP without efficient use for the compensation of the flexible metabolic fluxes and the metabolic regulation [60].

Acetate Overflow Metabolism

As the specific glucose consumption rate or the glycolysis flux increases, FBP concentration increases [68, 69]. The increased FBP allosterically enhances the activity of Pyk (and also Ppc) by feed-forward control (Fig. **7a**). The PEP concentration tends to be decreased due to activation of Pyk (and Ppc). PEP molecule allosterically inhibits Pfk activity by feed-back regulation, and the decrease in PEP concentration thus causes Pfk activity to be increased, and the glycolysis flux further increases, and in turn FBP concentration will also increase. On the other hand, the decrease in PEP and increase in PYR make PEP/PYR ratio to be decreased. This causes EIIA-P concentration to be decreased, and in turn less activates Cya, and thus cAMP is less formed. As a result, cAMP-Crp level decreases, which causes the decrease in *ptsG* gene expression, and this makes the glucose uptake rate to be repressed. This forms the negative feed-back loop for the initial increase in the glucose uptake rate as mentioned above (Fig. **7b**)

In addition to cAMP-Crp, Cra also plays important roles for catabolite regulation, where Cra detects FBP concentration, and Cra activity decreases with the increase of FBP concentration. In the above example, the increase in glucose uptake rate increases FBP concentration, and thus Cra activity decreases. This causes the increase of the expression of glycolysis genes such as *pfkA* and *pykF* genes, while it represses the expression of gluconeogenic pathway genes such as *ppsA* and *pckA* genes, which implies acceralation of increased glycolysis fluxes. The decrease in Cra activity also affects TCA cycle genes such that *icdA* and *aceA* gene expression is repressed (Appendix A), and thus TCA cycle is further repressed by this mechanism. The increase in glycolysis activity and the decrease in TCA cycle activity cause more acetate production. This is the overflow metabolism, which has been the subject of bacterial cultivation [66, 69, 70]. This acetate overflow metabolism is aslo discussed in Chapter 8 with computer simulation.

SYSTEMS ANALYSIS OF THE MAIN METABOLIC PATHWAY REGULATION

Let us further consider the above mechanism based on the mathematical formulation. For this, it may be of importance to extract the essential properties of the metabolic networks based on the simplified model. **Figure 9a** shows the schematic illustration of the simplified main metabolic pathways for the uptake of PTS and non-PTS carbohydrates. The state variables are GF6P (glucose 6-phosphate (G6P) + fructose 6-phosphate (F6P)), FBP, TP (triose phosphate), PEP, PYR, and EIIA/EIIA-P, where the PTS proteins such as EI, HPr, and EIIA and their phosphorylated state are lumped into EIIA and its phosphorylated state EIIA-P.

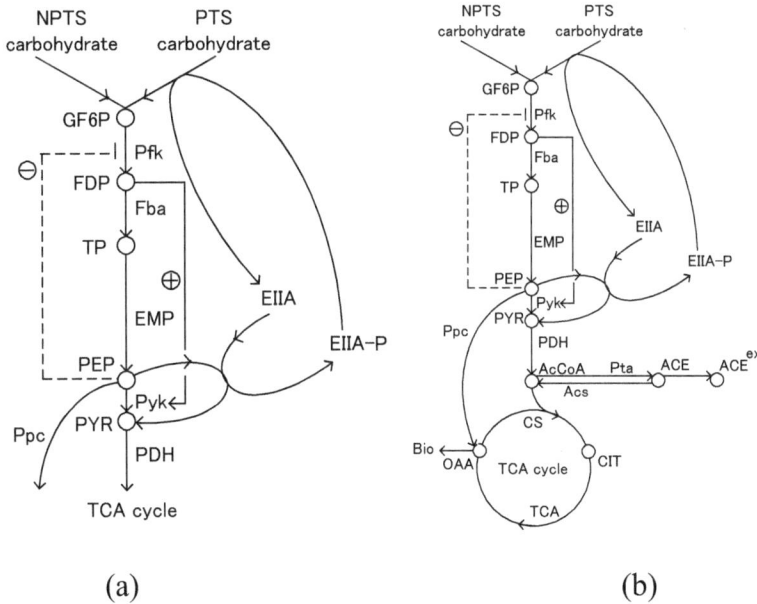

(a)   (b)

Figure 9: Schematic illustration of feed-forward and feed-back loops embedded in the main metabolism: (a) PTS, non-PTS and the glycolysis with feed-forward and feed-back loops, (b) Enzyme level and transcriptional regulation.

Since the reactions of PTS are very fast as compared to the glycolytic reactions or gene expression [71], the individual reactions of PTS can be lumped together [72]. Then the state equation for EIIA-P may be expressed as

$$\frac{d[EIIA - P]}{dt} = v_{PTS} - v_{PTS-up} \tag{1a}$$

The v_{PTS} is the reaction rate of the reversible PTS reaction, and v_{PTS-up} is the uptake rate of PTS carbohydrate. The v_{PTS} may be expressed as

$$v_{PTS} = k_{PTS}[PEP][EIIA] - k_{PTS}^-[PYR][EIIA - P] \tag{1b}$$

with the reaction rate constants k_{PTS} and k_{PTS}^-, where $[\bullet]$ denotes the concentration. Since PTS proteins are either phosphorylated or dephosphorylated, its total concentration $[EIIA]_0$ may be expressed as

$$[EIIA]_0 = [EIIA] + [EIIA - P] \tag{1c}$$

By substituting Eq. (1c) into Eq. (1b), and the resulting Eq. (1b) into Eq. (1a), and by assuming steady state, we have the following equation:

$$[EIIA - P] = \frac{\dfrac{[PEP]}{[PYR]}[EIIA]_0 - \dfrac{v_{PTS-up}}{k_{PTS}[PYR]}}{K_{PTS} + \dfrac{[PEP]}{[PYR]}} \tag{1d}$$

where K_{PTS} is an equilibrium constant defined as $K_{PTS} \equiv k_{PTS}^- / k_{PTS}$. Eq. (1d) indicates that as the PTS carbohydrate uptake rate increases, the negative term in the numerator of the RHS of Eq. (1d) becomes large and thus [EIIA-P] decreases. This also indicates that [E II A-P] is always smaller for PTS sugars than for non-PTS sugars, and the difference becomes eminent as the substrate uptake rate is increased.

Moreover, Eq. (1d) indicates that [E II A-P]/[E II A]₀ increases with respect to [PEP]/[PYR] as shown in Fig. **10a** [43]. Higher cell growth rates require higher glycolysis fluxes and lead to increased fluxes through Pyk. In the case without feed-forward regulation of Pyk by FBP, the PEP and PYR concentrations have to increase with respect to the cell growth rate to match this requirement (Fig. **10b**), where [PEP]/[PYR] is very sensitive to small fluctuations in the metabolite concentrations. On the other hand, if the feed-forward regulation of Pyk by FBP is active, higher fluxes can be realized by lowering the PEP concentration, where PYR concentration increases with increasing the cell growth rate [55]. The higher flux through Pyk is realized by the feed-forward activation by FBP (or equivalently G6P), where this is much less fragile to small fluctuations or uncertainties in the metabolite concentrations, since [PEP]/[PYR] decreases as the cell growth rate increases (Fig. **10c**) [55].

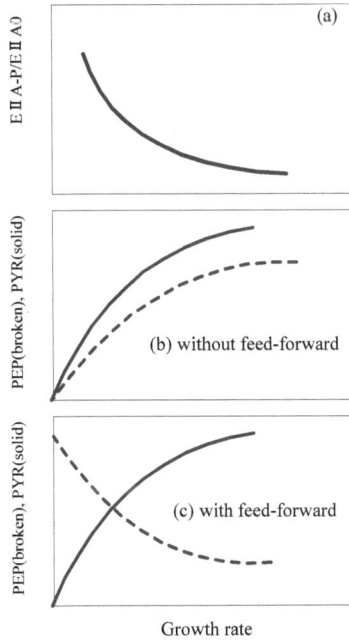

Figure 10: Behavior of E Ⅱ A-P with reapect to the cell growth rate (a), and PEP and PYR concentrations with reapect to the cell growth rate without feedforward regulation (b) and with feedforward regulation (c).

Consider the robustness (or sensitivity) of the glycolysis in view of feed-forward activation of Pyk by FBP. For this, consider the simple model, where the reversible reactions between FBP and PEP may be lumped into one reversible Michaelis-Menten reaction such as [56] (Fig. **11a**)

$$v_{E1} = v = \frac{v_{mE1}\left([FBP] - \dfrac{[PEP]}{K_{eq}}\right)}{K_{mFBPE1}\left(1 + \dfrac{[PEP]}{K_{mPEPE1}} + [FBP]\right)} \tag{2}$$

where E1 is the lumped enzyme catalyzing the reactions between FBP and PEP (Fig. **11a**). The Pyk reaction may correspond to the irreversible Michaelis-Menten kinetics such as

$$v_{Pyk} = v = \frac{v_{mPyk}[PEP]}{K_{mPEP,Pyk} + [PEP]} \tag{3}$$

Here the above simple model does not include any fluxes to be drained off to biosynthesis and any fluxes to enter from the pensose phosphate (PP) pathway. These effects may not be essential in the present analysis. The relationship between v *versus* [FBP] may become as shown in Fig. **11a**, where non-linearity is significant.

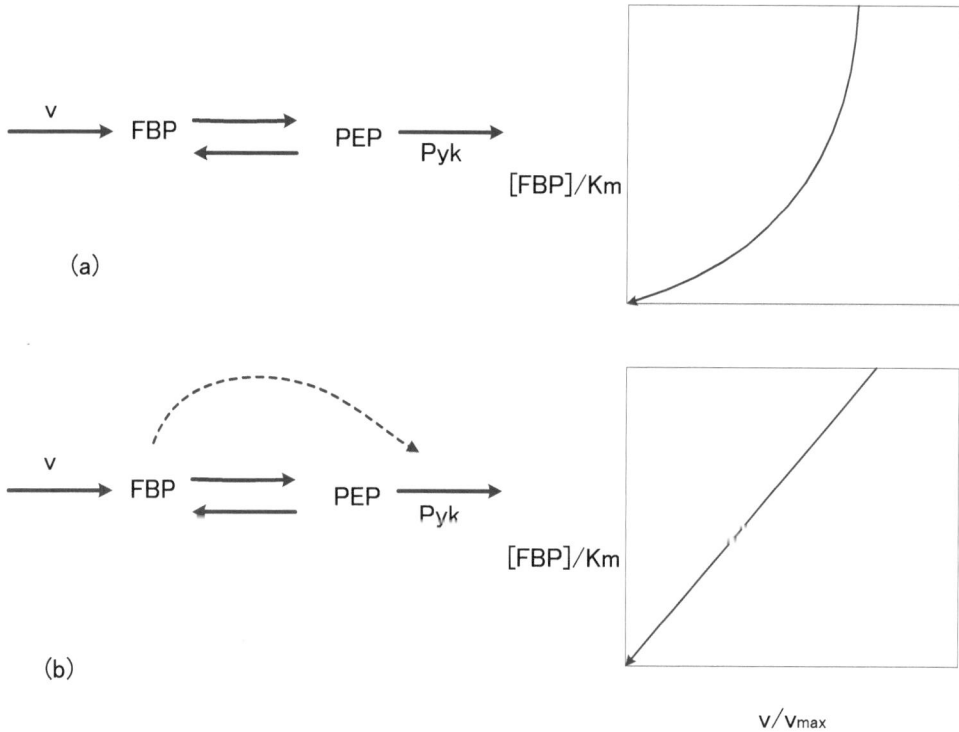

(a)

(b)

v/v$_{max}$

Figure 11: Effect of the feed-forward regulation on the robustness (linearization) (lower figure) from nonlinear kinetics (upper figure).

On the other hand, if the effect of feed-forward activation of Pyk by FBP is incorporated as shown in Fig. **11b**, the following Monod-Wyman-Changeux (MWC) equation may be obtained for Pyk [56]:

$$v_{Pyk} = v = \frac{v_{mPyk} \dfrac{[PEP]}{K_{mPEP,E2}} \left(1 + \dfrac{[PEP]}{K_{m,PEP,E2}}\right)^{n-1}}{L \left(1 + \dfrac{[FBP]}{K_{m,FBP,E2}}\right)^{-n} + \left(1 + \dfrac{[PEP]}{K_{m,PEP,E2}}\right)^{n}} \tag{4}$$

where L, n, and K_m denote allosteric equilibrium constant, cooperability constant, and affinity constant of FBP for Pyk, respectively [56]. As illustrated in Fig. **11b**, the relashionship between v and FBP concentration becomes more linear as compared to the case without such feed-forward regulation (Fig. **11b**). This makes the system robust against perturbations in the metabolite concentrations.

NITROGEN REGULATION

Next to carbon (C) source metabolism, nitrogen (N) metabolism is also important in understanding the metabolic regulation [73]. In *E. coli*, assimilation of N-source such as ammonia/ammonium (NH_4^+) using α-KG results in the synthesis of glutamic acid (Glu) and glutamine (Gln) (Fig. **12**). Glutamaine synthetase (GS) catalyzes the only pathway for glutamine biosynthesis. Glutamic acid can be synthesized by two pathways through the combined actions of GS and glutamate synthase (GOGAT) forming GS/GOGAT cycle, or by glutamate dehydrogenase (GDH) [74] (Fig. **12**).

When extracellular NH_4^+ concentration is low around 5 μM or less, ammonium enters into the cell *via* the transporter AmtB, and is converted to glutamine by GS, and UTase (uridylyl transferase) uridylylates both GlnK and GlnB [75] (Fig. **12**). When extracellular NH_4^+ concentration is more than 50 μM, UTase deuridylylates GlnK and GlnB. GlnK complexes with AmtB, thereby inhibiting the transport *via* AmtB, where GlnB interacts with NtrB and activates its phosphatase activity leading to dephosphorylation of NtrC, and NtrC-dependent gene expression ceases [75] as will be explained later in more detail. In this way, GlnK binds to AmtB and inhibits its activity under ammonium rich condition, while GlnK dissociates from AmtB at low ammonium concentration, thereby setting AmtB free to act for ammonium transport [76-78] (Fig. **12**).

Intracellular ammonium is assimilated into biomass in two steps: Namely, it is first captured in the form of glutamic acid using carbon skeleton of αKG *via* GS/GOGAT cycle. Then N-group in Glu is transferred to synthesize amino acids thus incorporating into biomass, while recycling the carbon skeleton back to αKG

[79]. The αKG pool, which integrates imbalance between the ammonium assimilation flux and the biomass incorporation flux activates AmtB *via* GlnK. If internal ammonium level drops, then the rate of ammonium assimilation will drop immediately. This results in αKG accumulation [80].

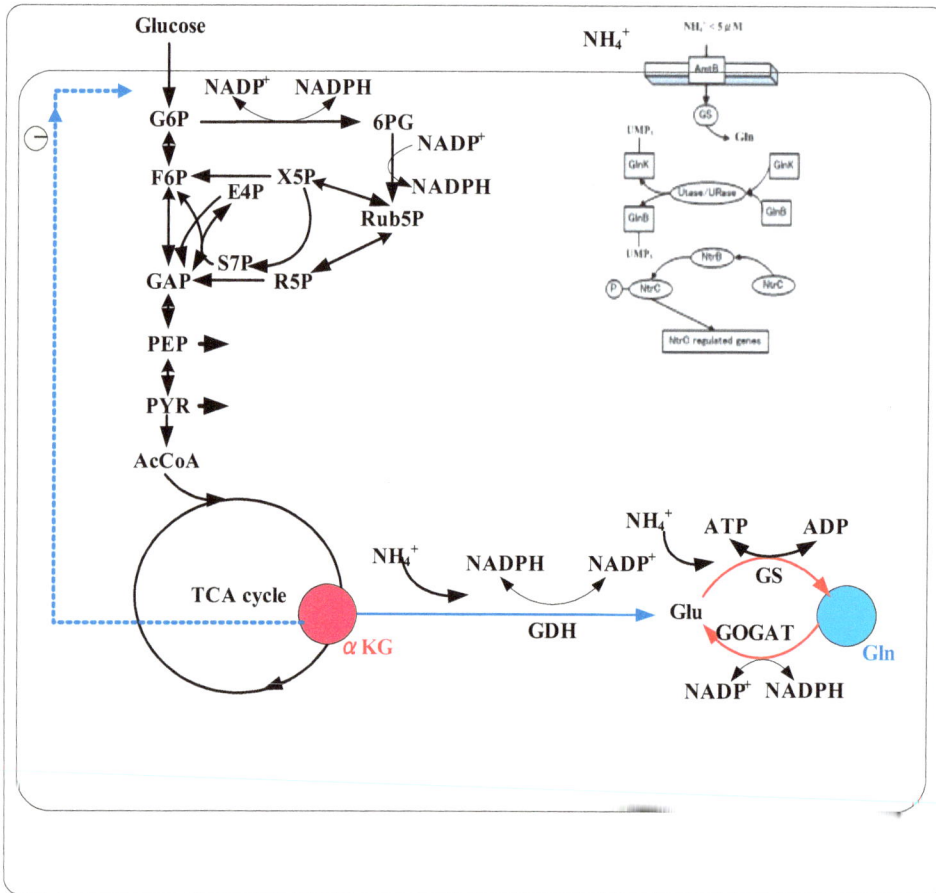

Figure 12: Ammonium transport and nitrogen regulation.

The central participants of Ntr response are NR_I (or NtrC) and NR_{II} (or NtrB), and RNA polymerase complexed to σ^{54}. NR_I is the transcriptional activator of σ^{54}-dependent promoters, while NR_{II} is a bifunctional protein that can either transfer phosphate to NR_I or control dephosphorylation of NR_I phosphate. N-limitation results in the phosphorylation of NR_I, which in turn stimulates the expression of *glnALG* operon.

Nac is involved in the transcriptional repression of *gdhA* gene under N-limitation [81], and it represses *gdhA* gene expression. NADPH is an important cofactor in GDH and (GS)-GOGAT activities. Under N limitation, the glutamate and glutamine synthetic pathways are expected to be repressed due to shortage of NH_3 for these reactions.

E. coli possesses two closely related P_{II} paralogues such as GlnB and GlnK, where GlnB is produced constitutively, and it regulates the NtrB (NR_{II})/NtrC (NR_I) two-component system [82]. The intracellular concentrations of NR_I and NR_{II} increase upon N-limitation [83-85]. The phosphorylated NtrC is an activator of various nitrogen-controlled genes such as *glnA* which codes for GS [86] and *glnK* encoding the second P_{II} paralogues [85]. The increased NR_I, presumably in the phosphorylated form such as NR_I-P activates the expression of *glnK* and *nac* promoters under N-limitation [87, 88]. The transcript levels of *glnK* and *nac* genes increases as C/N ratio increases, where *glnK* and *nac* promoters are sharply activated when ammonia is used up [88].

The Ntr system is composed of four enzymes (Fig. **13**): a uridylyltransferase/ uridylyl- removing enzyme (UTase/UR) encoded by *glnD* gene, a small trimeric protein, P_{II}, and the two-component system composed of NtrB and NtrC. GlnD controls the activity of GS by adenylylation/deadenylation through a bifunctional enzyme adenylyltransferase (ATase), the *glnE* gene product [89-91]. The activity of GlnK becomes high under N-limitation and contributes to the regulation of NtrC-dependent genes [92]. Upon GS adenylation, ATase activity is regulated by UTase/UR and P_{II} such that under nitrogen limitation, UTase covalently modifies P_{II} by addition of a UMP group at the specific residue and the resultant uridylylated form of P_{II} promotes deadenylylation of GS by ATase (Fig. **13**).

Conversely, under N-rich condition, the uridylyl removing activity of GlnD predominates, and the deuridylylated P_{II} promotes adenylation of GS by ATase. Adenylylation by ATase is promoted by deuridylylated P_{II} which is produced by UR action on P_{II} (UMP)$_3$ under higher N-concentration (low C/N ratio) (Fig. **13**). These indicate that UTase/UR and P_{II} acting together sense the intracellular nitrogen status [93]. The P_{II} signal transduction proteins such as GlnB and GlnK are uridylylated/deuridylylated in response to intracellular glutamine level, where low intracellular glutamine level, signalling N-limitation, leads to uridylylation of GlnB [93]. GlnB is shown to be allosterically regulated by α-KG, and thus GlnB may play a role in integrating signals of C/N status. The NtrB/NtrC two-component system and GlnE which adenylylates/deadenylylates GS are the receptors of GlnB signal transduction [92].

The phosphorylated NR_I/NtrC (NR_I/NtrC-P) activates transcription from N-regulated σ^{54}-dependent promoters by binding to the enhancers [82, 93-95]. P_{II} and the related GlnK control the phosphorylation state of NR_{II}/NtrB by stimulating the phosphatase activity of NR_{II}. The ability of GlnK and P_{II} to regulate the activities of NR_{II} is in turn regulated by C and N availability *via* allosteric control [82].

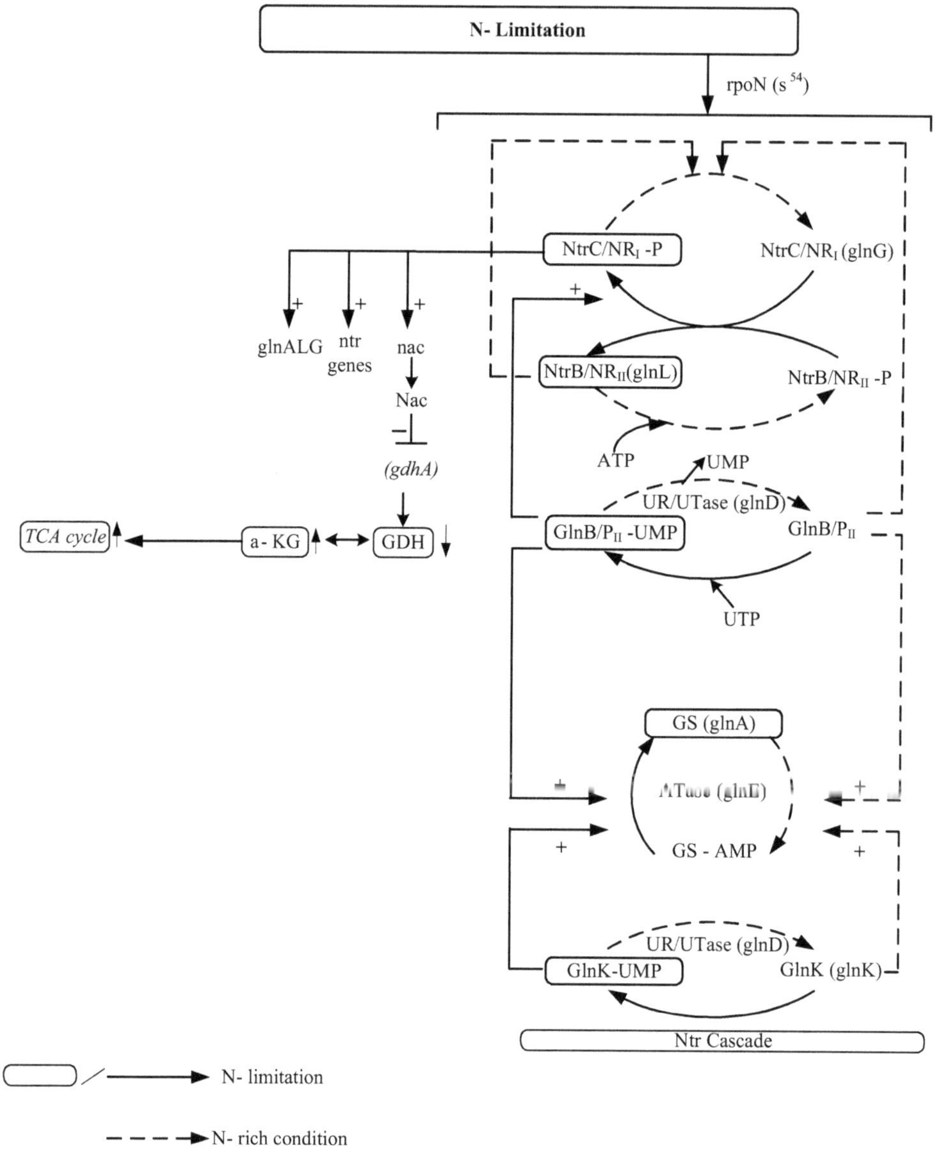

Figure 13: Overall mechanism of nitrogen regulation.

INTERACTION BETWEEN N-REGULATION AND C-REGULATION

During N-limitation, a sudden increase in nitrogen availability results in immediate increase in glucose uptake rate, and αKG plays an important role for this, where αKG directly blocks the glucose uptake under N-limitation by inhibiting EI of PTS [96] (Fig. **11**). This implies several things: (1) αKG inhibition of sugar uptake is for all PTS sugars by inhibiting EI but not carbohydrate specific E II, (2) this is performed without perturbing the concentrations of the glycolytic intermediates such as G6P, PEP, and PYR, (3) inhibition of EI by αKG leads to reduced amount of phosphorylated E II A^{Glc} and decreases cAMP level, where the effect of αKG on cAMP production is caused by the difference in E II A^{Glc} phosphorylation rather than a difference in substrate availability [96].

Among intracellular metabolites, α-keto acid such as αKG turns to be a mater regulator for catabolite regulation and co-ordination of different regulations [97]. Namely, when favored carbon source such as glucose was depleted, αKG level fall, and cAMP increases to stimulate other carbon catabolic machinery. Namely, when preferred carbon source such as glucose is abundant, the cell growth rate becomes higher with lower cAMP level, while if it became scarce, the cell growth rate declines with higher cAMP level. In particular, under nitrogen (N)- limitation, αKG accumulates due to decreased activity of glutamate dehydrogenase (GDH) and inhibits carbon assimilation, where there is less need for carbon-catabolic enzymes, and more demand for those involved in such nutrient assimilation. On the other hand, when anabolic nutrient such as ammonia is in excess, αKG concentration decreases due to activated GDH, producing glutamate from αKG, cAMP level increases, and carbon catabolic enzymes increases to accelerate carbon assimilation. Namely, αKG coordinates the catabolic (C)-regulation and N-regulation by inhibiting EI of PTS [96] or cAMP *via* Cya [98]. Moreover, the physiological function of cAMP signaling goes beyond simply enabling hierarchical utilization of carbon sources, but also controls the function of the proteome [97, 98]. In order to model this phenomenon, EI of PTS has to be expressed as the inhibition by αKG, or Cya is inhibited by keto acids such as OAA and PYR as well as αKG.

EFFECT OF OXYGEN LEVEL ON THE METABOLISM

Global regulators such as Fnr (fumarate nitrate reduction), Arc (anoxic respiration control) system, and Nar (nitrate reduction) are mainly responsible for the regulation under oxygen limitation and other electron acceptors in the culture environment, where Fnr directly senses molecular oxygen, and plays a role under anaerobic condition [99] in coordination with ArcA/B system, where Fnr activates *arcA* gene expression (Fig. **14**). Under oxygen limitation, Fnr binds a $[4Fe-4S]^{2+}$ cluster and becomes a transcriptionally active dimeric form. Molecular oxygen can oxidize the ion-sulfer cluster of the corresponding region, resulting in monomerization of the protein and subsequent loss of its ability to bind DNA [100]. The ArcA/B system plays a role under both anaerobic and micro-aerobic conditions [101, 102], where it is composed of ArcA, the cytosolic response regulator, and ArcB, the membrane bound sensor kinase. The ArcA/B two-component system responds to the redox state of the membrane-associated redox carriers such as quinones in the respiratory chain [103, 104]. The quinones pool decreases under oxygen limitation, and cause ArcB to be self-phosphorylated (ArcB-P), and then ArcB-P transphosphorylates ArcA (Fig. **14**) [105]. The ArcA-P then represses the expression of the TCA cycle and the glyoxylate shunt genes (Appendix A). Moreover, the genes which encode the primary dehydrogenases such as *glpD, lctPRD, aceE, F* and *lpdA* are also repressed by ArcA (Appendix A). The *cyoABCDE* operon is repressed by both ArcA and Fnr, while *cydAB* operon is activated by ArcA and repressed by Fnr (Fig. **14**) [106].

The expression of *pfl* genes which encode pyruvate formate lyase, Pfl is activated by ArcA and Fnr, whereas *aceE, F*, and *lpdA* which encode PDHc are repressed by ArcA under oxygen limitation (Fig. **14**). The formate can be excreted *via* Foc, or converted to hydrogen *via* formate dehydrogenase, FDH_N and formate hydrogen lyase, Fhl and deletion of FocAB, FDH_N and hydrogenase Hyd (Fig. **14**) [107, 108]. Moreover, the flux from PYR to AcCoA is blocked in *pfl* mutants ($\Delta pflA$ or $\Delta pflB$), and pyruvate exclusively goes to lactate formation *via* LDH reaction [109, 110]. Moreover, Fnr activates *frd* gene expression, while repressing *sdh* gene expression, resulting in branched pathways for TCA cycle under anaerobic condition.

As mentioned before, the TCA cycle activity is repressed as the glucose consumption rate was increased due to lower level of cAMP-Crp, which in turn causes acetate overflow metabolism. This also occurs by the higher redox ratio [63]. This phenomenon can be relaxed by activating TCA cycle by *arcA/B* genes knockout [101, 107, 111]. The activated TCA cycle produces more NADH, and allosterically inhibits CS and ICDH activities [62]. Thus, the NADH oxidation by the expression of *nox* gene coding for NADH oxidase, NOX in the *arcA* mutant further reduces the acetate formation, resulting in the increased recombinant protein production [63], while nicotinic acid and Na nitrate may also activate

TCA cycle [64]. The activation of the TCA cycle causes the decrease in the cell yield due to higher production of CO_2 in the TCA cycle.

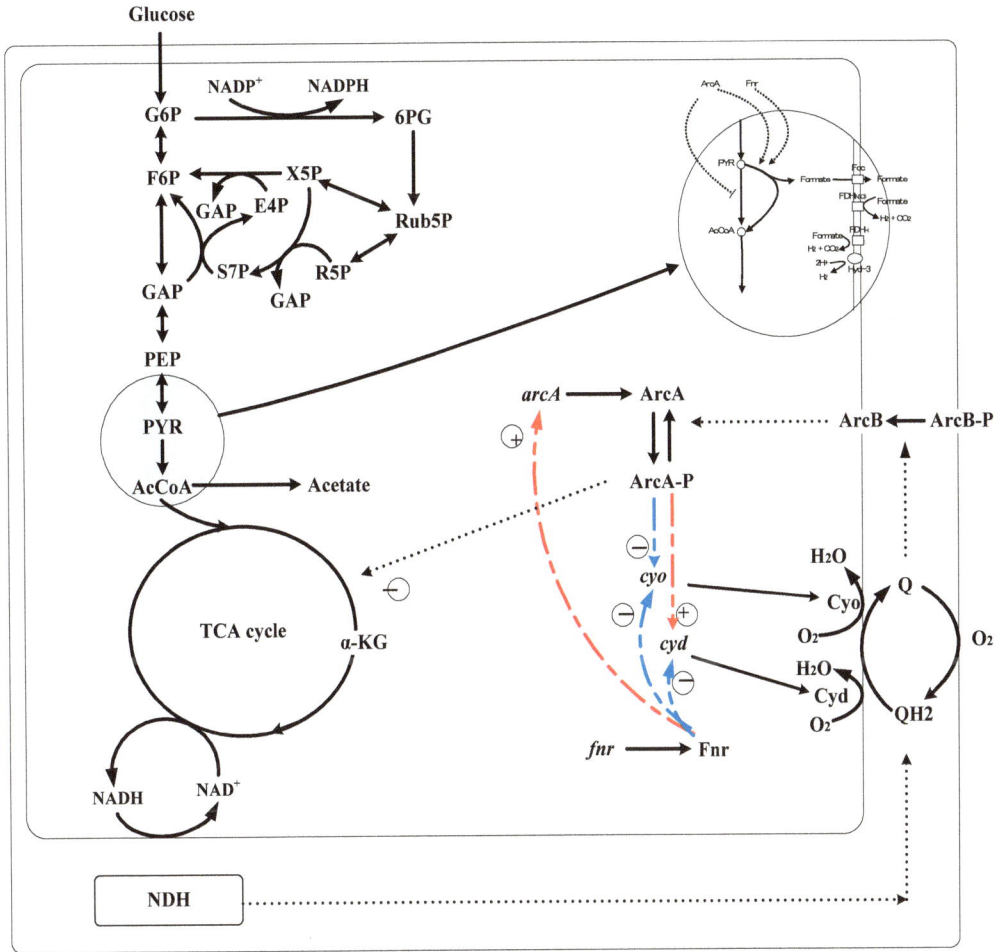

Figure 14: Metabolic regulation under reduced oxygen concentration.

Many bacteria utilize oxygen as the terminal electron acceptor, but they can switch to other acceptors such as nitrate under oxygen limitation. The reducing equivalents such as NADH are reoxidized in the respiratory chain, where oxygen, nitrate, fumarate, and dimethyl sulfoxide can be the electron acceptors. Nar plays a role when nitrate is present under oxygen limitation. Nar belongs to the two-component redox regulation systems, where it comprises a membrane sensor (NarX) that acts as a kinase causing phosphorylation of the regulator (NarL) under certain conditions [105]. The Nar system activates such genes as nitrate reduction encoding nitrate and nitrite reductases, and represses such genes as *frd* genes for fumarate reductase.

CONCLUDING REMARKS

Microbial production of biofuels and biochemicals from renewable resources or biomass has been paid recent attention from global sustainability and environmental protection points of view, and many attempts have been made for the cell design by metabolic engineering approach. However, the practical application is limited in many cases, and more innovative design of cell factories is desired. It is thus important to understand the metabolic regulation mechanism and develop appropriate models for the efficient design of cell factories. Here, the metabolic regulation mechanisms in response to nutrient limitations and culture conditions were explained focusing on *E.coli* metabolism. The transcriptional regulation is specific to the organism, while the enzyme level regulation mechanisms are conserved in many organisms, and thus the proper modeling on the latter regulation mechanisms will benefit to the development of many organisms.

NOTATION

PP pathway   = pentose phosphate pathway

TCA cycle   = tricarboxylic acid cycle

PTS   = phosphotransferase system

EI   = enzyme I

EII   – enzyme II

HPr   = histidine-phosphorylatable protein

Cya   = adenylate cyclase

CIT   = citrate

E4P   = erythrose-4-phosphate

FBP   = fructose-1,6-bisphosphate

F6P   = fructose-6-phosphate

G6P	=	glucose-6-phosphate
GAP	=	glycelaldehyde-3-phosphate
GLC	=	glucose
GOX	=	glyoxylate
ICI	=	isocitrate
KDPG	=	2-keto-3-deoxy-6-phosphogluconate
αKG	=	α-ketoglutarate
MAL	=	malate
OAA	=	oxaloacetate
PEP	=	phosphoenol pyruvate
6PG	=	6-phosphogluconate
PYR	=	pyruvate
R5P	=	ribose-5-phosphate
RU5P	=	ribulose-5-phosphate
S7P	=	sedoheptulose-7-phosphate
SUC	=	succinate
X5P	=	xylulose-5-phosphate
Ack	=	acetate kinase
Acs	=	acetyl coenzyme A synthetase
Adk	=	adenylate kinase
Ald	=	aldolase
CS	=	citrate synthase
Eno	=	enolase
Fum	=	fumarase

G6PDH	=	glucose-6-phosphate dehydrogenase
GAPDH	=	glyceraldehyde-3-phosphate dehydrogenase
ICDH	=	isocitrate dehydrogenase
Icl	=	isocitrate lyase
LDH	=	lactate dehydrogenase
MDH	=	malate dehydrogenase
Mez	=	malic enzyme
MS	=	malate synthase
Pck	=	phosphoenolpyruvate carboxykinase
PDH	=	pyruvate dehydrogenase
Pfk	=	phosphofructokinase
PGDH	=	6-phosphogluconate dehydrogenase
Pgi	=	phosphoglucose isomerase
Pgk	=	phosphoglycerate kinase
Ppc	=	phosphoenolpyruvate carboxylase
Pps	=	phosphoenolpyruvate synthase
Pta	=	phosphotransacetylase
Pyk	=	pyruvate kinase
Ru5P	=	ribulose phosphate epimerase
R5PI	=	ribose phosphate isomerase
SDH	=	succinate dehydrogenase
Tal	=	transaldolase
Tkt	=	transketolase

APPENDIX A

Effect of global regulators on the metabolic pathway gene expression

Global regulator	Metabolic pathway genes
Cra	+ : *aceBAK, cydB, fbp, icdA, pckA, pgk, ppsA* − : *acnB, adhE, eda, edd, pfkA, pykF, zwf*
Crp (cAMP-Crp)	+ : *aceEF,acnAB,acs, focA, fumA, gltA, lpdA, malT, manXYZ, mdh, mlc, pckA,* *pdhR, pflB, pgk, ptsG, sdhCDAB, sucABCD, ugpABCEQ,* − : *cyaA, lpdA, rpoS*
ArcA/B	+ : *cydAB, focA, pflB* − : *aceBAK, aceEF, acnAB, cyoABCDE, fumAC, gltA, icdA, lpdA, mdh,* *nuoABCDEFGHIJKLMN, pdhR, sdhCDAB, sodA, sucABCD*
Mlc	− : *crr, manXYZ, malT, ptsG, ptsHI*
PdhR	− : *aceEF, lpdA*
CsrA	+ : *eno,pfkA,pgi,pykF,tpiA* − : *fbp,glgC, glgA, glgB,pgm, ppsA,pckA,*
Fur	− : *entABCDEF, talB,sodA*

RpoS	+ : *acnA, acs, ada, appAR, appB, argH, aroM, dps, bolA, fbaB, fumC, gabP, gadA,*
	     gadB, katE, katG, ldcC, narY, nuv, pfkB, osmE, osmY, poxB, sodC,talA, tktB,ugpE,C,
	     xthA, yhgY,
	− : *ompF*

SoxR/S	+ : *acnA, cat, fumC, fur, sodA, sox, zwf*
OxyR	+; *ahpC, ahpF, katG:*
PhoR/B	+ : *phoBR, phoA-psiF, asr, pstSCAB-phoU* − : *phoH, phnCHN, ugpA, argP*
Fnr	+ : *acs, focA, frdABCD, pflB, yfiD* − : *acnA, cyoABCDE, cydAB, fumA, fnr, icdA, ndh, nuoABCEFGHIJKLMN,* *sdhCDAB, sucABCD*

REFERENCES

(Introduction)

[1]   Seshasayee AS, Bertone P, Fraser GM, Luscombe NM. Transcriptional regulatory networks in bacteria: from input signals to output responses. Curr Opin Microbiol 2006; 9: 511-519.

[2]   Kotte O, Zaugg JB, Heinemann M. Bacterial adaptation through distributed sensing of metabolic fluxes. Mol Syst Biol 2010; 6: 355.

[3]   Shimizu K. Regulation systems of bacteria such as *Escherichia coli* in response to nutrient limitation and environmental stresses. Metabolites 2014; 4:1-35.

[4]   Matsuoka Y, Shimizu K. Metabolic regulation in *Escherichia coli* in response to culture environments *via* global regulators. Biotechnol J 2011; 6:1330-1341.

[5]   Chuvukov V, Gerosa L, Kochanowski K, Sauer U. Coordination of microbial metabolism. Nature Reviews 2014; 12:327-340

[6]   Gunnewijk MG, van den Bogaard PT, Veenhoff LM, Heuberger EH, de Vos WM, Kleerebezem M, Kuipers OP, Poolman B. Hierarchical control *versus* autoregulation of carbohydrate utilization in bacteria. J Mol Microbiol Biotechnol 2001; 3: 401-413.

[7]   Deutscher J, Francke C, Postma PW. How phosphotransferase system-related protein phosphorylation regulates carbohydrate metabolism in bacteria. Microbiol Mol Biol Rev 2006; 70: 939-1031.

[8]   Postma PW, Lengeler JW, Jacobson GR. Phosphoenolpyruvate: carbohydrate phosphotransferase systems. In: Neidhardt FC, Curtiss III R, Ingraham JL, Lin ECC, Low KB, Magasanik B, Reznikoff WS, Riley M, Schaechter M, Umbarger HE, Eds. *Escherichia coli* and *Salmonella*: Cellular and Molecular Biology, 2nd edition. Washington DC, ASM Press 1996; pp. 1149-1174.

[9]   Tchieu JH, Norris V, Edwards JS, Saier MH Jr. The complete phosphotransferase system in *Escherichia coli*. J Mol Microbiol Biotechnol 2001; 3: 329-346.

[10]   Chou CH, Bennett GN, San KY. Effect of modulated glucose uptake on high-level recombinant protein production in a dense *Escherichia coli* culture. Biotechnol Prog 1994; 10: 644-647.

[11]   Gosset G. Improvement of *Escherichia coli* production strains by modification of the phosphoenolpyruvate: sugar phosphotransferase system. Microb Cell Fact 2005; 4:14.

[12]   Lunin VV, Li Y, Schrag JD, Iannuzzi P, Cygler M, Matte A. Crystal structures of *Escherichia coli* ATP-dependent glucokinase and its complex with glucose. J Bacteriol 2004; 186: 6915-6927.

[13]   Saier MH Jr, Ramseier TM. The catabolite repressor/ activator (Cra) protein of enteric bacteria. J Bacteriol 1996; 178: 3411-3417.

[14]   Kornberg HL. Fructose transport by *Escherichia coli*. Philos Trans R Soc Lond B Biol Sci 1990; 326: 505-513.

[15]   Kornberg HL. Routes for fructose utilization by *Escherichia coli*. J Mol Microbiol Biotechnol 2001; 3: 355-359.

[16]   Yao R, Kurata H, Shimizu K. Effect of *cra* gene mutation on the metabolism of *Escherichia coli* for a mixture of multiple carbon sources. Adv Biosci Biotechnol 2013; 4: 477-486.

[17]   Crasnier-Mednansky M, Park MC, Studley WK, Saier Jr MH. Cra-mediated regulations of *Escherichia coli* adenylate cyclase. Microbiology 1997; 143:785-792

[18]   Griffith JK, Baker ME, Rouch DA, Page MG, Skurray RA, Paulsen IT, Chater KF, Baldwin SA, Henderson PJ. Membrane transport proteins: implications of sequence comparisons. Curr Opin Cell Biol 1992; 4:684-695

[19]   Sumiya M, Davis EO, Packman LC, McDonald TP, Henderson PJ. Molecular genetics of a receptor protein for d-xylose, encoded by the gene *xylF*, in *Escherichia coli*. Receptors Channels 1995; 3:111-128

[20]   Song S, Park C. Organization and regulation of the d-xylose operons in *Escherichia coli* K-12: XylR acts as a transcriptional activator. J Bacteriol 1997; 179:7025-7032

[21]   Vasudevan P, Briggs M. Biodiesel production−current state of the art and challenges. J Ind Microbiol Biotechnol 2008; 35: 421–430.

[22]   Dharmadi Y, Murarka A, Gonzalez R. Anaerobic fermentation of glycerol by Escherichia coli: A new platform for metabolic engineering. Biotechnol & Bioeng 2006; 94: 821–829.

[23]   Clomburg JM, Gonzalez R. Anaerobic fermentation of glycerol: a platform for renewable fuels and chemicals, Trends in Biotechnol 2013; 31 (1): 20-28

[24]   Almeida JRM, Fávaro LCL, Betania F Quirino BF. Biodiesel biorefinery: opportunities and challenges for microbial production of fuels and chemicals from glycerol waste. Biotechnol for Biofuels 2012; 5: 48

[25]   Martínez-Gómez K, Flores N, Castañeda HM, Martínez-Batallar G, Hernández-Chávez G, Ramírez OT *et al*. New insights into *Escherichia coli* metabolism: carbon scavenging, acetate metabolism and carbon recycling responses during growth on glycerol. Micob Cell Fact 2012; 11:46

[26] Oh MK, Liao JC. Gene expression profiling by DNA microarrays and metabolic fluxes in *Escherichia coli*. Biotechnol Prog 2000; 16:278–286.

[27] Peng, L, Shimizu, K. Global metabolic regulation analysis for *Escherichia coli* K12 based on protein expression by 2-dimensional electrophoresis and enzyme activity measurement. Appl Microbiol Biotechnol 2003; 61:163-178

[28] Applebee MK, Joyce AR, Conrad TM, Pettigrew DW, Palsson BO. Functional and metabolic effects of adaptive glycerol kinase (GLPK) mutants in *Escherichia coli*. J Biol Chem 2011; 286:23150-23159

[29] Cheng KK, Lee BS, Masuda T, Ito T, Ikeda K, Hirayama A *et al*. Global metabolic network reorganization by adaptive mutations allows fast growth of *Escherichia coli* on glycerol. Nat Commun 2014; 5:3233

[30] Monod J. Recherches sur la Croissance de cultures Bacteriennes [thesis], Hermann et Cie, Paris, France, 1942.

[31] Magasanik B. Catabolite repression. Cold Spring Harb Symp Quant Biol 1961; 26: 249-256.

[32] Liu M, Durfee T, Cabrera JE, Zhao K, Jin DJ, Blattner FR. Global transcriptional programs reveal a carbon source foraging strategy by *Escherichia coli*. J Biol Chem 2005; 280: 15921-15927.

[33] Blencke HM, Homuth G, Ludwig H, Mäder U, Hecker M, J. Stülke J. Transcriptional profiling of gene expression in response to glucose in *Bacillus subtilis*: regulation of the central metabolic pathways. Metab Eng 2003; 5: 133-149.

[34] Moreno MS, Schneider BL, Maile RR, Weyler W, Saier MH Jr. Catabolite repression mediated by the CcpA protein in *Bacillus subtilis*: novel modes of regulation revealed by whole genome analyses. Mol Microbiol 2001; 39: 1366-1381.

[35] Görke B, Stülke J. Carbon catabolite repression in bacteria: many ways to make the most out of nutrients. Nat Rev Microbiol 2008; 6: 613-624.

[36] Plumbridge J. Expression of *ptsG*, the gene for the major glucose PTS transporter in *Escherichia coli*, is repressed by Mlc and induced by growth on glucose. Mol Microbiol 1998; 29: 1053-1063.

[37] De Reuse H, Danchin A. The *ptsH*, *ptsI*, and *crr* genes of the *Escherichia coli* phosphoenol pyruvate-dependent phosphotransferase system: a complex operon with several modes of transcription. J Bacteriol 1988; 170: 3827-3837.

[38] Bettenbrock K, Fischer S, Kremling A, Jahreis K, Sauter T, Gilles ED. A quantitative approach to catabolite repression in *Escherichia coli*. J Biol Chem 2006; 281: 2578-2584.

[39] Hogema BM, Arents JC, Bader R, Eijkemans K, Yoshida H, Takahashi H *et al*. Inducer exclusion in *Escherichia coli* by non-PTS substrates: the role of the PEP to pyruvate ratio in determining the phosphorylation state of enzyme IIA^{Glc}. Mol Microbiol 1998; 30: 487-498.

[40] Park YH, Lee BR, Seok YJ, Peterkofsky A. *In vitro* reconstitution of catabolite repression in *Escherichia coli*. J Biol Chem 2006; 281: 6448-6454.

[41] Tanaka Y, Kimata K, Aiba H. A novel regulatory role of glucose transporter of *Escherichia coli*: membrane sequestration of a global repressor Mlc. EMBO J 2000; 19: 5344-5352.

[42] Lee SJ, Boos W, Bouché JP, Plumbridge J. Signal transduction between a membrane-bound transporter, PtsG, and a soluble transcription factor, Mlc of *Escherichia coli*. EMBO J 2000; 19: 5353-5361.

[43] Bettenbrock K, Sauter T, Jahreis K, Kremling A, Lengeler JW, Gilles ED. Correlation between growth rates, $EIIA^{Crr}$ phosphorylation, and intracellular cyclic AMP levels in *Escherichia coli* K-12. J Bacteriol 2007; 189: 6891-6900.

[44] Winkler HH, Wilson TH. Inhibition of β-galactoside transport by substrates of the glucose transport system in *Escherichia coli*. Biochim Biophys Acta 1967; 135: 1030-1051.

[45] Hogema BM, Arents JC, Bader R, Postma PW. Autoregulation of lactose uptake through the LacY permease by enzyme IIA^{Glc} of the PTS in *Escherichia coli* K-12. Mol Microbiol 1999; 31: 1825-1833.

[46] Nelson SO, Wright JK, Postma PW. The mechanism of inducer exclusion. Direct interaction between purified III^{Glc} of the phosphoenol pyruvate: sugar phosphotransferase system and the lactose carrier of *Escherichia coli*. EMBO J 1983; 2: 715-720.

[47] Smirnova I, Kasho V, Choe JY, Altenbach C, Hubbell WL, Kaback HR. Sugar binding induces an outward facing conformation of LacY. Proc Natl Acad Sci USA 2007; 104: 16504-16509.

[48]   Misko TP, Mitchell WJ, Meadow ND, Roseman S. Sugar transport by the bacterial phosphotransferase system. Reconstitution of inducer exclusion in *Salmonella typhimurium* membrane vesicles, J Biol Chem 1987; 262: 16261-16266.
[49]   Nanchen A, Schicker A, Revelles O, Sauer U. Cyclic AMP-dependent catabolite repression is the dominant control mechanism of metabolic fluxes under glucose limitation in *Escherichia coli*. J Biotechnol 2008; 190: 2323-2330.
[50]   Moat AG, Foster JW, Spector MP, Eds. Microbiology, 4th edition. NewYork: Wiley-Liss 2002.
[51]   Saier MH Jr, Ramseier TM, Reizer J. Regulation of carbon utilization. In: Neidhardt FC, Curtiss III R, Ingraham JL, Lin ECC, Low KB, Magasanik B, Reznikoff WS, Riley M, Schaechter M, Umbarger HE, Eds. In *Escherichia coli* and *Salmonella*: Cellular and Molecular Biology, 2nd edition. Washington DC, ASM Press 1996; pp.1325-1343.
[52]   Shimada T, Yamamoto K, Ishihama A. Novel members of the Cra regulon involved in carbon metabolism in *Escherichia coli*. J Bacteriol 2011; 193 (3), 649-659
[53]   Sarkar D, Shimizu K. Effect of *cra* gene knockout together with other genes knockouts on the improvement of substrate consumption rate in *Escherichia coli* under microaerobic condition. Biochem Eng J 2008; 42: 224-228.
[54]   Sarkar D, Siddiquee KAZ, Araúzo-Bravo MJ, Oba T, Shimizu K. Effect of *cra* gene knockout together with *edd* and *iclR* genes knockout on the metabolism in *Escherichia coli*. Arch Microbiol 2008; 190: 559-571.

(Enzyme level and transcriptional regulation in the central metabolism)

[55]   Kremling A, Bettenbrock K, Gilles ED. A feed-forward loop guarantees robust behavior in *Escherichia coli* carbohydrate uptake. Bioinformatics 2008; 24:704-710
[56]   Kochanowski K, Volkmer B, Gerosa L, Haverkorn van Rijsewijk, BR, Schmidt A, Heinemann M. Functioning of a metabolic flux sensor in *Escherichia coli*. PNAS USA 2013; 110:1130-1135
[57]   Huberts DH, Niebel B, Heinemann M. A flux-sensing mechanism could regulate the switch between respiration and fermentation. FEMS Yeast Res 2012; 12(2), 118-128
[58]   Christen S, Sauer U. Intracellular characterization of aerobic glucose metabolism in seven yeast species by 13C flux analysis and metabolomics. FEMS Yeast Res 2011; 11: 263-272.
[59]   Boels E, Hollenberg CP. The molecular genetics of hexose transport in yeasts. FEMS Microbiol Rev 1997; 21:85-111.
[60]   Yang C, Hua Q, Baba T, Mori H, Shimizu K. Analysis of *Escherichia coli* anaprelotic metabolism and its regulation mechanisms from the metabolic responses to altered dilution rates and phosphoenolpyruvate carboxykinase knockout. Biotechnol & Bioeng 2003; 84:129-144
[61]   Lee B, Yen J, Yang L, Liao JC. Incorporating qualitative knowledge in enzyme kinetic models using fuzzy logic. Biotechnol Bioeng 1999; 63 (6): 722-729
[62]   Nizam SA, Zhu JF, Ho PY, Shimizu K. Effects of *arcA* and *arcB* genes knockout on the metabolism in *Escherichia coli* under aerobic condition, Biochem Eng J 2009; 44:240-250
[63]   Vemuri GN, Eiteman MA, Altman E. Increased recombinant protein production in *Escherichia coli* strains with overexpressed water-forming NADH oxidase and a deleted ArcA regulatory protein. Biotechnol & Bioeng 2006a; 94:538-542
[64]   Nizam SA, Shimizu K. Effects of *arcA* and *arcB* genes knockout on the metabolism in *Escherichia coli* under anaerobic and microaerobic conditions. Biochem Eng J 2008; 42: 229-236
[65]   Vemuri GN, Altman E, Sangurdekar DP, Khodursky AB, Eiteman MA. Overflow metabolism in *Escherichia coli* during steady-state growth: transcriptional regulation and effect of the redox ratio. Appl Environ Microbiol 2006b; 72:3653-3661
[66]   Wolfe AJ. The acetate switch. Microbiol Mol Biol Reviews 2005; 69, 12 – 50
[67]   Xu YF, Amador-Noguez D, Reaves ML, Feng XJ, Rabinowitz JD. Ultrasensitive regulation of anapleurosis *via* allosteric activation of PEP carboxylase. Nat Chem Biol 2012; 8:562-568
[68]   Schaub J, Reuss M. In vivo dynamics of glycolysis in E.coli shows need for gowth-rate dependent metabolome analysis. Biotechnol Prog 2008; 24, 1402-1407
[69]   Valgepea K, Adamberg K, Nahku R, Lahtvee PJ, Arike L, Vilu R. Systems biology approach reveals that overflow metabolism of acetate in *Escherichia coli* is triggered by carbon catabolite repression of acetyl-CoA synthetase. BMC Syst. Biol 2010; 4: 166.

[70]   Majewski RA, Domach MM. Simple constrained-optimization view of acetate overflow in *Escherichia coli*. Biotechnol & Bioeng 1990; 35: 732-738

[71]   Kremling A, Fischer S, Sauter T, Bettenbrock K, Gilles ED. Time hierarchies in the *Escherichia coli* carbohydrate uptake and metabolism. BioSystems 2004; 73, 57-71.

[72]   Kremling A, Bettenbrock K, Gilles ED. Analysis of global control of *Escherichia coli* carbohydrate uptake. BMC Syst Biol 2007; 1:42.

[73]   Van Heeswijk WC, Westerhoff HV, Boogerd FC. Nitrogen assimilation in Escherichia coli: putting molecular data into a systems perspective. Microbiol Mol Biol Rev 2013; 77 (4), 628-695.

[74]   Yan D. Protection of the glutamate pool concentration in enteric bacteria. PNAS USA 2007; 104: 9475-9480.

[75]   Ninfa AJ, Jiang P, Atkinson MR, Peliska JA. Integration of antagonistic signals in the regulation of nitrogen assimilation in *Escherichia coli*. Curr Top Cell Regul 2000; 36: 31-75.

[76]   Gruswitz F, O'Connell J 3rd, Stroud RM. Inhibitory complex of the transmembrane ammonia channel, AmtB, and the cytosolic regulatory protein, GlnK, at 1.96 A. *Proc Natl Acad Sci USA* 2007; 104: 42-47.

[77]   Radchenko MV, Thornton J, Merrick M. Control of AmtB-GlnK complex formation by intracellular levels of ATP, ADP, and 2-oxoglutarate. *J Biol Chem* 2010; 285: 31037-31045.

[78]   Truan D, Huergo LF, Chubatsu LS, Merrick M, Li XD, Winkler FK. A new P(II) protein structure identifies the 2-oxoglutarate binding site. *J Mol Biol* 2010; 400: 531-539.

[79]   Kim M, Zhang Z, Okano H, Yan D, Groisman A, Hwa T. Need-based activation of ammonium uptake in *Escherichia coli*. Mol Syst Biol 2012; 8: 616.

[80]   Yuan J, Doucette CD, Fowler WU, Feng XJ, Piazza M, Rabitz HA, Wingreen NS, Rabinowitz JD. Metabolomics-driven quantitative analysis of ammonia assimilation in *E. coli*. Mol Syst Biol 2009; 5: 302

[81]   Camarena L, Poggio S, García N, Osorio A. Transcriptional repression of *gdhA* in *Escherichia coli* is mediated by the Nac protein. FEMS Microbiol Lett 1998; 167: 51-56.

[82]   Ninfa AJ, Atkinson MR. PII signal transduction proteins. Trends Microbiol 2000; 8: 172-179.

[83]   Reitzer LJ, Magasanik B. Expression of *glnA* in *Escherichia coli* is regulated at tandempromoters. Proc Natl Acad Sci USA 1985; 82: 1979-1983.

[84]   Atkinson MR, Ninfa AJ. Mutational analysis of the bacterial signal-transducing protein kinase/phosphatase nitrogen regulator II (NR(II) or NtrB). J Bacteriol 1993; 175: 7016-7023.

[85]   Atkinson MR, Blauwkamp TA, Ninfa AJ. Context dependent functions of the PII and GlnK signal transduction proteins in *Escherichia coli*. J Bacteriol 2002; 184: 5364-5375.

[86]   Blauwkamp TA, Ninfa AJ. Physiological role of the GlnK signal transduction protein of *Escherichia coli*: survival of nitrogen starvation. Mol Microbiol 2002; 46: 203-214.

[87]   van Heeswijk WC, Hoving S, Molenaar D, Stegeman B, Kahn D, Westerhoff HV. An alternative P(II) protein in the regulation of glutamine synthetase in *Escherichia coli*. Mol Microbiol 1996; 21: 133-146.

[88]   Pahel G, Rothstein DM, Magasanik B. Complex *glnA-glnL-glnG* operon of *Escherichia coli*. J Bacteriol 1982; 150: 202-213.

[89]   Shapiro BM, Stadtman ER. Glutamine synthetase deadenylylating enzyme. Biochem Biophys Res Commun 1968; 30: 32-37.

[90]   Stadtman ER. Discovery of glutamine synthetase cascade. Methods Enzymol 1990; 182: 793-809.

[91]   Jaggi R, van Heeswijk WC, Westerhoff HV, Ollis DL, Vasudevan SG. The two opposing activities of adenylyl transferase reside in distinct homologous domains, with intra-molecular signal transduction. EMBO J 1997; 16: 5562-5571.

[92]   Maheswaran M, Forchhammer K. Carbon-sourcedependent nitrogen regulation in *Escherichia coli* is mediated through glutamine-dependent GlnB signalling. Microbiology 2003; 149: 2163-2172.

[93]   Merrick MJ, Edwards RA. Nitrogen control in bacteria. Microbiol Rev 1995; 59: 604-622.

[94]   Jiang P, Peliska JA, Ninfa AJ. The regulation of *Escherichia coli* glutamine synthetase revisited: role of 2-ketoglutarate in the regulation of glutamine synthetase adenylylation state. Biochemistry 1998; 37: 12802-12810.

[95]   Kustu S, Santero E, Keener J, Popham D, Weiss D. Expression of sigma 54 (*ntrA*)-dependent genes is probably united by a common mechanism. Microbiol Rev 1989; 53: 367-376.

(Interaction between N-regulation and C-regulation)

[96]   Doucette CD, Schwab DJ, Wingreen NS, Rabinowitz JD. Alpha-ketoglutarate coordinates carbon and nitrogen utilization *via* enzyme I inhibition. *Nat Chem Biol* 2011; **7**: 894-901.
[97]   Rabinowitz J, Silhavy TJ Metabolite turns master regulator. Nature 2012; 500: 283-284
[98]   You C, Okano H, Hui S, Zhang Z, Kim M, Gunderson CW *et al*. Coordination of bacterial proteome with metabolism by cyclic AMP signaling. Nature 2013; 500: 301-306.
[99]   Gunsalus RP. Control of electron flow in *Escherichia coli*: coordinated transcription of respiratory pathway genes. J Bacteriol 1992; 174 (22): 7069–7074
[100]   Salmon K, Hung SP, Mekjian K, Baldi P, Hatfield GW, Gunsalus RP. Global gene expression profiling in *Escherichia coli* K12: Theeffects of oxygen availability and FNR. J Biol Chem 2003; 278 (32): 29837–29855.
[101]   Alexeeva S, Hellingwerf KJ, de Mattos MJT. Requirement of ArcA for redox regulation in *Escherichia coli* under microaerobic but not anaerobic or aerobic conditions. J Bacteriol 2003; 185 (1): 204–209.
[102]   Zhu, J, Shalel-Levanon S, Bennett G, San KY. Effect of the global redox sensing/ regulation networks on *Escherichia coli* and metabolic flux distribution based on C-13 labeling experiments. Metab Eng 2006; 8 (6): 619–627.
[103]   Georgellis D, Kwon O, Lin ECC. Quinones as the redox signal for the arc two-component system of bacteria. Science 2001; 292 (5525): 2314–2316.
[104]   Malpica R, Franco B, Rodriguez C, Kwon O, Georgellis D. Identification of a quinone-sensitive redox switch in the ArcB sensor kinase, PNAS USA 2004; 101 (36): 13318–13323.
[105]   Constantinidou C, Hobman JL, Grifiths L, Patel MD, Penn CW, Cole JA *et al*. A reassessment of the FNR regulon and transcriptomic analysis of the effects of nitrate, nitrite, NarXL, and NarQP as *Escherichia coli* K12 adapts from aerobic to anaerobic growth. J Biol Chem 2006; 281 (8): 4802-4815
[106]   Kessler D, Knappe J. Anaerobic dissimilation of pyruvate, in *E. coli* and *Salmonella*: Cellular and Molecular Biology, Neidhardt FC, Curtiss R, Ingraham JI *et al*., Eds., vol. 1, pp. 199–205, ASM Press,Washington, DC, USA, 2nd edition, 1996.
[107]   Toya Y, Nakahigashi K, Tomita M, Shimizu K. Metabolic regulation analysis of wild-type and *arcA* mutant *Escherichia coli* under nitrate conditions using different levels of omics data. Mol Biosyst 2012; 8: 2593-2604.
[108]   Maeda T, Sanchez-Torres V, Wood TK. Enhanced hydrogen production from glucose by metabolically engineered *Escherichia coli*. Appl Microbiol Biotechnol 2007; 77: 879-890
[109]   Zhu J, Shimizu K. The effect of *pfl* genes knockout on the metabolism for optically pure d -lactate production by *Escherichia coli*. Appl Microbiol Biotechnol 2004; 64; 367-75
[110]   Zhu J, Shimizu K. Effect of a single-gene knockout on the metabolic regulation in *E. coli* for d -lactate production under microaerobic conditions. Metabolic Eng 2005; 7; 104-15.
[111]   Valgepea K, Adamberg K, Seiman A, Vilu R. *Escherichia coli* achieves faster growth by increasing catalytic and translation rates of proteins. Mol BioSys 2014; 9, 2344-2358

CHAPTER 8

Modeling and Computer Simulation for the Main Metabolism

Abstract: Modeling of the main metabolic pathways such as glycolysis, TCA cycle, pentose phosphate (PP) pathway, and anaplerotic pathways is considered, where the fluxes obtained by computer simulation can be used to compute the specific ATP production rate, the specific CO_2 production rate, and the specific NAD(P)H production rate. The specific ATP production rate thus computed can be used for the estimation of the specific growth rate. The model can be further extended for catabolite regulation by incorporating the effects of transcription factors such as Cra and cAMP-Crp, where acetate overflow metabolism and co-consumption of multiple sugars can be clarified by this modeling approach. Modeling for anaerobic fermentation is also considered, where aerobic/ anaerobic switch can be made by incorporation of the roles of ArcA/B and Fnr. Modeling for NH_3 assimilation pathways is then considered, where this may be combined with the main metabolism to simulate nitrogen regulation at various C/N ratios. Finally, amino acid synthetic pathways are considered for lysine production.

Keywords: Modeling, kinetics, computer simulation, dynamic simulation, gene knockout mutant, *Escherichia coli*, catabolite regulation, co-consumption of multiple carbon sources, nitrogen metabolism, anaerobic fermentation, transcription factors, lysine fermentation.

INTRODUCTION

It is quite important to understand the complex and highly interrelated cellular behavior quantitatively. This may be achieved with the help of modeling and computer simulation by integrating different levels of ever-increasing amount of experimental data using biological knowledge. The ultimate goal of systems biology is to develop *in silico* models of a whole cell or cellular processes that can predict cellular phenotypes in response to culture environment and/or genetic perturbation. If this could be made by appropriate modeling, the cell design can be made more efficiently without conducting exhaustive experiments, and one can screen out the promising candidates, followed by the verification of only some of the candidates by experiments. It is important to model the main metabolic pathways, since energy generation and cell synthesis may be estimated for the phenotypic growth characteristics based on the intracellular fluxes of the main metabolic pathways.

Some of the mathematical models which can describe the dynamic behavior of the intracellular metabolite concentrations of the central metabolic pathways have

Kazuyuki Shimizu and Yu Matsuoka

been developed for *Saccharomyces cerevisiae* [1-3]. The intracellular metabolite concentrations were measured for the pulse addition of glucose during continuous culture at low dilution rate, and the time profile was compared to the dynamic behavior predicted by the computer simulation with model parameter identification [4-7]. The dynamic behavior of *Escherichia coli* cell has also been investigated by fast sampling system [8]. The kinetic models for the glycolysis and pentose phosphate (PP) pathway have been developed for *E.coli* to simulate the transient data obtained by the fast sampling system [9]. This model does not contain tri carboxylic acid (TCA) cycle and fermentative pathways, and thus cannot simulate the typical batch and continuous cultures. This can be achieved by considering almost all the kinetic equations for the TCA cycle and anaplerotic pathways as well as glycolysis and PP pathway [10].

A wealth of information is available on the molecular biology, biochemistry, and physiology of cellular metabolism in response to culture environment, and some attempts have been made for the modeling and simulation as stated above. It is important to make modeling based on the integrated information from gene level to flux level [11, 12]. Hardiman [13] considered the topology of the global regulatory network in *E. coli* for time-series metabolic flux analysis from the cell growth phase to the nutrient-limited stationary phase in the fed-batch culture. The catabolic regulation together with PTS has been modeled by several researchers [14-19].

In the computer simulation, the intracellular metabolite concentrations can be estimated by solving the mass balances with enzymatic equations. The model parameters such as K_s and K_i may be used as those given in the literature obtained by *in vitro* measurement, while the parameter values of v_{max} may be better tuned by the experimental data of batch and continuous cultures [9, 10]. This is based on the fact that most of K_s and K_i values do not much affect the growth phenotype under typical growth condition, and it is not appropriate to identify those parameter values based on the experimental data in the typical batch culture. It must be, however, careful for the case of allosteric regulation, where the corresponding parameters may become important, and must be identified by the experimental data.

In the present chapter, the kinetic modeling of the main metabolic pathways are explained for the computer simulation of batch and continuous cultures, where the cell growth rate may be estimated in relation to ATP production rate. It is important to incorporate the effects of the transcription factors on the metabolic reactions into the model to simulate the phenotypic behavior under genetic and/ or environmental perturbations. The modeling for the nitrogen regulation is then explained. The modeling and simulation for the peripheral metabolism such as lysine synthetic pathways are also explained. The idea of modeling the effect of oxygen concentration on the metabolic changes will also be explained.

MODELING FOR THE MAIN METABOLISM

Brief Summary for the Modeling of the Main Metabolism

For the modeling of the main metabolism, consider the metabolic pathways as given in Fig. **1** for the glycolysis, TCA cycle, PP pathway, anaplerotic pathways, and gluconeogenic pathways. For the proper modeling, it is important to take into account enzyme level regulation such as allosteric regulation as well as transcriptional regulation *via* global regulators or transcription factors. Referring to Fig. **1**, the mass balance equations are expressed as follows:

a) Glycolysis

$$\frac{d[Glc]}{dt} = v_{NPTS} - v_{Glk} - v_{PTS-up} - \mu[Glc] \tag{1a}$$

$$\frac{d[G6P]}{dt} = v_{PTS-up} - v_{Pfk} - v_{G6PDH} - v_{Bio,G6P} - v_{Gpm} - \mu[G6P] \tag{1b}$$

$$\frac{d[F6P]}{dt} = v_{Pgi} + v_{Fbp} + v_{Tkb} + v_{Tal} - v_{Pfk} - \mu[F6P] \tag{1c}$$

$$\frac{d[FDP]}{dt} = v_{Pfk} - v_{Ald} - v_{Fbp} - \mu[FDP] \tag{1d}$$

$$\frac{d[DHAP]}{dt} = v_{Ald} - v_{Tpi} - \mu[DHAP] \tag{1e}$$

$$\frac{d[GAP]}{dt} = v_{Ald} + v_{Tpi} - v_{GAPDH} + v_{Tka} + v_{Tkb} - v_{TAl} - v_{Bio,GAP} - \mu[GAP] \tag{1f}$$

$$\frac{d[1,3BPG]}{dt} = v_{GAPDH} - v_{Pgk} - \mu[1,3BPG] \tag{1g}$$

$$\frac{d[3PG]}{dt} = v_{Pgk} - v_{Pgm} - v_{Bio,3PG} - \mu[3PG] \tag{1h}$$

$$\frac{d[2PG]}{dt} = v_{Pgm} - v_{Eno} - \mu[2PG] \tag{1i}$$

$$\frac{d[PEP]}{dt} = v_{Eno} + v_{Pck} + v_{Pps} - v_{PykF} - v_{Ppc} - v_{PTS} - v_{Bio,PEP} - \mu[PEP] \tag{1j}$$

$$\frac{d[PYR]}{dt} = v_{Pyk} + v_{Mez} + v_{PTS} - v_{PDH} - v_{PpsA} - v_{Bio,PYR} - \mu[PYR] \tag{1k}$$

PDH and Acetate Formation Pathways

$$\frac{d[AcCoA]}{dt} = v_{PDH} + v_{Acs} - v_{CS} - v_{Pta} - v_{MS} - v_{Bio,AcCoA} - \mu[AcCoA] \tag{2a}$$

$$\frac{d[AcP]}{dt} = v_{Pta} - v_{Ack} - \mu[AcP] \tag{2b}$$

$$\frac{d[Ace]}{dt} = v_{Ack} - v_{Acs} - \mu[Ace] \tag{2c}$$

TCA Cycle

$$\frac{d[CIT]}{dt} = v_{CS} - v_{Acn} - \mu[CIT] \tag{3a}$$

$$\frac{d[ICIT]}{dt} = v_{Acn} - v_{ICDH} - v_{Icl} - \mu[ICIT] \tag{3b}$$

$$\frac{d[\alpha KG]}{dt} = v_{ICDH} - v_{KGDH} - v_{GDH} - v_{Bio,\alpha KG} - \mu[\alpha KG] \tag{3c}$$

$$\frac{d[SCoA]}{dt} = v_{IKGDH} - v_{SCS} - \mu[SCoA] \tag{3d}$$

$$\frac{d[SUC]}{dt} = v_{SCS} + v_{Icl} - v_{SDH} - \mu[SUC] \tag{3e}$$

$$\frac{d[FUM]}{dt} = v_{SDH} - v_{Fum} - \mu[FUM] \tag{3f}$$

$$\frac{d[MAL]}{dt} = v_{Fum} + v_{MS} - v_{MDH} - v_{Mez} - \mu[MAL] \tag{3g}$$

$$\frac{d[OAA]}{dt} = v_{MDH} + v_{Ppc} - v_{CS} - v_{Pck} - v_{Bio,OAA} - \mu[OAA] \tag{3h}$$

Glyoxylate Pathway

$$\frac{d[GOX]}{dt} = v_{Icl} - v_{MS} - \mu[GOX] \tag{4a}$$

PP Pathway

$$\frac{d[PGL/6PG]}{dt} = v_{G6PDH} - v_{6PGDH} - \mu[PGL/6PG] \tag{5a}$$

$$\frac{d[RU5P]}{dt} = v_{6PGDH} - v_{Rpe} - v_{Rpi} - \mu[RU5P] \tag{5b}$$

$$\frac{d[R5P]}{dt} = v_{Rpi} - v_{TktA} - v_{Bio,R5P} - \mu[R5P] \tag{5c}$$

$$\frac{d[X5P]}{dt} = v_{Rpe} + v_{Xyk} - v_{TktA} - \mu[X5P] \tag{5d}$$

$$\frac{d[S7P]}{dt} = v_{TktA} - v_{Tal} - \mu[S7P] \tag{5e}$$

$$\frac{d[E4P]}{dt} = v_{Tal} - v_{TktB} - v_{Bio,E4P} - \mu[E4P] \tag{5f}$$

where μ is the specific growth rate. The suffix "Bio" means biosynthetic flux. The kinetic rate equations for the enzymatic reactions are given in Chapter 4. In the above equations, the specific rates as shown by $v_{\bullet}$ are functions of substrate, product, and some specific metabolite concentrations. In fact, $v_{\bullet}$ are also functions of transcription factors as will be explained later in this chapter. The terms associated with μ account for the dilution of the intracellular metabolites caused by the cell growth. [ ] represents the concentration of metabolite, where some were lumped together.

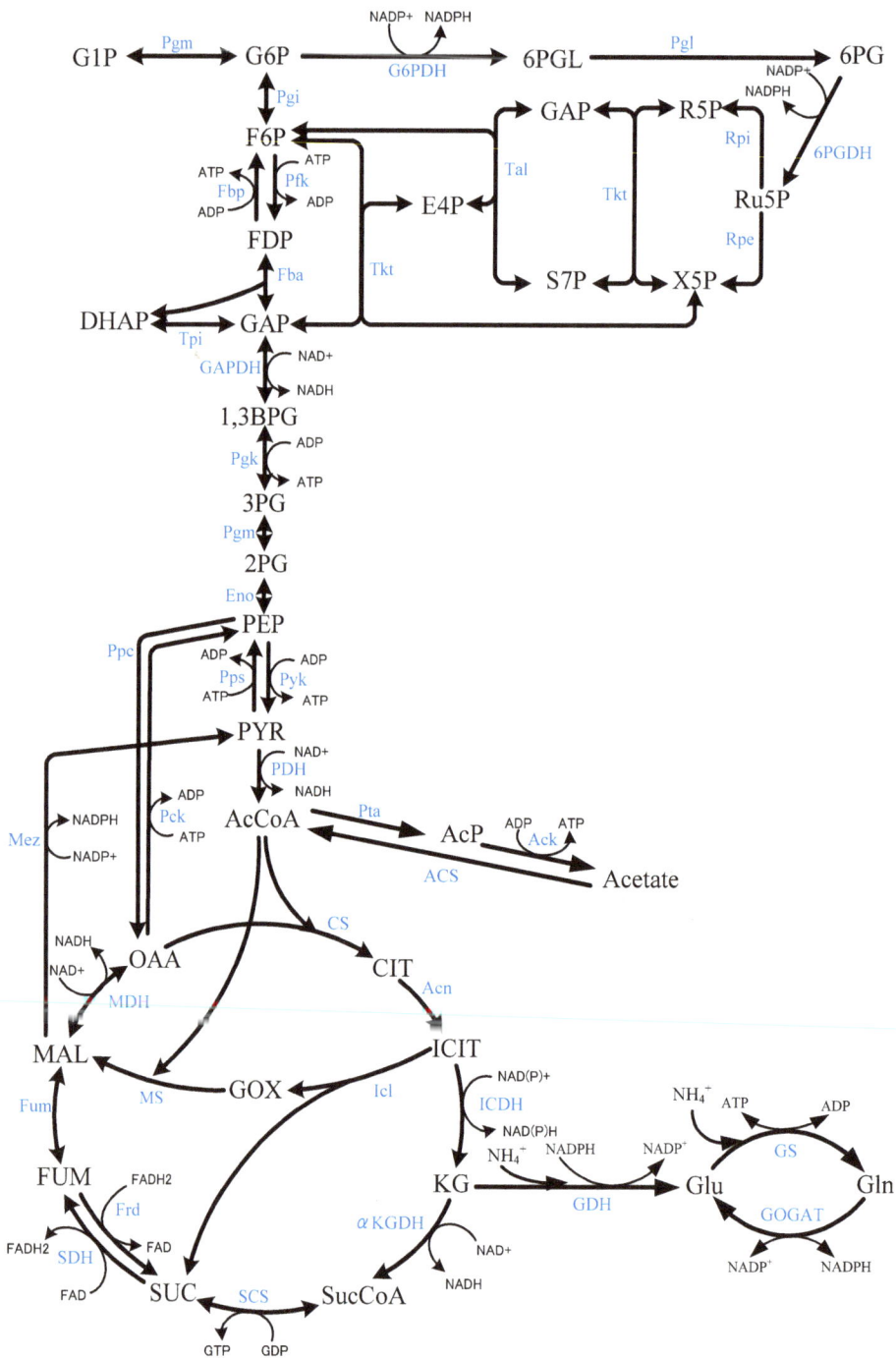

Figure 1: Main metabolic pathways for the simulation of *E. coli.*

Once the metabolic fluxes of the main metabolic pathways were obtained by computer simulation, the specific ATP production rate, the specific CO_2 production rate, and the specific NAD(P)H production rate can be also estimated. ATP is produced either by the substrate level phosphorylation or oxidative phosphorylation, where the reducing equivalents such as NADH and $FADH_2$ can contribute in generating ATP *via* oxidative phosphorylation under aerobic condition. The pathways involved in the electron transfer and the oxidative phosphorylation have variable stoichiometry depending on the use of different NADH dehydrogenases such as NDH-I and NDH-II, and different cytochromes such as Cyo and Cyd in *E.coli*. The number of H^+ translocated from periplasm into cytosol by the membrane bound H^+-ATPase to phosphorylate ADP to form ATP has been estimated to be 2-4 under aerobic condition. Referring to Fig. **1**, the specific ATP production rate may be estimated by

$$v_{ATP} = OP_{NADH} + OP_{FADH_2} + v_{Pgk} + v_{Pyk} + v_{Ack} + v_{SCS} - v_{Pfk} - v_{Pck} \qquad (6)$$

where OP_{NADH} and OP_{FADH_2} are the specific ATP production rates *via* oxidative phosphorylation, which may be estimated by

$$OP_{NADH} = [v_{GAPDH} + v_{PDH}(+v_{ICDH}) + v_{KGDH} + v_{MDH}] \times (P/O) \qquad (7a)$$

and

$$OP_{FADH_2} = v_{SDH} \times (P/O)' \qquad (7b)$$

where (P/O) and $(P/O)'$ are the P/O ratios for NADH and $FADH_2$, respectively, and most likely to be 2.5 and 1.5, respectively, under aerobic condition. Likewise, the specific CO_2 production rate may be estimated by

$$v_{CO_2} = v_{PGDH} + v_{PDH} + v_{ICDH} + v_{KGDH} + v_{Mez} + v_{Pck} - v_{Ppc} \qquad (8)$$

and the specific NADPH production rate may be estimated by

$$v_{NADPH} = v_{G6PDH} + v_{PGDH}(+v_{ICDH}) + v_{Mez} \qquad (9)$$

where ICDH is considered to be NADH-forming or NADPH-forming enzyme depending on the type of organisms. Since NADPH plays an important role for detoxification of reactive oxygen species (ROSs) generated under oxidative stress condition, some prokaryotic microorganisms such as *E.coli* produce NADPH at ICDH in the TCA cycle together with the reactions at G6PDH, 6PGDH, and possibly at Mez [20].

Modeling for the Cell Growth Rate

The estimation for the cell growth rate is by far important. The most typical equation for the cell growth rate is expressed by the Monod type model such as

$$\mu(S) = \frac{\mu_m S}{K_s + S} \tag{10a}$$

where S is the substrate concentration and μ_m and K_s are the model parameters. The drawback of this equation is that the cell keeps growing with the specific growth rate at μ_m as far as $S \gg K_s$ where K_s is usually quite small for the case of using glucose as the substrate. This does not reflect the fermentation characteristics in practice. Therefore, the following empirical equation may be sometimes considered in practice for batch cultivation:

$$\mu(S, X) = \mu_m \left(1 - \frac{X}{X_m}\right)^n \frac{S}{K_s + S} \tag{10b}$$

where X is the cell concentration and X_m is the final value of X in the batch culture, and n is the model parameter. X_m may be set as $S(0)Y_{X/S}$ where $S(0)$ is the initial substrate concentration, and $Y_{X/S}$ is the yield coefficient. Although this equation may be used to fit the dynamic behaviour in the batch culture, Eq. (10b) is the empirical equation, and may not be able to express the cell growth rate for the cells growing under various environmental conditions and for the mutants.

The cell growth rate is determined in principle by the cell synthesis from the precursors formed in the main metabolism (anabolism) and ATP formation (catabolism) in the typical growth condition. Consider the latter case for the modeling of the cell growth rate. Fig. **2a** shows the linear relationship between the specific growth rate and the specific ATP production rate [10, 21]. As stated before, ATP is generated by either the substrate level phosphorylation or oxidative phosphorylation from NADH and FADH$_2$ at the respiratory chain. NDH-I transports 2 H$^+$/e$^-$, while NDH-II transports 0 H$^+$/e$^-$. Then the quinol oxidases such as cytochromes Cyt bO$_3$ and Cyt bd transport 2 H$^+$/e$^-$ and 1 H$^+$/e$^-$, respectively [22]. Then the number of H$^+$ transported into the cytosol by the membrane bound H$^+$-ATPase to phosphorylate ADP to form ATP has been estimated to be 2-4, where 2.5 may be the most likely under aerobic condition [23]. Here, instead of assuming such values, consider P/O ratio to be the model parameter to be tuned by the experimental data. Then the cell growth rate may be further modified as

$$\mu(S, X, v_{ATP}) = \mu_m \left(1 - \frac{X}{X_m}\right)^n \frac{S}{K_s + S} k_{ATP} v_{ATP}(\bullet) \tag{11}$$

where $v_{ATP}(\bullet)$ is the specific ATP production rate computed from the fluxes, and k_{ATP} is the adjustable model parameter.

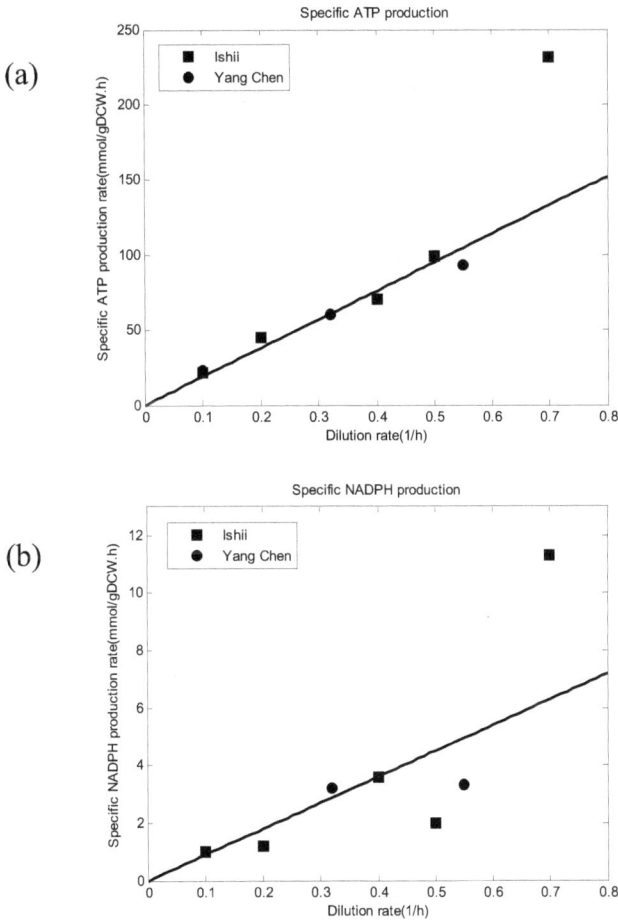

Figure 2: The specific ATP and NADPH production rates with respect to dilution rate in the continuous culture: (a) the specific ATP production rate *vs.* dilution rate; (b) the specific NADPH production rate *vs.* dilution rate. The rectangle symbols are the experimental data from [24] while circle symbols are the experimental data from [25].

Fig. **3** shows the batch cultivation result for the wild-type *E.coli*, where the lines are the simulation result, while the symbols are the experimental data [10]. Fig. **3a** shows the effect of k_{ATP}, while Fig.**3b** shows the effect of (P/O) ratio on the cultivation characteristics, where k_{ATP} and (P/O) ratio are the model parameters as shown in Eqs. (7a), (7b) and (11). Fig. **3c** shows the simulation result where k_{ATP} and (P/O) ratio were fitted to the experimental data, where $k_{ATP} = 0.10$ and $(P/O) = 2.5$.

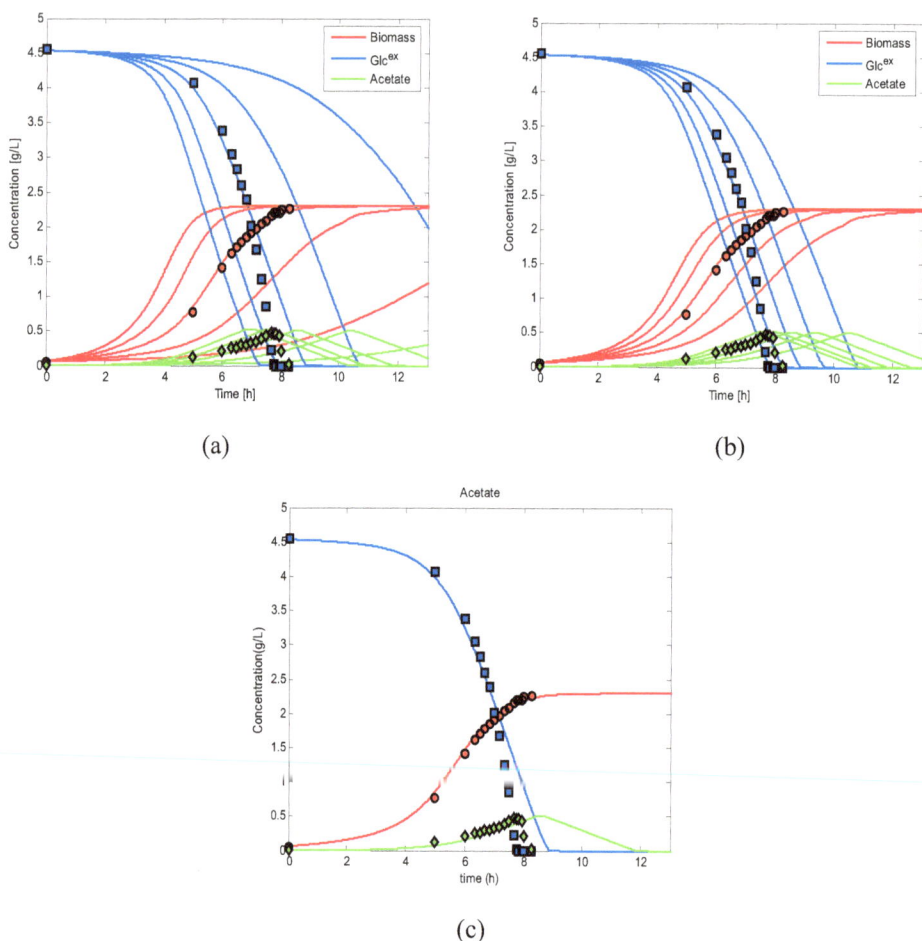

(a)

(b)

(c)

Figure 3: Effect of k_{ATP} and P/O ratio on the cell growth characteristics: (a) Effect of k_{ATP} ; (b) effect of P/O ratio; (c) cell growth characteristics for the best fitted value where $k_{ATP} = 0.10$ and P/O ratio = 2.5 [10].

Some of the initial concentrations of the intracellular metabolites such as OAA, PEP and PYR affect the cultivation characteristics as shown by Fig. **4** [10]. Since those will change depending on the pre-cultivation, those may be also considered to be adjustable parameters for the simulation of batch culture, but it is of minor importance.

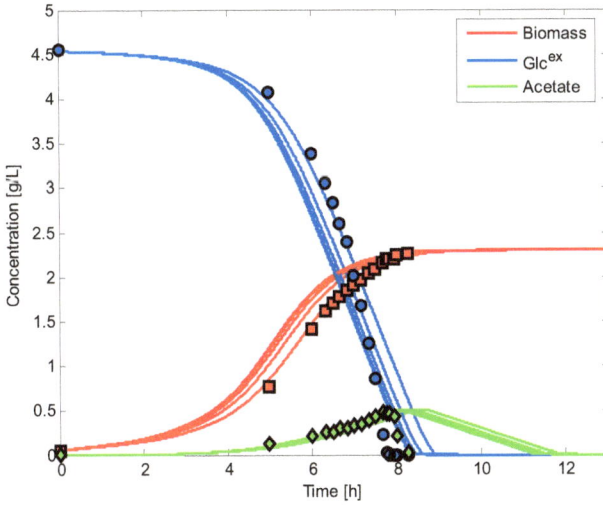

(a)

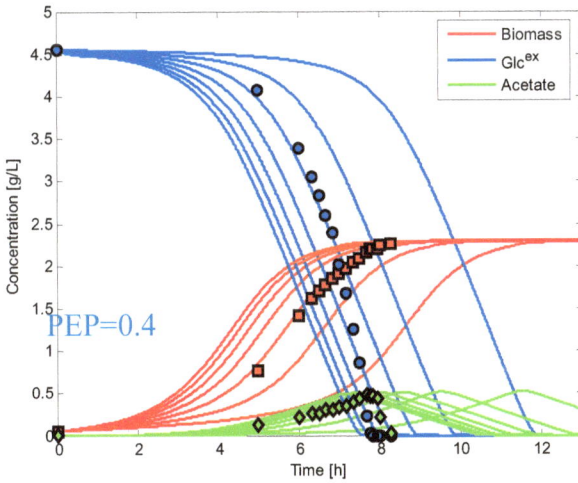

(b)

Figure 4: Effect of initial OAA concentrations (0.01, 0.016, 0.011, 0.16, 0.21) (a) and initial PEP concentrations (0.04, 0.1, 0.16, 0.22, 0.28, 0.34, 0.4) on the fermentation chacacteristics [10].

Moreover, Fig. **2b** shows the relationship between the specific growth rate μ and the specific NADPH production rate, v_{NADPH} computed based on the fluxes obtained by ^{13}C-labeling experiments [10, 21], where it indicates the linear relationship such that

$$v_{NADPH} = k_{NADPH}\mu \tag{12}$$

Once we identified k_{NADPH} value from Fig. **2b**, we may compute v_{NADPH} after μ was given. Then, the flux of the oxidative PP pathway may be estimated to provide the same v_{NADPH} value.

Simulation of the Specific Pathway Knockout Mutant

From the practical application point of view, the primary interest is the predictability of the model for the specific gene knockout. Fig. **5a** shows the simulation result of estimating the cell growth rate by the above method for the case of Ppc knockout (broken lines) as compared to the wild type (solid lines), where the filled symbols are the experimental data for the wild type, and the open symbols are those for Ppc mutant [10], where some rule-based knowledge is incorporated in the simulation. Fig. **5a** shows reasonably good predictability of the simulation result. Upon Ppc knockout, PEP concentration increases, which in turn activates Pyk activity as well as PTS activity, and thus PYR concentration increases. Although AcCoA concentration increases, v_{CS} became lower. This is due to lower OAA concentration caused by Ppc knockout, indicating that Ppc mutant has difficulty in growing in synthetic media [26]. The simulation result indicates the importance of anapleorotic route of Ppc to backup OAA. The significant decrease in OAA concentration caused significant decrease in TCA cycle fluxes as well as intermediate metabolite concentrations, and this caused lower ATP production (Fig. **5b**), which then caused lower cell growth rate. However, the overall cell yield or the final cell concentration for Ppc mutant was higher as compared to the wild type, consistent with experimental data (Fig. **5a**). The change in the specific CO_2 production rate (q_{co_2}) is also shown in Fig. **5c**, where the lower CO_2 production rate is due to lower TCA cycle activity, and this contributes to the increase in the cell yield [10].

Modeling for Catabolite Regulation by Incorporating the Effects of TFs

Catabolite regulation is typically observed in many microorganisms when different carbohydrates are present in a medium, where glucose is preferentially consumed, and the uptake of other carbon sources is repressed [27]. The central players for catabolite regulation in *E. coli* are the phosphoenol pyruvate (PEP):

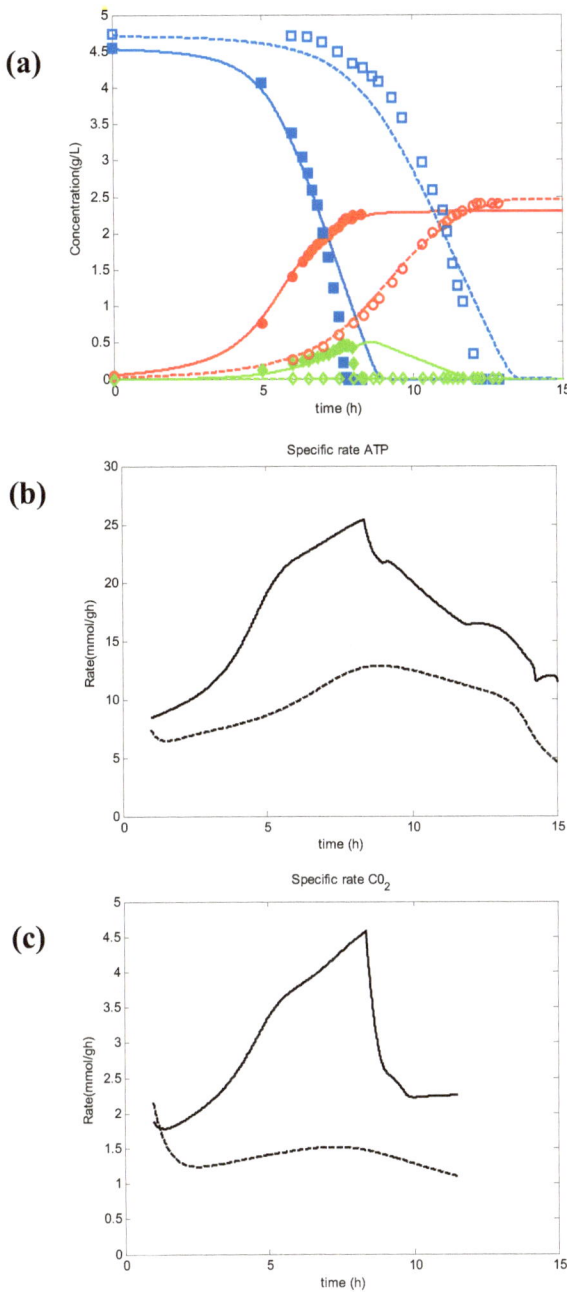

Figure 5: Effect of Ppc knockout on fermentation characteristics, where solid lines represent the simulation result for the wild type, and dotted lines represent the simulation result for Ppc knockout mutant. The filled symbols represent experimental data for wild type, open symbols are for Ppc knockout mutant: (a) fermentation characteristics; (b) specific ATP production rate; (c) specific CO_2 production rate [10].

carbohydrate phosphotransferase systems (PTSs) and cAMP-Crp as explained in Chapter 7, where cAMP-Crp activates *ptsG* gene expression as well as TCA cycle gene expression (Appendix A of Chapter 7). These PTS systems have been modeled by several researchers [14-17], where the model equations for PTS are given in Appendix A.

In addition to cAMP-Crp, which acts depending on the level of glucose concentration, the catabolite repressor/activator protein (Cra) originally characterized as the fructose repressor (FruR) plays an important role in the control of carbon flow in *E. coli* as explained in Chapter 7. The carbon uptake and glycolysis genes such as *ptsHI*, *pfkA*, *pykF*, *zwf* and *edd-eda* are repressed, while gluconeogenic pathway genes such as *ppsA*, *fbp*, *pckA*, *icdA* and *aceBAK* are activated by Cra in *E.coli* [28] (Appendix A of Chapter 7).

Under glucose limitation, cAMP is generated from ATP by Cya, where Cya is activated by the phosphorylated E Ⅱ A (E Ⅱ A-P) as explained in Chapter 7. The following equation may be considered for v_{Cya} as a function of [EIIA-P]:

$$v_{Cya} = \frac{v_{Cya}^{\max}[EIIA-P]}{[EIIA-P]+K_{Cya}} \tag{13}$$

Since cAMP-Crp activates the expression of *ptsG* gene, which encodes EIIBC^{Glc}, the specific glucose consumption rate v_{PTS-up} may be expressed as

$$v_{PTS-up} = v_{PTS-up}(\bullet)\frac{[cAMP-Crp]}{[cAMP-Crp]+K_{cAMP-Crp}} \tag{14}$$

where $v_{PTS-up}(\bullet)$ is the kinetic rate equation as given in Chapter 4. Since *acs* is activated by cAMP-Crp (Appendix A of Chapter 7), v_{Acs} may be also expressed as

$$v_{Acs} = v_{Acs}(\bullet)\left(\frac{[cAMP-Crp]}{[cAMP-Crp]+K_{acs-cAMP-Crp}}\right) \tag{15a}$$

where $v_{Acs}(\bullet)$ is the kinetic rate equation as given in Chapter 4. CS is encoded by *gltA*, and it is under control of cAMP-Crp such that

$$v_{CS} = v_{CS}(\bullet)\left(\frac{[cAMP-Crp]}{[cAMP-Crp]+K_{gltA-cAMP-Crp}}\right) \tag{15b}$$

In the same way, v_{TCA} may be a function of cAMP-Crp and Cra such that

$$v_{TCA} = v_{TCA}(\bullet)\left(\frac{[cAMP-Crp]}{[cAMP-Crp]+K_{tca-cAMP-Crp}}\right)\left(\frac{[Cra]}{[Cra]+K_{tca-Cra}}\right) \tag{15c}$$

where v_{TCA} is the lumped flux of TCA cycle other than CS for simplicity, and the above equation reflects the fact that *icdA* is activated by Cra, while *icdA*, *sucABCD*, *sdhCDAB*, *fumA*, *mdh* are activated by cAMP-Crp, but care must be taken for the modeling where the regulation of PDH is rather complex. Note that *lpdA* which encodes a subunit of PDH is also under control of cAMP-Crp (Appendix A of Chapter 7).

Noting that Cra activity is repressed by FBP, let the reaction rate for Cra be expressed as

$$v_{Cra} = k_{Cra}\left(1 - \frac{[FBP]}{[FBP]+K_{Cra}}\right) \tag{16}$$

Since Cra activates gluconeogenic pathway genes, while it represses glycolytic gene expression, v_{Pfk} and v_{Pyk} may be expressed as functions of [Cra] such that

$$v_{Pfk} = v_{Pfk}(\bullet)\left(1 - \frac{[Cra]}{[Cra]+K_{pfk-Cra}}\right) \tag{17a}$$

$$v_{Pyk} = v_{Pyk}(\bullet)\left(1 - \frac{[Cra]}{[Cra]+K_{pyk-Cra}}\right) \tag{17b}$$

while v_{Pps}, v_{Fbp}, and v_{Pck} may be expressed as

$$v_{Pps} = v_{Pps}(\bullet)\left(\frac{[Cra]}{[Cra]+K_{pps-Cra}}\right) \tag{18a}$$

$$v_{Fbp} = v_{Fbp}(\bullet)\left(\frac{[Cra]}{[Cra]+K_{fbp-Cra}}\right) \tag{18b}$$

$$v_{Pck} = v_{Pck}(\bullet)\left(\frac{[Cra]}{[Cra]+K_{pck-Cra}}\right) \tag{18c}$$

where $v_{\bullet}\,(\bullet)$ is the reaction rate as given in Chapter 4.

Model Equations for the Uptake of Carbon Sources

In the case of glucose uptake by non PTS, the state equations for the glucose concentration may be expressed as

$$\frac{d[Glc]}{dt} = v_{NPTS} - v_{Glk} - v_{PTS4'} - \mu[Glc] \tag{19}$$

where the suffix NPTS denotes the transport other than glucose-PTS. v_{NPTS} and v_{Glk} may be expressed as

$$v_{NPTS} = \frac{k_{npts}[NPTS][Glc^{ex}]}{K_S + \left(1 + \dfrac{[EIIA]}{K_I}\right)\left[Glc^{ex}\right]} \tag{20a}$$

$$v_{Glk} = \frac{v_{Glk}^{max}[Glc]}{\left(1 + \dfrac{[G6P]}{K_I}\right)\left(K_S + [Glc]\right)} \tag{20b}$$

$$v_{PTS4'} = \frac{\alpha_{EIICB}k_4[EIICB]_0[EIIA \sim P][Glc]}{\left(K_{EIIA\sim P} + [EIIA \sim P]\right)\left(K_{Glc} + [Glc] + \dfrac{K_{Glc}\left[Glc^{ex}\right]}{K_{Glc^{ex}}}\right)} \tag{20c}$$

In addition to glucose-PTS, glucose is also transported by other transporters such as mannose PTS and the galactose transport system (Man, MglBAC and GalP). The synthesis of these proteins is also under control of cAMP-Crp. Here, all of these transport systems are assumed to be represented by NPTS, and the mass balance equation for NPTS may be expressed as

$$\frac{d[NPTS]}{dt} = k_{max}\left(\frac{1}{K_{npts}} + \frac{[Crp][cAMP]}{\alpha K_2 K_{npts} K_b}\right)[D_{npts}] - \mu[NPTS] \tag{21}$$

where $[D_{npts}]$ denotes the NPTS concentration of DNA template of *npts*, where NPTS is assumed to be encoded by the artificially defined *npts* gene. In the case of *ptsG* mutation, the concentration of the phosphorylated E Ⅱ A (E Ⅱ A-P) increases, where the phosphate is transferred from PEP to E Ⅱ A by the cascade of PTS, since the phosphorylated E Ⅱ A cannot transfer phosphate to E Ⅱ CB encoded by *ptsG*. The increase in E Ⅱ A-P activates Cya, and in turn cAMP level increases. This causes the activation of non-PTS transporter.

In the typical biofuels production from lignocellulosic biomass, the main carbon source is the mixture of glucose and xylose. In the case of xylosc assimilation, the following equations may be considered, where these are xylose transport (XT), xylose isomerase (Xyi), and xylose kinase (Xyk) as also mentioned in Chapter 4 [29]

$$v_{XT} = \frac{k_{cat}[XT][XYL^{ex}]}{K_{XYL^{ex}} + [XYL^{ex}]} \tag{22a}$$

$$v_{Xyi} = \frac{v_{Xyi}^{max}[XYL]}{K_{XYL} + [XYL]} \tag{22b}$$

$$v_{Xyk} = \frac{v_{Xyk}^{max}[XYLU][ATP]}{K_{XYLU}K_{ATP} + K_{ATP}[XYLU] + K_{XYLU}[ATP] + [XYLU][ATP]} \tag{22c}$$

Note that XT is under control of cAMP-Crp, and thus this may be also expressed as a function of cAMP-Crp.

COMPUTER SIMULATION FOR CATABOLITE REGULATION FOR OVERFLOW METABOLISM AND FOR THE CASE OF USING MULTIPLE CARBON SOURCES

Consider the fermentation characteristics of *E. coli* in the continuous and batch culture based on the computer simulation using the model as mentioned above, where it is able to express the catabolite regulation mechanism by incorporating the roles of TFs. Note that here we assume that enzymatic reaction is affected by TFs, since gene expression is in fact under control of TFs from the molecular biology point of view.

Fig. **6** shows the effect of the specific glucose consumption rate on the intracellular metabolite concentrations, TFs, and the intracellular metabolic fluxes. The intracellular PEP concentration decreases, while PYR concentration increases (Fig. **6c**), and thus [PEP]/[PYR] ratio decreases, resulting in the decrease in EIIA-P, and increase in EIIA, and cAMP-Crp concentration decreases (Fig. **6a**) as the specific glucose consumption rate increases or equivalently as the dilution rate increases in the continuous culture. This is consistent with the experimental data, except for the case at low glucose consumption rate or at low dilution rate in the continuous culture due to nutrient starvation [30]. Fig. **6d** shows the change in *crp* and *cra* gene expressions with respect to the dilution rate [21], where the trends are the similar as compared to the simulation result. Note that the dilution rate may be converted to the specific glucose consumption rate with constant yield coefficient, $Y_{X/S}$ ($\doteqdot 0.5$). Fig. **6b** shows that the TCA cycle fluxes such as CS, ICDH, KGMAL and MDH all decreases, which corresponds to the decrease in cAMP-Crp, where KGMAL denotes the lumped flux from αKG to MAL in the simulation. The experimental ^{13}C-metabolic fluxes of the TCA cycle show the similar trend. Moreover, acetate concentration increases as the specific glucose consumption rate increases (Fig. **6c**). This is also consistent with the experimental data [21, 31]. As the specific glucose consumption rate increases, FBP concentration increases (Fig. **6c**), which represses Cra (Fig. **6a**), which in turn repressed TCA cycle (Fig. **6f**), and activates glycolytic pathways such as Pfk and EMP (as denoted by Emp) etc. (Fig. **6e**). Namely, the activation of glycolysis and the repression of TCA cycle activity cause acetate production to be increased as the specific glucose consumption rate increases.

Fig. **7** shows the simulation result for the batch cultivation of wild-type *E. coli* using a mixture of glucose and xylose (non-PTS carbohydrate) as a carbon source [19], where symbols are the experimental data. It indicates that the ratio of PEP/PYR is low resulting in the lower EIIA-P level, and thus the levels of Cya, cAMP, and cAMP-Crp are low during the initial 6 hours, where glucose is preferentially consumed, and acetate is kept producing, while little xylose is consumed due to catabolite repression. Moreover, FBP concentration is high, and thus Cra activity is low, and Pfk and Pyk avtivities are higher during this period. After glucose was depleted, the ratio of PEP/PYR becomes higher, and EIIA-P concentration increases, and in turn Cya, cAMP, and cAMP-Crp all increase. Then acetate and xylose are both consumed, where Acs, Icl, MS, and Pps are activated. Upon depletion of glucose, FBP concentration decreases, and Cra is activated, and Pfk and Pyk activities become lower. Upon increase in cAMP-Crp and Cra, the gluconeogenic pathways as well as the acetate uptake pathway and the glyoxylate pathway are activated.

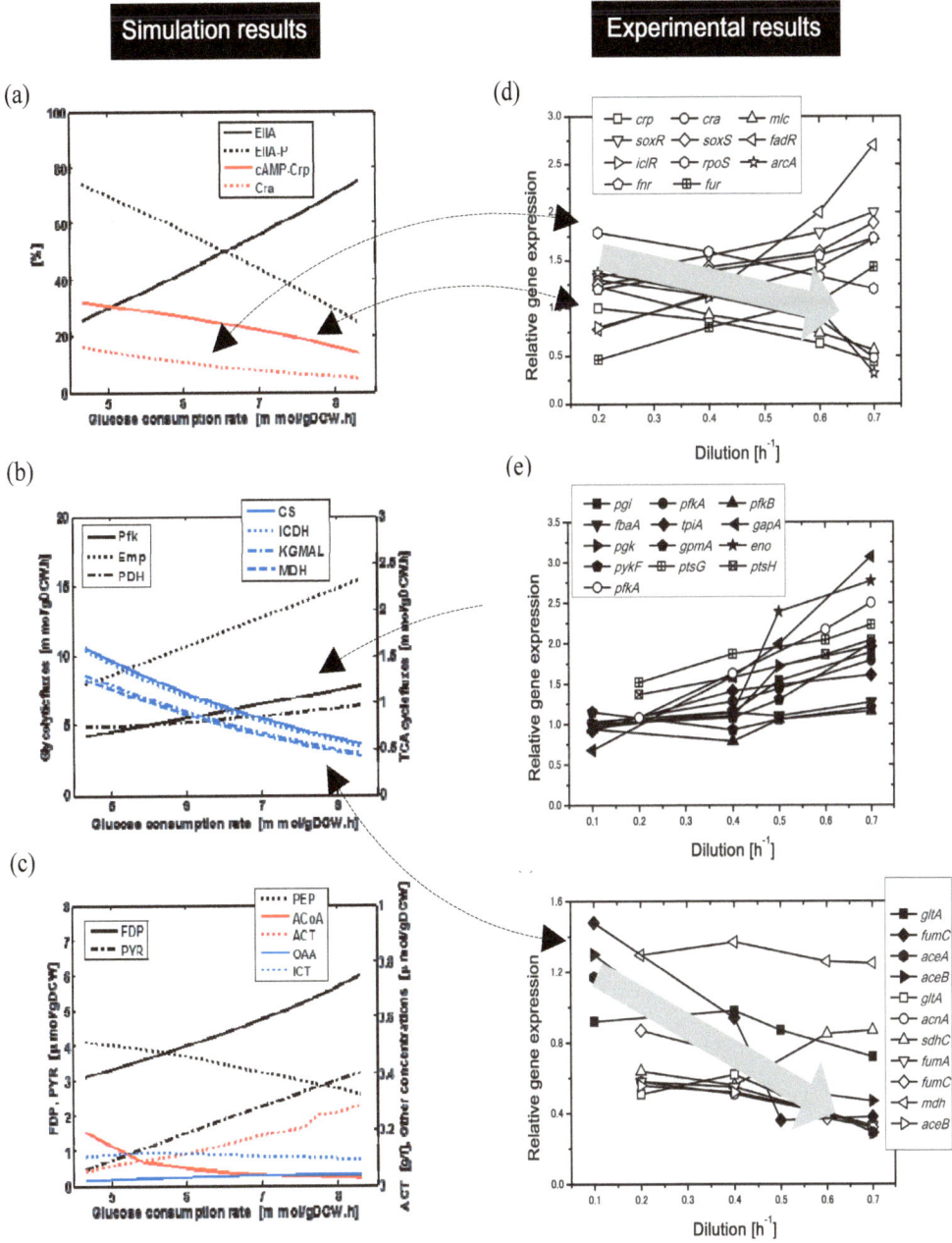

Figure 6: Effect of the specific glucose consumption rate on the intracellular metabolite concentrations, TFs, fluxes, and gene expressions of wild-type *E. coli*: (a) simulation result for cAMP-Crp, Cra, and E Ⅱ A with and without P, (b) simulation result for fluxes, (c) simulation result for metabolite concentrations [19], (d) experimental data for the TFs, (e) experimental data for the glycolysis, (f) experimental data for the TCA cycle, [21, 24].

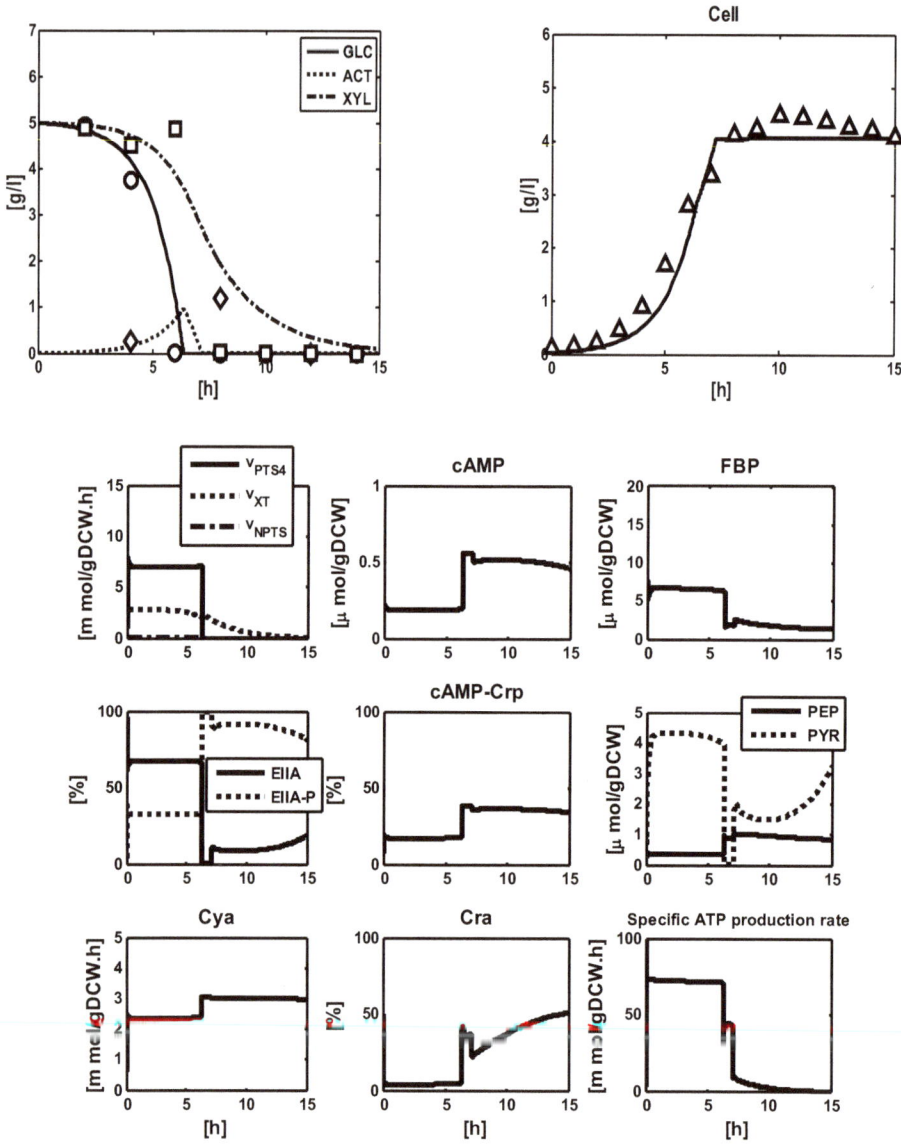

Figure 7: Batch culture of wild-type *E. coli* where a mixture of glucose and xylose is used as a carbon source, where the lines are simulation results and the symbols are the experimental data: ○, glucose concentration; □, xylose concentration; ◇, acetate concentration; △, cell concentration [19].

Fig. **8** shows the simulation result for the batch cultivation of *ptsG* gene knockout (Δ*ptsG*) mutant for the case of using a mixture of glucose and xylose, where the symbols are the experimental data [19]. In consistent with the experimental data, simultaneous consumption of glucose and xylose can be made. However, the consumption rates are lower as compared to the case of wild type. One of the reasons for the slow consumption of glucose is due to utilization of less efficient Mgl and GalP together with Glk, where the former is lumped together as NPTS in the present modeling approach. Since PEP/PYR is relatively higher in Δ*ptsG* strain, EIIA-P, Cya, cAMP, and cAMP-Crp are relatively higher from the beginning of the culture as compared to the case of using wild-type strain. Note that FBP is lower and Cra is higher as compared to the case of wild type. The higher Cra activity causes the repression of Pfk and Pyk activities, which in turn results in the lower glucose consumption rate for Δ*ptsG* strain. Note also that little acetate is formed in the Δ*ptsG* strain. It is consistent with the experimental data.

In the end, this indicates that co-consumption of multiple carbon sources can be made by *ptsG* mutant, and it can be made clear for its mechanism based on the intracellular information obtained by computer simulation. The present model can be used to predict the fermentation characteristics of other mutant such as Pyk mutant [19]. The important feature of the present modeling approach is that unlike experimental data analysis and/or ^{13}C-metabolic flux analysis, the detailed intracellular information can be seen to analyze the regulation mechanism for the phenotypic characteristics. This is quite important to get insight into the significant improvement for metabolic engineering in practice. It should be kept in mind that the model has to be corrected or improved by modification if some of the trends in the metabolite concentrations are different from the experimentally measured data. By repeatedly doing this, the model can be updated, and it gives some indication on which data should be further checked by experiments.

MODELING FOR NITROGEN ASSIMILATION

As explained in Chapter 7, next to carbon regulation, nitrogen regulation is important in practice, where Fig. **9** shows the metabolic pathways for ammonium assimilation, which involve glutamate dehydrogenase (GDH), glutamine synthetase (GS) and glutamate synthase (GOGAT) pathways. Glutamate and glutamine play key roles in cellular metabolism and serve as precursors of protein synthesis. Glutamate can be synthesized from two different pathways, where one is formed from αKG by the reaction catalyzed by GDH, and the other is synthesized from glutamine by GS/GOGAT as explained in Chapter 4. GS is active during low ammonium concentration, while GDH is active at higher ammonium concentration since GS has a higher affinity than GDH for ammonia. The activity of GS is controlled by P_{II} which acts in response to the concentration of glutamine and αKG.

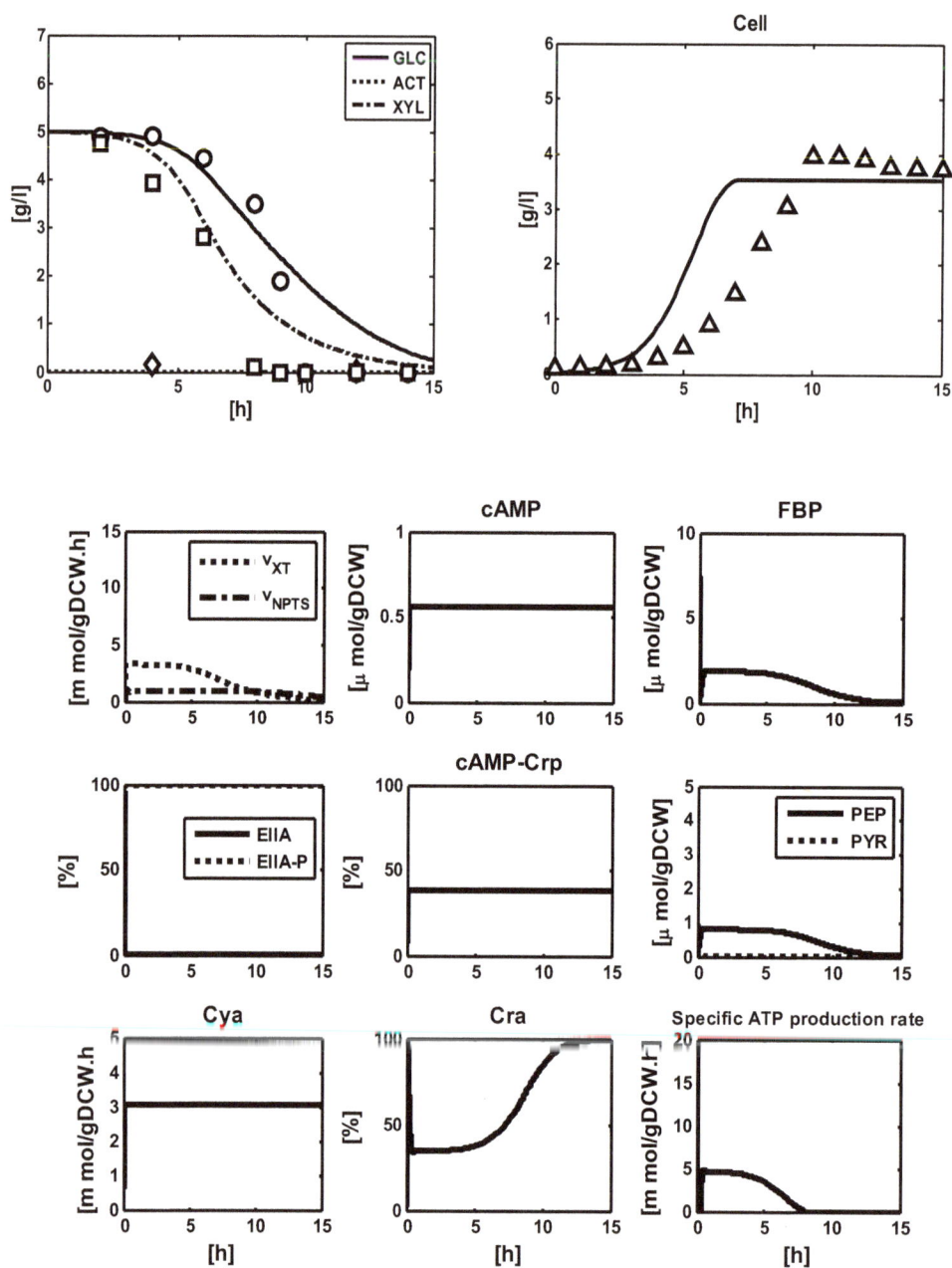

Figure 8: Batch culture of *ptsG* mutant *E. coli* where a mixture of glucose and xylose is used as a carbon source, where the lines are simulation results and the symbols are the experimental data: ○, glucose concentration; □, xylose concentration; ◇, acetate concentration; △, cell concentration [19].

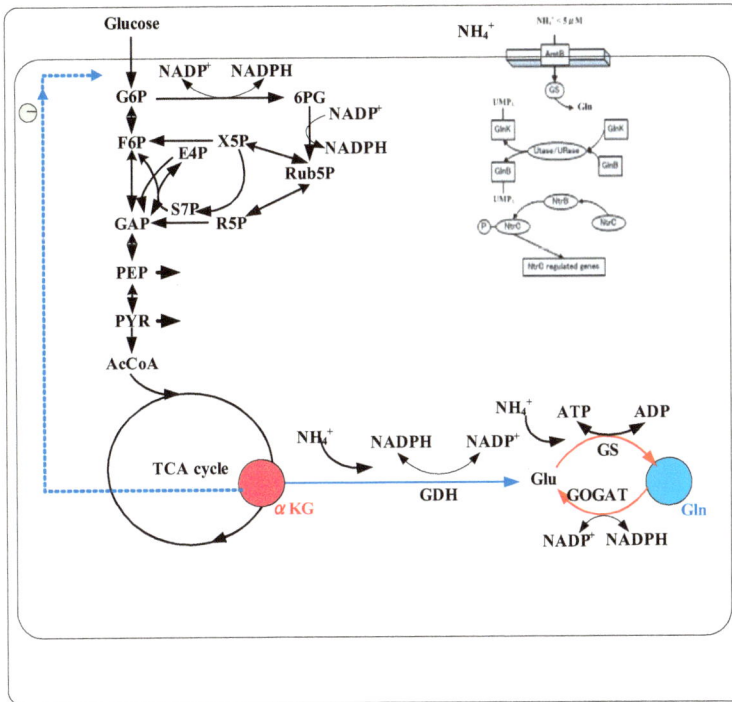

Figure 9: Nitrogen metabolism and its regulation.

Several kinetic models have been proposed for ammonium assimilation [32-34].

The mass balance equations may be expressed for the metabolites such as glutamate, glutamine and aspartate as well as such proteins as GS and P II .

The activation (adenylylation) and inactivation (deadenylylation) of GS depends on C/N ratio (ratio of αKG (C) concentration to glutamine (N) concentration). In the case where C/N ratio is higher (N-limitation), GS is adenylylated. Sensing of the C/N ratio for GS adenylylation involves the protein P_{II}, which presents in two forms; urydylylated P_{II}-UMP and deurydylylated P_{II}. Urydylylation/deurydylylation of P_{II} catalyzed by UT/UR enzymes are promoted by glutamine. UT and UR as well as AT and AR may be expressed as the appropriate equations [35].

It should be noted that the combination of the models incorporating catabolite regulation mechanism with nitrogen regulation for the main metabolism is useful to study how the metabolic regulation is made under multiple nutrient limitations.

Namely, under nitrogen limitation, αKG accumulates and such keto-acid inhibits EI of PTS, thus reducing the glucose consumption rate simultaneously [36].

MODELING FOR AMINO ACID (LYSINE) SYNTHETIC PATHWAYS

It is strongly desirable from the practical application point of view for the modeling to extend the pathways from the main metabolism to the peripheral metabolism such as amino acid synthetic pathways, since such model can be used for the metabolic engineering to improve the amino acid fermentation.

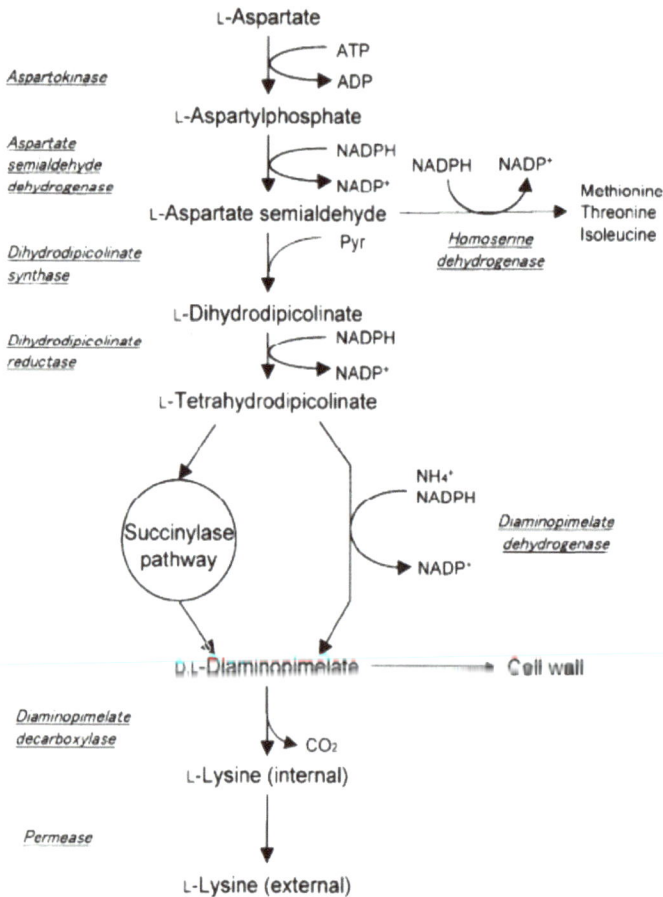

Figure 10: Metabolic pathways for lysine synthesis.

As explained in Chapter 4, the bacterial synthesis of lysine takes place *via* the diaminopimelate (DAP) pathway as shown in Fig. **10**. Aspartate (ASP) is formed from OAA in the TCA cycle by transamination. ASP is then activated *via* phosphorylation by aspartokinase (Ask) and reduced to give aspartate semialdehyde (ASA) in the first two steps. L-Aspartate semialdehyde is a branch point to enter into either threonine, methionine and isoleucine syntheses or lysine synthesis. Dihydrodipicolinate synthase (DHPS) and dihydrodipicolinate reductase (DHPR) catalyze the third and fourth steps in the lysine biosynthetic pathway, respectively, and they are the enzymes which commit flux to the biosynthesis of DAP and lysine. The synthesis of DAP from L-tetrahydrodipicolinate (THDP) is accomplished by three separate routes: the succinylase and acetylase pathways, in which N-succinylated or N-acetylated intermediates are generated, and the infrequently encountered dehydrogenase pathway. In *C. glutamicum*, the dehydrogenase and succinylase pathways have been shown to operate simultaneously [37, 38]. The synthesized meso-DAP can be either used for cell wall synthesis or decarboxylation to L-lysine catalyzed by diaminopimelate decarboxylase (DAPDC). With the accumulation of internal lysine, a secretion system functions and lysine is excreted into the medium.

The relevant mathematical equations may be described as [39, 40]

$$\frac{d[ASA]}{dt} = v_{Ask} - v_{DHPS} - \mu[ASA] \tag{23a}$$

$$\frac{d[THDP]}{dt} = v_{DHPS} - v_{DAPDH} - v_{Succinilase} - \mu[THDP] \tag{23b}$$

$$\frac{d[DAP]}{dt} = v_{DAPDH} - v_{Succinilase} - v_{DAPDC} - \mu[DAP] \tag{23c}$$

$$\frac{d[Lys^{in}]}{dt} = v_{DAPDC} - v_{Permease} - \mu[Lys^{in}] \tag{23d}$$

$$\frac{d[Lys^{out}]}{dt} = v_{Permease} \cdot V \cdot X \tag{23e}$$

where v. represents the rate of the enzyme-catalyzed reaction or transport process, and μ is the specific growth rate. The terms associated with μ account for the dilution of the intracellular metabolites caused by the cell growth. The V in Eq. (23e) is the specific cell volume, which may be assumed to be 1.9 ml/g DCW [41]. X is the dry cell weight. [•] represents the concentration of metabolite.

Although some intermediates in the lysine biosynthetic pathway (*e.g.*, DAP and lysine) participate in the biosynthesis, consumption of these intermediates is included due to the low specific growth rate (generally less than 0.1 h^{-1}) in the lysine production phase. Relevant calculations suggest that during this phase, less than 0.5 mM/h of intracellular lysine is used for biosynthesis (that is, less than 0.5% of the overall flux through this pathway) using the data that approximately 0.33 mmol lysine is needed for synthesizing 1 g biomass [42]. For intracellular DAP, this consumption for anabolism is much less.

Fig. **11** shows the batch fermentation result for lysine fermentation, where this is the homoserine auxotroph, and thus lysine is accumulated after homoserine was depleted [39].

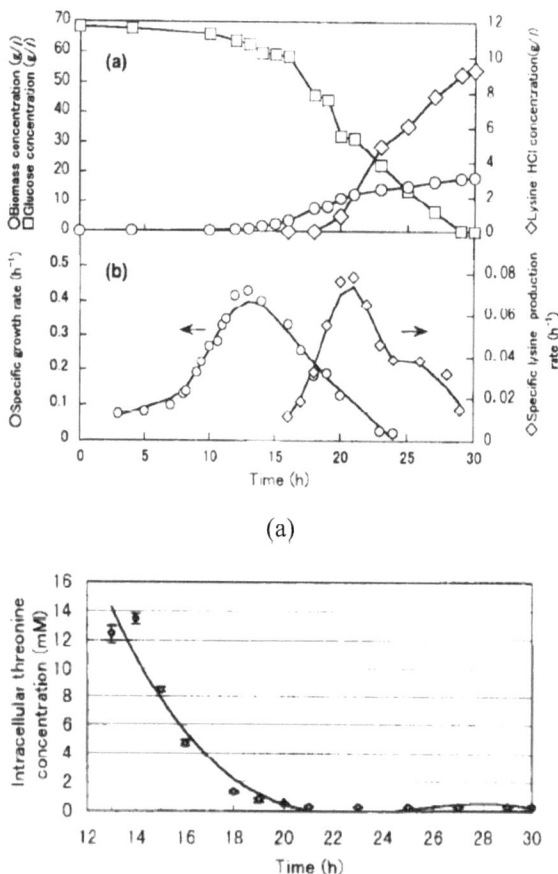

Figure 11: Batch cultivation of *Corynebacterium glutamicum* for lysine production: (a) biomass (○), glucose (□), L-lysine hydrochloride (◇) concentrations, (b) specific growth rate (○) and specific L-lysine production rate (◇), (c) intracellular L-threonine concentration [39].

Fig. **12** shows the comparison of the simulation result with experimentally measured data, where the model parameters are given in Table **1** in Chapter 4. Fig. **12** indicates satisfactory fitting of the experimental data to the simulation result [39]. This model can be used in practice. For example, the rate limiting pathway for lysine production can be clarified by applying MCA for the present model as explained in Chapter 5.

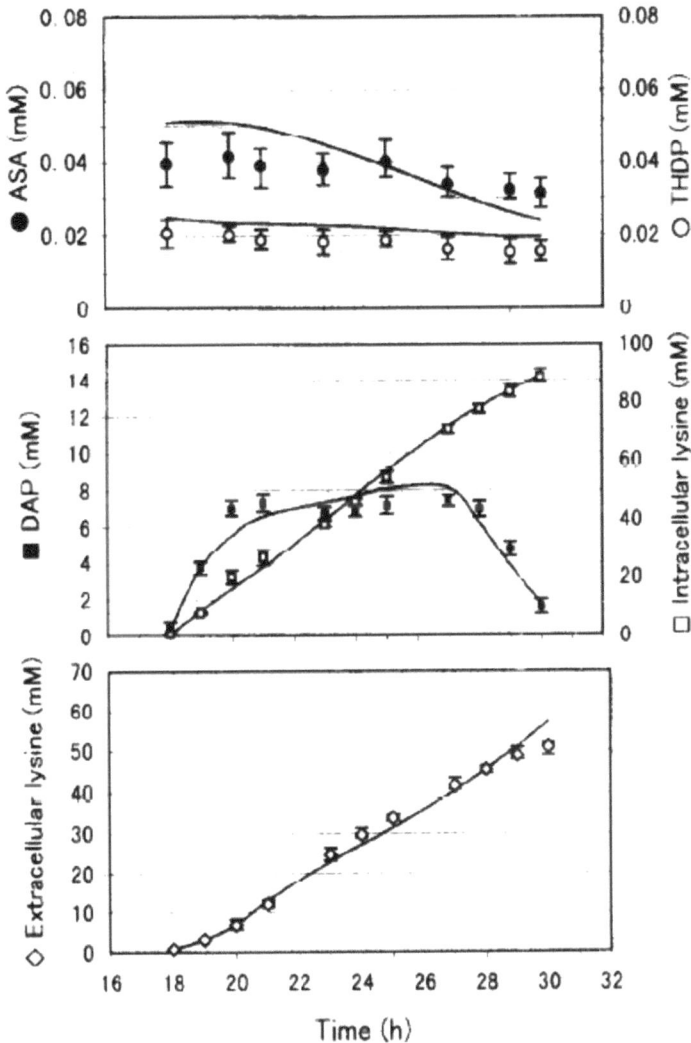

Figure 12: Comparison of simulation result and the experimental data for lysine production using *Corynebacterium glutamicum*, where the lines are simulation results and the symbols are the experimental data: ●, ASA concentration; ○, THDP concentration; ■, DAP concentration; □, intracellular L-lysine concentration; ◇, extracellular L-lysine concentration [39].

MODELING FOR THE METABOLISM UNDER ANAEROBIC CONDITION

In addition to nutrient sources, oxygen level is also quite important from the metabolic regulation and metabolic engineering points of view, where most of the useful metabolite production is made under anaerobic condition in practice, while the cell growth rate is enhanced under aerobic condition. It is, thus, of practical interest to properly model the cell metabolism under both aerobic and anaerobic conditions. For the modeling of the metabolism under anaerobic conditions, it is important to incorporate the effects of TFs such as ArcA/B and Fnr, where these detect oxygen level or redox state, and regulate a set of gene expression to control the metabolism. Fnr plays important roles under anaerobic condition, while ArcA/B play roles under both anaerobic and micro-aerobic conditions as mentioned in Chapter 7.

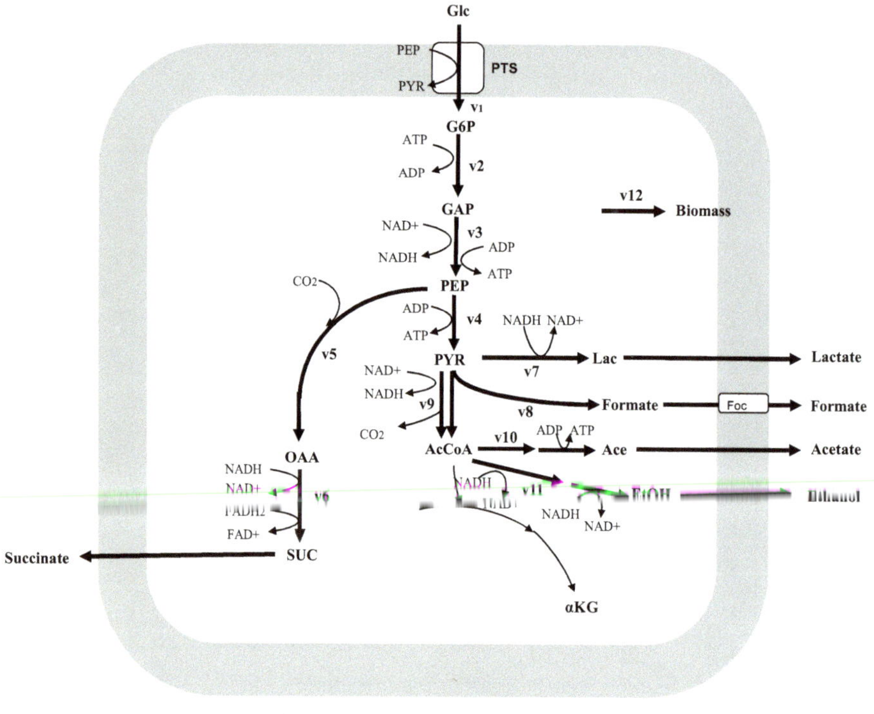

Figure 13: Metabolic pathways under anaerobic condition.

Under anaerobic conditions, the cell such as *E. coli* excretes lactate, ethanol, succinate, and formate (CO_2 and H_2 as well) as well as acetate, where the relative production rates for these metabolites are governed by the demand for redox neutrality (Fig. **13**). The succinate is formed from PEP *via* Ppc, where the *frd* genes are activated by Fnr, while *sdh* genes are repressed by ArcA, and thus the fluxes of the left half of the TCA cycle are reversed forming branched pathways disconnecting at KGDH pathway under anaerobic condition (Fig. **13**).

Under micro-aerobic condition, pyruvate serves as a common substrate for pyruvate formate-lyase (Pfl) and pyruvate dehydrogenase complex (PDHc), and this branch point involves the cleavage of PYR. The expression of *pfl* genes which encode Pfl is activated by ArcA and Fnr, and it becomes higher at lower oxygen concentrations or at higher $NADH/NAD^+$ ratio, whereas *aceE,F* which encode α and β subunits of PDHc is repressed by ArcA under oxygen limited condition. The AcCoA, product of both Pfl and PDHc reactions, is converted to acetate and ethanol or subsequently undergo further oxidation in the TCA cycle (Fig. **13**). The ArcA then represses the expression of the genes involved in the TCA cycle and the glyoxylate shunt.

Alexeeva *et al.* [43] investigated the effect of different oxygen supply rates on the metabolism in *arc*A mutant, where this gives some hint for the modeling of the metabolism under both aerobic and anaerobic conditions. Namely, the activities of ArcA/B and Fnr must be the decreasing functions with respect to dissolved oxygen concentration such that

$$v_{ArcA/B} = k_{ArcA/B}\left(1 - \frac{[O_2{}^*]}{[O_2{}^*] + K_{ArcA/B}}\right) \tag{24a}$$

$$v_{Fnr} = k_{Fnr}\left(1 - \frac{[O_2{}^*]}{[O_2{}^*] + K_{Fnr}}\right) \tag{24b}$$

where $[O_2{}^*]$ is the dissolved oxygen concentration. Then some of the rate equations as given in Chapter 4 may be expressed as functions of ArcA and Fnr such that

$$v_{Pfl} = v_{Pfl}(\bullet)\left(\frac{ArcA}{ArcA + K_{pfl-ArcA}}\right)\left(\frac{Fnr}{Fnr + K_{pfl-Fnr}}\right) \tag{25a}$$

$$v_{PDH} = v_{PDH}(\bullet)\left(1 - \frac{ArcA}{ArcA + K_{pdh-ArcA}}\right) \tag{25b}$$

$$v_{Frd} = v_{Frd}(\bullet)\left(\frac{Fnr}{Fnr + K_{frd-Fnr}}\right) \tag{25c}$$

$$v_{CS} = v_{CS}(\bullet)\left(1 - \frac{ArcA}{ArcA + K_{gltA-ArcA}}\right) \tag{25d}$$

$$v_{TCA} = v_{TCA}(\bullet)\left(1 - \frac{ArcA}{ArcA + K_{tca-ArcA}}\right) \tag{25e}$$

Note that in predicting the distribution of the fermentation products, $NADH/NAD^+$ ratio must be properly incorporated, or the $NADH/NAD^+$ balance has to be closed.

CONCLUDING REMARKS

For the proper modeling, it is important to incorporate the essential feature of metabolic regulation, where the gene level and enzyme level regulations must be appropriately incorporated, where the roles of transcription factors must be included for the former. Quantitative model which can predict the experimental data is important in the practical application point of view.

In particular, in the case of biofuels production from biomass, it is important to simulate the catabolite regulation for the assimilation of multiple carbon sources obtained by the hydrolysis of the cellulosic biomass in practice.

Moreover, it is also of practical importance to develop the model applicable under both aerobic and anaerobic conditions. This may be attained by incorporating the effects of transcription factors such as Fnr and ArcA/B on the metabolic pathway enzymes. Although the specific production rate of ATP and NAD(P)H can be estimated from the fluxes of the central metabolism, it is not easy to estimate their concentrations, where NADH concentration is important for the fermentative reaction pathways under anaerobic conditions.

Although the computer simulation using the kinetic models may be limited to the main metabolism, the coordinative regulation between carbon and nitrogen regulations can be simulated by combining both catabolite and nitrogen regulations into the model. In particular, the coordination may be simulated *via* the change in the concentrations of α-keto acids such as αKG acting on PTS and cAMP [36, 44].

NOTATION

Acs	=	acetyl coenzyme A (AcCoA) synthetase
Ald	=	aldolase
ASA	=	aspartate semialdehyde
Ask	=	aspartate kinase
ASP	=	aspartate
ATase	=	adenylyl transferase
CCR	=	carbon catabolite repression
CIT	=	citrate
CS	=	citrate synthase
CyaA	=	adenylate cyclase
DAP	=	diaminopimelate
DAPDC	=	diaminopimelate decarboxylase
DHPS	=	dihydodipicolinate
E4P	=	erythrose-4-phosphate
EI	=	enzyme I
EII	=	enzyme II
Eno	=	enolase
F6P	=	fructose-6-phosphate
FBP	=	fructose-1,6-bisphosphate
Fum	=	fumarase
G6P	=	glucose-6-phosphate

G6PDH   = glucose-6-phosphate dehydrogenase

GAP   = glycelaldehyde-3-phosphate

GAPDH   = glyceraldehyde-3-phosphate dehydrogenase

GF6P   = glucose-6-phosphate and fructose-6-phosphate

GLC   = glucose

GOX   = glyoxylate

HPr   = histidine-phosphorylatable protein

ICDH   = isocitrate dehydrogenase

ICI   = isocitrate

Icl   = isocitrate lyase

KDPG   = 2-keto-3-deoxy-6-phosphogluconate

αKG   = α-ketoglutarate

LDH   = lactate dehydrogenase

MAL   = malate

MDH   = malate dehydrogenase

Mez   = malic enzyme

MS   = malate synthase

OAA   = oxaloacetate

Pck   = phosphoenolpyruvate carboxykinase

PDH   = pyruvate dehydrogenase

PEP   = phosphoenol pyruvate

Pfk   = phosphofructokinase

6PG   = 6-phosphogluconate

PGDH	=	6-phosphogluconate dehydrogenase
Pgi	=	phosphoglucose isomerase
Pgk	=	phosphoglycerate kinase
Ppc	=	phosphoenolpyruvate carboxylase
PP pathway	=	pentose phosphate pathway
Pps	=	phosphoenolpyruvate synthase
Pta	=	phosphotransacetylase
PTS	=	phosphotransferase system
Pyk	=	pyruvate kinase
PYR	=	pyruvate
R5P	=	ribose-5-phosphate
R5PI	=	ribose phosphate isomerase
RU5P	=	ribulose-5-phosphate
S7P	=	sedoheptulose-7-phosphate
SDH	=	succinate dehydrogenase
SUC	=	succinate
Tal	=	transaldolase
TCA cycle	=	tricarboxylic acid cycle
THDP	=	tetrahydrodipicolinate
TktA	=	transketolase I
TktB	=	tranketolase II
UTase/ UR	=	uridyl transferase/ uridylyl-removing enzyme
X5P	=	xylulose-5-phosphate

APPENDIX A MODELING FOR PTS

For the glucose phosphorylation step of PTS, a unidirectional random mechanism may be considered [16], where glucose and phosphorylated $EIIA^{Glc}$ bind to $EIICB^{Glc}$. The following reaction steps may be considered for the cascade of phosphorylation in PTS [17]:

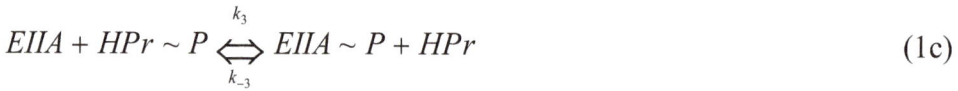

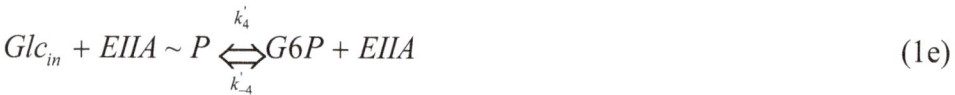

$$EI + PEP \underset{k_{-1}}{\overset{k_1}{\Longleftrightarrow}} EI \sim P + PYR \tag{1a}$$

$$EI \sim P + HPr \underset{k_{-2}}{\overset{k_2}{\Longleftrightarrow}} HPr \sim P + EI \tag{1b}$$

$$EIIA + HPr \sim P \underset{k_{-3}}{\overset{k_3}{\Longleftrightarrow}} EIIA \sim P + HPr \tag{1c}$$

$$Glc + EIIA \sim P \underset{k_{-4}}{\overset{k_4}{\Longleftrightarrow}} G6P + EIIA \tag{1d}$$

$$Glc_{in} + EIIA \sim P \underset{k_{-4}'}{\overset{k_4'}{\Longleftrightarrow}} G6P + EIIA \tag{1e}$$

The equations for the phosphorylated and unphosphorylated states of EI, HPr, and EIIA may then be derived [16]. Here, we assume EI and HPr to be in equilibrium, and consider only EIIA and phosphorylated EIIA, EIIA~P in relation to PEP, PYR, Glc and G6P as

$$\frac{d[EIIA]}{dt} = -v_{PTS1} + v_{PTS4} \tag{2a}$$

$$\frac{d[EIIA \sim P]}{dt} = v_{PTS1} - v_{PTS4} \tag{2b}$$

where v_{PTS1} and v_{PTS4} may be expressed as [17]

$$v_{PTS1} = k_1[PEP][EIIA] - k_{-1}[PYR][EIIA \sim P] \tag{3a}$$

$$v_{PTS4} = \frac{k_4 [EIICB]_0 [EIIA \sim P][Glc]}{(K_{EIIA \sim P} + [EIIA \sim P])(K_{Glc} + [Glc])} \qquad (3b)$$

Moreover, the reaction rates $v_{PTS4'}$ in Eq. (1e) may be expressed as [17]

$$v_{PTS4'} = \frac{\alpha_{EIICB} k_4 [EIICB]_0 [EIIA \sim P][Glc]}{(K_{EIIA \sim P} + [EIIA \sim P])\left(K_{Glc} + [Glc] + \dfrac{K_{Glc} [Glc^{ex}]}{K_{Glc^{ex}}} \right)} \qquad (3c)$$

Further detailed analysis may be found elsewhere [16,17].

REFERENCES

[1]   Rizzi M, Theobald U, Querfurth E, Rohrhirsch T, Baltes M, Reuss M. *In vivo* investigations of glucose transport in *Saccharomyces cerevisiae*. Biotechnol Bioeng 1996; 49: 316-327.

[2]   Rizzi M, Baltes M, Theobald U, Reuss M. *In vivo* analysis of metabolic dynamics in *Saccharomyces cerevisiae* II. Mathematical model. Biotechnol Bioeng 1997; 55: 592-608.

[3]   Vaseghi S, Baumeister A, Rizzi M, Reuss M. *In vivo* dynamics of the pentose phosphate pathway in *Saccharomyces cerevisiae*. Metab Eng 1999; 1: 128-140.

[4]   Theobald U, Mailinger W, Reuss M, Rizzi M. *In vivo* analysis of glucose induced fast changes in yeast adenine nucleotide pool applying a rapid sampling technique. Anal Biochem 1993; 214: 31-37.

[5]   Theobald U, Mailinger W, Baltes M, Rizzi M, Reuss M. *In vivo* analysis of metabolic dynamic in *Saccharomyces cerevisiae*: I. Experimental observations. Biotechnol Bioeng 1997; 55: 305-316.

[6]   Buziol S, Bashir I, Baumeister A, Claassen W, Rizzi M, Mailinger W, Reuss M. A new bioreactor coupled rapid stopped-flow-sampling-technique for measurement of transient metabolites in time windows of milliseconds. *In Proceeding of the Fourth International Congress on Biochemical Engineering. Stuggart*, Fraunhofer IRB, 2000: 79-83.

[7]   Buziol S, Bashir I, Baumeister A, Claassen W, Rizzi M, Mailinger W, Reuss M. A new bioreactor coupled rapid stopped-flow sampling technique for measurement of metabolite dynamics on a subsecond time scale. Biotechnol Bioeng 2001; 80: 632-636.

[8]   Hoque MA, Ushiyama H, Tomita M, Shimizu K. Dynamic responses of the intracellular metabolite concentrations of the wild type and *pykA* mutant *Escherichia coli* against pulse addition of glucose or NH$_3$ under those limiting continuous cultures. Biochem Eng J 2005; 26: 38-49.

[9]   Chassagnole C, Noisommit-Rizzi N, Schmid JW, Mauch K, Reuss M. Dynamic modeling of the central carbon metabolism of *Escherichia coli*. Biotechnol Bioeng 2002; 79: 53-73.

[10]   Kadir TA, Mannan AA, Kierzek AM, McFadden J, Shimizu K. Modeling and simulation of the main metabolism in *Escherichia coli* and its several single-gene knockout mutants with experimental verification. Microb Cell Fact 2010; 9: 88.

[11]   Kotte O, Zaugg JB, Heinemann M. Bacterial adaptation through distributed sensing of metabolic fluxes. Mol Syst Biol 2010; 6: 355.

[12]   Usuda Y, Nishio Y, Iwatani S, Van Dien SJ, Imaizumi A, Shimbo K, Kageyama N, Iwahata D, Miyano H, Matsui K. Dynamic modeling of *Escherichia coli* metabolic and regulatory systems for amino-acid production. J Biotechnol 2010; 147: 17-30.

[13]   Hardiman T, Lemuth K, Keller MA, Reuss M, Siemann-Herzberg M. Topology of the global regulatory network of carbon limitation in *Escherichia coli*. J Biotechnol 2007; 132: 359-374.

[14]   Kremling A, Jahreis K, Lengeler JW, Gilles ED. The organization of metabolic reaction networks: A signal-oriented approach to cellular models. Metab Eng 2000; 2: 190-200.

[15]   Kremling A, Gilles ED. The organization of metabolic reaction networks. II. Signal processing in hierarchical structured functional units. Metab Eng 2001; 3: 138-150.

[16]   Kremlng A, Fischer S, Sauter T, Bettenbrock K, Gilles ED. Time hierarchies in the *Escherichia coli* carbohydrate uptake and metabolism. BioSystems 2004; 73: 57-71.

[17]   Bettenbrock K, Fischer S, Kremling A, Jahreis K, Sauter T, Gilles ED. A quantitative approach to catabolite repression in *Escherichia coli*. J Biol Chem 2006; 281: 2578-2584.

[18]   Nishio Y, Usuda Y, Matsui K, Kurata H. Computer-aided rational design of the phosphotransferase system for enhanced glucose uptake in *Escherichia coli*. Mol Syst Biol 2008; 4: 160.

[19]   Matsuoka Y, Shimizu K. Catabolite regulation analysis of *Escherichia coli* for acetate overflow mechanism and co-consumption of multiple sugars based on systems biology approach using computer simulation. J Biotechnol 2013; 168: 155-173.

[20]   Mailloux RJ, Bériault R, Lemire J, Singh R, Chénier RR, Hamel RD, Appanna VD, The tricarboxylic acid cycle, an ancient metabolic network with a novel twist. Plos One 2007; 2: e690.

[21]   Yao R, Hirose Y, Sarkar D, Nakahigashi K, Ye Q, Shimizu K. Catabolic regulation analysis of *Escherichia coli* and its *crp*, *mlc*, *mgsA*, *pgi* and *ptsG* mutants. Microb Cell Fact 2011; 10: 67.

[22]   Gennis RB, Stewart V. Respiration. In: Neidhardt FC, Curtiss III R, Ingraham JL, Lin ECC, Low KB, Magasanik B, Reznikoff WS, Riley M, Schaechter M, Umbarger HE, Eds. *Escherichia coli* and *Salmonella*: Cellular and Molecular Biology, 2nd edition. Washington DC, ASM Press 1996; pp. 217-261.

[23]   Harold FM, Maloney PC. Energy transduction by ion currents. In: Neidhardt FC, Curtiss III R, Ingraham JL, Lin ECC, Low KB, Magasanik B, Reznikoff WS, Riley M, Schaechter M, Umbarger HE, Eds. *Escherichia coli* and *Salmonella*: Cellular and Molecular Biology, 2nd edition. Washington DC, ASM Press 1996; pp. 283-306.

[24]   Ishii N, Nakahigashi K, Baba T, Robert M, Soga T, Kanai A *et al.* Multiple high-throughput analyses monitor the response of *E. coli* to perturbations. Science 2007; 316: 593-597.

[25]   Yang C, Hua Q, Baba T, Mori H, Shimizu K. Analysis of *Escherichia coli* anaprelotic metabolism and its regulation mechanisms from the metabolic responses to altered dilution rates and phosphoenolpyruvate carboxykinase knockout. Biotechnol Bioeng 2003; 84: 129-144.

[26]   Peng L, Arauzo-Bravo MJ, Shimizu K. Metabolic flux analysis for a *ppc* mutant *Escherichia coli* based on ^{13}C-labeling experiments together with enzyme activity assays and intracellular metabolite measurements. FEMS Microbiol Lett 2004; 235: 17- 23.

[27]   Görke B, Stülke JR. Carbon catabolite repression in bacteria: many ways to make the most out of nutrients. Nat Rev Microbiol 2008; 6: 613-624.

[28]   Shimada T, Yamamoto K, Ishihama A. Novel members of the Cra regulon involved in carbon metabolism in *Escherichia coli*, J Bacteriolo 2011; 193(3):649-659.

[29]   Altintas MM, Eddy CK, Zhang M, McMillan JD, Kompala DS. Kinetic modeling to optimize pentose fermentation in *Zymomonas mobilis*. Biotechnol Bioeng 2006; 94: 273-295.

[30]   Bettenbrock K, Sauter T, Jahreis K, Kremling A, Lengeler JW, Gilles ED. Correlation between growth rates, EIIA^{crr} phosphorylation and intracellular cyclic AMP levels in *Escherichia coli* K-12, J Bacteriol 2007; 189: 6891-6900.

[31]   Valgepea K, Adamberg K, Nahku R, Lahtvee PJ, Arike L, Vilu R. Systems biology approach reveals that overflow metabolism of acetate in *Escherichia coli* is triggered by carbon catabolite repression of acetyl-CoA synthetase. BMC Syst Biol 2010; 4: 166.

[32]   Bruggeman FJ, Boogerd FC, Westerhoff HV. The multifarious short-term regulation of ammonium assimilation of *Ecsherichia coli*: dissection using an *in silico* replica. FEBS J, 2005; 272: 1965-1985.

[33]   Ma H, Boogerd FC, Goryanin I. Modelling nitrogen assimilation of *Escherichia coli* at low ammonium concentration. J Biotech 2009; 144: 175-183.

[34]   Lodeiro A, Melgarejo A. Robustness in *Escherichia coli* glutamate and glumamine synthesis studied by a kinetic model. J Biol Phys 2008; 34: 91-106.

[35]   Yuan J, Doucette D, Fowler WU, Feng XJ, Piazza M, Rabits HA, Wingreen NS, Rabinowitz JD. Metabolomics-driven quantitative analysis of ammonium assimilation in *E. coli*. Mol Syst Biol 2009; 5: 302.

[36]   Doucette CD, Schwab DJ, Wingreen NS, Rabinowitz JD. α-ketoglutarate coordinates carbon and nitrogen utilization *via* Enzyme I inhibition. Nat Chem Biol 2011; 7: 894-901.

[37]   Schrumpf B, Schwarzer A, Kalinowski J, Ptihler A, Eggeling L, Sahm H. A functionally split pathway for lysine synthesis in *Corynebacterium glutamicum*. J Bacteriol 1991; 174: 4510-4516.

[38]   Sonntag K, Eggeling L, de Graaf AA, Sahm H. Flux partitioning in the split pathway of lysine synthesis in *Corynebacterium glutamicum*. Quantification by ^{13}C- and ^{1}HNMR spectroscopy. Eur J Biochem 1993; 213: 1325-1331.

[39]   Yang C, Hua Q, Shimizu K. Development of a kinetic model for L-lysine biosynthesis in *Corynebacterium glutamicum* and its application to metabolic control analysis. J Biosci Bioeng 1999; 88: 393-403.

[40]   Hua Q, Yang C, Shimizu K. Metabolic control analysis for lysine synthesis using *Corynebacterium glutamicum* and experimental verification. J Biosci Bioeng 2000; 90: 184-192.

[41]   Bröer S, Eggeling L, Kramer R. Strains of *Corynebacterium glutamicum* with different lysine productivities may have different lysine excretion systems. Appl Environ Microbial 1993; 59: 316-321.

[42]   Vallino JJ, Stephanopoulos G. Metabolic flux distributions in *Corynebacterium glutamicum* during growth and lysine overproduction. Biotechnol Bioeng 1993; 41: 633-648.

[43]   Alexeeva S, Hellingwerf KJ, Teixeira de Mattos MJ. Requirement of ArcA for redox regulation in *Escherichia coli* under microaerobic but not anaerobic or aerobic conditions. J Bacteriol 2003; 185: 204-209.

[44]   You C, Okano H, Hui S, Zhang Z, Kim M, Gunderson CW, Wang YP, Lenz P, Yan D, Hwa T. Coordination of bacterial proteome with metabolism by cyclic AMP signalling. Nature 2013; 500: 301-306.

CHAPTER 9

Concluding Summary

Biotechnological modeling is the extraction and reconstruction of the phenomena occurring in the living organisms and will be of immense value to

1) biotechnologists aiming to improve fermentation performance by optimal operation,

2) metabolic engineers aiming to design *de novo* recombinant strains able to capture the available nutrients, degrade pollutants, vaccinate human and or animals, and produce next generation biofuels,

3) basic scientists aiming to understand the metabolic regulation mechanism in living organisms, and its alternative realization by synthetic biology, and

4) systems biologists aiming to advance the science of modeling.

Although the main advantage of modeling is its predictability, and it must contain essential feature of the living systems. Completeness of the model that reflects the detailed phenomena occurring in the living cells may not be necessary for it to improve predictions or rationalizations. The complexity of the model and the accuracy of its prediction depends on the purposes of using it, where it is discriminated from others depending on the underlying assumptions, structure of the model, and the available data.

For the proper modeling, the structure of the model must be first determined, followed by model (parameter) identification, sensitivity analysis, optimization, model validation by the experimental data, and experimental design. Some model reduction may be considered if the available data are limited.

Systems biology or systems analysis gives foundation for modeling, where basic linear systems and its dynamic responses with and without feed-forward and feed-back control systems, graph theoretic approaches for the network analysis, and data analysis are briefly explained in Chapter 2.

The modeling may start with understanding the principles which govern the living organisms on earth, where the law of conservation palys an essential role toward the notion of balance. Such basic principles are given based on the theory of transport phenomena in Chapter 3. Moreover, several simple or unstructure models as well as the artificial neural networks (ANNs) based on only the input-output data are given in Chapter 3.

In Chapter 4, the kinetic or enzymatic modeling of the main metabolic pathways is given, where the central metabolism is the hub for the synthesis of cell constituents, and generates energy (catabolism) and synthesizes the precursor metabolites for the cell synthesis (anabolism). Thus it is critical to construct the accurate model for the central metabolism. Chapter 4 gives detailed kinetic expression for each pathway in the central metabolism as well as anaplerotic, gluconeogenetic, and fermentation pathways. In relation to ammonia assimilation and nitrogen regulation, the glutamate/ glutamin synthetic pathways are also explained. Moreover, the modeling for the peripheral metabolic pathways such as lysine synthetic pathways is also given in Chapter 4.

In Chapter 5, model (parameter) identification with sensitivity analysis is briefly explained together with local and global optimization methods, where the local search method such as gradient method may be combined with global search method such as genetic algorithm for model identification. The optimal control problem is also explained for the non-linear system, where the model must include the effect of culture environmental conditions such as temperature and pH etc. to be optimized with respect to time.

In Chapter 6, the typical methods for the analysis on the steady-state and dynamic characteristics are briefly explained. The system is inherently non-linear, and the complicated phenomena observed in nature or in living organisms, such as oscillatory behavior and chaotic behavior, are originated from the non-linearity. As mentioned in the text, the simple non-linear system equation can give complicated chaotic behavior depending on the parameter values, and this implies that the model may not necessarily be complex.

In Chapter 7, the enzyme level and transcriptional regulation mechanisms are explained in relation to catabolite regulation and nitrogen regulation focusing on *E.coli* cell metabolism. Enzyme level regulation mechanism is conserved in many organisms, while the transcriptional regulation may be specific to the organism, where the time scale for the former is on the order of second or minutes, while that of the latter is on the order of hours, and those must be taken into account into the modeling.

In Chapter 8, some computer simulation is explained based on the kinetic modeling with the rate equations as explained in Chapter 4 together with transcriptional (catabolite) regulation as explained in Chapter 7, where the attention was focused on the acetate overflow metabolism in the continuous culture and co-consumption of multiple sugars in the batch culture of *E.coli*. The power of computer simulation based on the appropriate modeling is clearly shown. Namely, it is quite easy to understand the fermentation mechanism by looking at the behavior of the intracellular metabolite concentrations and the activities of transcription factors. Moreover, many fruitful predictions in response to genetic and/ or environmental perturbations can be obtained, and thus the metabolic engineering can be made more systematic and efficient way with minimum number of experiments.

Finally, it may be pointed out the importance of the relevant experimental data for the model identification and model validation. The computer simulation must be validated by the additional experiments, where the additional experimental data can improve the model structure, and thus both experiments and appropriate modeling are important for biotechnological development or innovation.

Subject Index

www.ingramcontent.com/pod-product-compliance
Lightning Source LLC
Chambersburg PA
CBHW050806220326
41598CB00006B/128